Superconductor Detectors

Masataka Ohkubo

Superconductor Detectors

Fundamentals and Analytical Applications

Springer

Masataka Ohkubo
Department of Electronics
and Manufacturing
National Institute of Advanced Industrial
Science and Technology (AIST)
Tsukuba, Ibaraki, Japan

ISBN 978-3-031-84192-7 ISBN 978-3-031-84193-4 (eBook)
https://doi.org/10.1007/978-3-031-84193-4

This Springer imprint is published by the registered company Springer Nature Switzerland AG
The registered company address is: Gewerbestrasse 11, 6330 Cham, Switzerland

Preface

It is no exaggeration to claim that detector development has been pioneering new science and technology. New findings have been achieved by new innovative detectors and analytical instruments rather than new concepts in most cases. Our five senses are extraordinary systems that seamlessly combine analog amplification with digital signal processing, making them impossible to replicate with man-made technology. However, the five senses cannot cover some of physical quantities to be measured. The quantities are also called measurands. Man-made detectors expand our vision to nature. This book treats of measurands such as energy and flux of particles. The term "particle" is divided into two classes: massless particles of photons and nonzero-mass particles such as atoms and molecules. The measurands include the number of particles and their kinetic energies, forming so-called spectra. Other important measurands are arrival times and locations of particles. The man-made detectors play a pivotal role in measurement and analysis across a wide range of fields, including scientific research, industrial development, pollution control, environmental monitoring, and forensic science. Collectively, these fields are referred to as analytical science. The term "analytical sciences" is also often used, but the singular form "analytical science" is used here to represent the entire area of this field without dividing it into multiple fields. Medical diagnosis is also an important field, but we leave it to the literature. The history indicates that new detectors have brought us new insight and innovation in science and technology. Analytical science naturally requires analytical instruments; however, the importance of detector technology is often buried in the complex system of analytical instruments including analyzing software tools and databases in addition to hardware. This book turns the spotlight on the detector hardware technology and how superconductor detectors (or cryogenic detectors) have revolutionized total performance of analytical instruments and produced unconventional analytical data beyond the fundamental limits of conventional instruments.

The purpose of this book is to describe operation principles and to clarify the importance of superconductor detectors in analytical science. I expand the conventional understanding of semiconductor detectors as well as superconductor detectors and reveal unsolved problems for future research. It is based on my experience starting from 1993 at Electrotechnical Laboratory (ETL), after moving from Toyota Central R&D labs. I hope that this book can contribute to deepening the reader's understanding of detector science, expanding the area of applications, advancing and pioneering science and technology, and finally solving social problems.

This book is arranged such that one can trace a difference between quantum and thermal detectors; advantages and disadvantages, and similarity and dissimilarity between semiconductor and superconductor detectors. Before describing the latest research results, I review the pioneer's ideas and achievements far ahead of the times. I recommend students, researchers, and engineers, who plan to join this field, to read the book from the beginning. Even experts may expand new insights into well-known detector physics. There are many unsolved questions, for example, long-range proximity effect, which affects both quantum and thermal superconductor detectors, and excess quasiparticle density saturated at low temperature, which is common to superconductor detectors and qubits for quantum computing. They are stated as unsolved or open questions for future research.

Appendix A describes the definition of energy resolution and the mathematical derivation of the theoretical energy resolution, which are the basics before reading the text. Appendix B explains characteristic X-ray emission lines and excitation inside X-ray detector materials, which are important for not only detector operation but also material analysis. Appendix C deals with terminology and nomenclature for superconductor detectors. The international standards are indispensable for exchanging ideas among researchers and engineers. We use the standard abbreviations for the superconductor detector types, the terms, and the graphical symbols, considering the recommendation of international electrotechnical commission (IEC) standards. For example, we use the term of "superconductor detectors" instead of "superconducting detectors," because they are not always in a superconducting state (See Appendix C.3).

Tsukuba, Japan Masataka Ohkubo
December 2024

Acknowledgments I deeply appreciate S. Noguchi and A. Ohta of Toyohashi University of Technology; T. Hioki, Y. Kido, and J. Kawamoto of Toyota Central R&D Labs Inc.; my family, my parents, my sister, and the people at my home village Ohno for making the Ph.D. degree and the subsequent research life possible. The research direction in this book was guided by N. Kobayashi, I. Sakamoto, and N. Hayashi of Electrotechnical Lab. (ETL); and M. Kurakado of RIKEN as one of the pioneers in the LTD community.

It is impossible to list all of my collaborators at ETL and National Institute of Advanced Industrial Science and Technology (AIST) and those who contributed to this research and development from other institutes worldwide: the partial list can be found in a related paper: https://doi.org/10.1063/5.0151581. Without them, this book could not be completed. They deserve my deep appreciation for their experimental and theoretical assistance, encouragement, suggestion, discussion, and proposals. Finally, special thanks are due to Kenneth Howell for inviting me to write this book, and to the staff at Springer Nature for their patience with my slow writing and their thoughtful editorial support.

Competing Interests There are no competing interests to be declare at any locations of this book.

Ethics Approval This book contains no contents that require ethics approval for human participants and animal studies.

Contents

About the Author

Masataka Ohkubo is a native of Kochi prefecture in Japan, graduated from National Institute of Technology, Kochi College, in 1979, and received an M.S. degree in electrical and electronic engineering from Toyohashi University of Technology in 1983.

He worked with Toyota Central R&D Labs., Inc. from 1983 to 1993. In the middle of that period, high-temperature superconductors (HTSs) were discovered. He used techniques such as co-evaporation and pulsed laser deposition to make YBCO epitaxial thin films for superconductor electronics. Ion beam analysis was one of the analytical techniques to characterize the single-crystalline films. With this work, he received a Ph.D. degree from Toyohashi University of Technology in 1991. In 1993, he joined the quantum radiation division at Electrotechnical Laboratory (ETL). Between 1995 and 1997, he was a guest researcher at Karlsruhe Research Center (Karlsruhe Institute of Technology) and a visiting researcher at University of Tübingen. He pursued a possibility for utilizing the intrinsic Josephson effect in BSCCO thin films to develop superconductor X-ray detectors. It was unsuccessful; however, he learned BSCCO crystal growth and electrical properties. After returning to ETL, his research topics included X-ray spectroscopy and mass spectrometry with superconductor tunnel junction detectors (STJDs), transition edge sensor detectors (TESDs), and superconductor strip particle detectors (SSPDs). In 2001, the ETL was merged into the National Institute of Advanced Industrial Science and Technology (AIST). After that, he was pursuing innovations in analytical science with superconductivity.

In 2013, he established a clean room facility called CRAVITY for analog and digital superconductivity in AIST. He was appointed to visiting professorships at Japanese universities and research organizations. His current duties include the convener of International Electrotechnical Commission—Technical Committee 90/Working Group 14 (IEC-TC90/WG14) for superconductor electronic devices, standardizing measurement methods, graphical symbols such as Josephson junctions, and terminology; and an associate editor of IEEE superconductivity news forum (SNF). He is now a Researcher Emeritus of AIST.

Abbreviations

$2\Delta_g$	Energy gap in superconducting state/energy difference between the top of quasihole states and the bottom of quasielectron states.
a	$\varepsilon = aE_g + b$
AFM	Atomic force microscope
ASC	Applied Superconductivity Conference
AV	Abrikosov vortex
b	$\varepsilon = aE_g + b$
BSA	Bovine serum albumin
CAD	Collisionally activated dissociation
CCD	Charge coupled device
CID	Collision-induced dissociation
Δ	Order parameter
Δ_g	Half of superconducting energy gap $2\Delta_g$
Da	Dalton (mass unit identical to unified atomic mass unit)
ΔE	Energy resolution in FWHM
ΔE_{tun}	Fluctuation due to multiple quasiparticle tunneling in FWHM
DFPT	Density functional perturbation theory
DFT	Density functional theory
DOS	Density of states
DR	Electron-ion dissociative recombination
DSP	Digital signal processing
DUT	Device under test
E	Detected energy/energy relative to the bottom of the conduction band for electrons/energy relative to the top of the valence band for holes
ε	Average energy consumed for creating one e–h pair in solid state detectors
e–h	Electron-hole pair
E_0	Incident particle energy
EDS	Energy dispersive spectroscopy
E_F	Fermi energy

E_g	Band gap or Energy gap in semiconductors and insulators
EI	Electron impact
E_k	Quasiparticle excitation energy from the superconducting ground state
EM	Electron-multiplier
EPMA	Electron probe microanalyzer
ESI	Electrospray ionization
ETD	Electron transfer dissociation
ETF	Electrothermal feedback
EUCAS	European Conference on Applied Superconductivity
F	Fano factor
FAB	Fast atom bombardment
FWHM	Full width at half maximum
G	Relative standard deviation of tunnel uncertainty
GM	Geiger-Müller
H	Heat of formation
HTS	High-temperature superconductor
I–U	Current–voltage
I_b	Bias current
I_c	Critical current
ICMC	International Cryogenic Materials Conference
IEC	International Electrotechnical Commission
IEEE	Institute of Electrical and Electronics Engineers
IgG	Immunoglobulin G
IgM	Immunoglobulin M
ISD	In-source decay
ISEC	International Superconductive Electronics Conference
I_{sub}	Quasiparticle tunnel current in a bias region below $2\Delta_g$
J_c	Critical current density
j_c	Local critical current density
JV	Josephson vortex
KER	Kinetic energy release
LA phonon	Longitudinal acoustic phonon
LO phonon	Longitudinal optical phonon
LSI	Large-scale integration
LTD	Low temperature detector
LTSEM	Low temperature scanning electron microscope
LTSSM	Low temperature scanning synchrotron microscope
MALDI	Matrix-assisted laser desorption/ionization
MAR	Multiple Andreev reflection
MCP	Microchannel plate
MKID	Microwave kinetic inductance detector
MS	Mass spectrometry
MS/MS	Tandem mass spectrometry
NIS	Normal-metal insulator superconductor

$\overline{N}$	Average number of e–h pairs in semiconductors
$\overline{N}_{qp}$	Average number of quasiparticle pairs in superconductors
$\overline{N}_{tun}$	Average number of quasiparticle tunnel times
PIXE	Particle induced X-ray emission
PMT	Photomultiplier tube
p_s	Probability of subsystem being in a microstate of E_s
PSD	Post-source decay
P_{tun}	Quasiparticle tunneling probability
R_d	Dynamic resistance in a voltage bias region below $2\Delta_g/e$ in superconductor tunnel junctions
r_{ee}	Electron-electron scattering rate/e–h pair generation rate
r_{ep}	Phonon emission rate of electrons (τ_{ep}: corresponding time)
RF	Radio frequency
RMS	Root mean square
r_{pb}	Cooper pair breaking rate of phonons
RRR	Residual resistivity ratio
σ	Standard deviation/Ionization cross-section
σ_{tun}	Standard deviation due to multiple quasiparticle tunneling
S-N transition	Superconducting–normal transition
S/N ratio	signal to noise ratio
SDD	Silicon drift detector
SEM	Scanning electron microscope
SIS	Superconductor insulator superconductor
SPT	Superconductor phase transition
SQUID	Superconductor quantum interference device
SR	Synchrotron radiation
SS	Superconductor strip
SSPD	Superconductor strip particle detector/Superconductor strip photon detector
STJD	Superconductor tunnel junction detector
T	Temperature
τ_0	Material-dependent characteristic time
τ_{AV}	Quasiparticle loss time due to AV trapping
τ_{eff}	Effective quasiparticle lifetime/Effective decay time of output pulses ($= 1/\gamma_{loss}$)
τ_r	Quasiparticle lifetime due to recombination into Cooper pairs
TA phonon	Transverse acoustic phonon
TC	Technical Committee
T_c	Critical temperature
TEM	Transmission electron microscope
TESD	Transition edge sensor detector
TO phonon	Transverse optical phonon
U_b	Bias voltage
VLSI	Very large-scale integration

V_m	Product of critical current and normal-sate resistance in superconductor tunnel junctions
ω	Angular frequency/Radiative X-ray emission yield
W	Average energy consumed for creating one electron-ion pair in gaseous detectors
Ω_D	Debye energy
WDS	Wave dispersive spectroscopy
WG	Working group
XAFS	X-ray absorption fine structure
XES	X-ray emission spectroscopy
ξ	Coherence length
ξ_k	Normal-state single-particle kinetic energy of the state k relative to E_F
ℓ	Electron mean free path
Φ_0	Flux quantum ($2.067\,834 \times 10^{-15}$ Wb)

Chapter 1
Historical Overview

Abstract Chapter 1 provides a short sketch of this book, a history of the low temperature detector community, the particle types to be detected, and introduction to all the superconductor detector types since the discovery of superconductivity. A detailed history of each superconductor detector type is covered in the subsequent Chapters. Finally, the superconductor materials used for detectors are listed.

Keywords History · Particle types · Superconductor materials

1.1 Introduction

Detectors play indispensable roles in analytical science. They convert physical quantities to be measured to electric signals. This book deals with single particle detection, not covering wave detection. There are two types of particles: massless quanta of photons and nonzero-mass particles (massive particles) such as atoms and molecules. Photons in this book range from far infrared to γ-rays, covering ~ 0.1 eV to ~ MeV. The nonzero-mass particles with kinetic energies in a range from a few eV to a few MeV have a wide mass range from electrons with a mass of 9.1×10^{-31} kg to macromolecules with ~ 2×10^{-21} kg.

Common detectors for analytical instruments employ, for example, alkaline metals for external photoelectric effect or secondary particle emission. They operate at room temperature. Other examples are semiconductors for internal photoelectric effect or particle-induced ionization, normally operating at a low temperature due to liquid nitrogen or Peltier cooling. These detectors based on the quantum effects are called "quantum detectors." The external photoelectric effect covers a low photon-energy range down to infrared rays near 1.2 eV, which corresponds to a wavelength of 1 μm. This is a typical limit of conventional technology for detecting electromagnetic waves as single quanta. The external photoelectric effect normally results in the emission of one secondary electron, which is multiplied at a factor of ~ 10^6 with electron multiplication; for example, in photomultiplier tube (PMT) for photon counting in photometry and electron-multiplier tube (EMT) for molecule counting in mass

M. Ohkubo, *Superconductor Detectors*, https://doi.org/10.1007/978-3-031-84193-4_1

spectrometry (MS). The external effect generally provides a single particle counting capability only, not measuring particle energies. Therefore, we need to place a spectroscopic analyzer in front of a detector. On the other hand, the internal effect is generally utilized for measuring the energies of single particles. They are sometimes called solid-state ionization chambers, which is an analogue of gaseous ionization chambers. The number of electron–hole (e–h) pairs created beyond a semiconductor bandgap represents the absorbed kinetic energy. The practical lower limit for this energy measurement with semiconductors is approximately a photon energy of 50 eV in a range of soft X-rays.

Superconductor quantum detectors that operate at a temperature well below 1 K have a capability of measuring single photon energies of infrared rays, which is never realized by any other detectors. They also expand the single photon detection limit to ~ 0.3 eV (a wavelength of ~ 4 μm stepping into a far-infrared region), which is far beyond the lower limit of photon counting with the external photoelectric effect. In a higher photon energy of X-rays, the energy resolution of superconductor detectors is far superior to that of semiconductor detectors. Superconductor detectors outperform conventional detectors in the two aspects: (1) expansion of the area for detecting electromagnetic waves as single quanta and (2) remarkably-high energy resolution, which revolutionize analytical science. Appendix A.1 describes the definition of energy resolution in detail. Note that we use the term of "superconductor detectors" instead of "superconducting detectors," because they are not always in a superconducting state (See Appendix C.3).

Other detection scheme in contrast to quantum detection is thermal detection, which consists of an absorber and a thermometer. Conventionally, it is known as bolometry, which measures temperature rise due to continuous particle incidence. In other word, bolometers measure particle flux in most cases. Thermistors and superconductor thermal detectors changed that situation by adding a capability of single-particle energy measurement. Especially, superconductor thermometers that have a sharp normal-superconducting transition enable measurement of temperature rise due to single-particle incident events. The superconductor thermal detectors outperform both superconductor quantum detectors and thermistor-based thermal detectors in energy resolution. The superconductor thermal detectors can also measure single-photon energy down to ~ 0.3 eV.

Superconductor detectors naturally use a superconductor at least for one of the three detector components: absorption, electric signal generation, and amplification/conversion in readout circuitry. In conventional analytical instruments, the signal generation devices normally utilize a semiconductor. Unconventional instrumental performance due to superconductivity overcomes the inherent problems of conventional analytical instruments. The state-of-art superconductor electronics including analog and digital circuits appears in Ref. [1]. We are not insisting that superconductors replace semiconductors in the entire field of analytical science; operating superconductor detectors in synergy with semiconductors advances analytical science. Currently, analog amplification and digital signal processing are in most cases performed by semiconductor technology after a superconductor detector converts measurands to electrical signals. Superconductor sensors/detectors as well as other superconductor devices engaged large attention in the superconductivity centennial conference combining three biennial conferences—EUCAS-ISEC-ICMC in 2011 [2] as well as biennial ASC.

Superconductor detectors are classified into the quantum detection, in which an electric signal is directly produced after absorbing a particle, and the thermal detection, in which temperature rise due to particle incidence is measured by a thermometer. Both superconductor quantum and thermal detectors rely on superconductivity at one or more segments of the detector components; therefore, all superconductor detector types can be also categorized as quantum detection. Nevertheless, we adopt the categorization based on the electric signal creation mechanisms, following the categorization widely accepted in the field of radiation detectors [3].

This book begins with an introduction to the low temperature detector (LTD) field. LTDs do not exclusively refer to superconductor detectors; however, most LTDs exploit superconductivity. After mentioning a brief history of low temperature detectors in Chap. 1, the general operation principles of quantum and thermal detection are explained briefly in Chap. 2. Contrastive descriptions of quantum detection versus-thermal detection in Chaps. 3 and 4 help the readers to understand the underlaying detector physics of superconductor detectors. In both Chapters, we describe the theoretical and experimental efforts to achieve the predicted high performance of superconductor detectors. The readers will find fundamental problems on the classical Fano factor, the relation between energy gap and mean carrier-generation energy, and the intermediate state in the middle of superconducting transition, followed by engineering hurdles on small sensitive area and spatially inhomogeneous response. Nevertheless, enormous efforts have matured superconductor detectors and brought outstanding performance, which enables us to solve the inherent problems of conventional analytical instruments. In Chap. 5, the performance figures of the different types of superconductor detectors are listed and compared to find suitable applications. Chapter 6 begins with the basics of analytical instruments. This introduction to instruments is followed by the examples of unconventional analytical results obtained by using superconductor detectors, focusing on X-ray spectroscopy and mass spectrometry with three detector types: superconductor tunnel junction detector (STJD), transition edge sensor detector (TESD), and superconductor strip particle detector (SSPD). Note that the term of "mass spectrometry" instead of "mass spectroscopy" is recommended by the MS community. Microwave kinetic inductance detector (MKID) and metallic magnetic calorimetric detector (MMCD) are described in Chaps. 3 and 4. The analytical results contribute to a wide range of fields such as dentistry, molecular biology, semiconductors for energy-saving society, atmosphere escape from planets, and prebiotic organic molecules in the interstellar space.

1.2 Low Temperature Detectors (LTDs)

The research field of low temperature detectors has a history of 38 years since the first international workshop on low temperature detectors (LTD-1) was held in 1987 with about 40 participants, although there were sporadic pioneering meetings before LTD-1. The initial purpose of the LTD community was to devise methods to detect neutrinos and dark matter, both of which are difficult to detect even these days. Topical reviews summarized the status in 1996 [4, 5]. Detectors operating at a cryogenic temperature would have a chance to achieve an ultimately high sensitivity for detecting very small energy transferred to an absorber via interaction with particles. Many detector concepts were proposed at the beginning of a series of LTD workshops, followed by detector fabrication and proof-of-concept experiments. Some of them were successful and some others failed. The successful LTDs developed further and matured for practical use. One of the important LTD applications in addition to elementary particle physics and astronomy is analytical science, which is the subject of this book. Other recent trends of superconductor detectors include quantum information fields such as quantum cryptography and quantum computing.

Figure 1.1 shows the plot of kinetic energy of particles as a function of mass. Massless photons are located at zero on the x-axis, separated from nonzero-mass particles such as neutrinos, dark matter, electrons, neutrons, and molecules. Usefulness of LTDs in many fields resulted in the expansion of the LTD community. For example, the participants of LTD-11 in 2005 increased to 225 from 40 in LTD-1 [6]. In addition to the initial purpose of neutrino detection and dark matter search, the objects to be measured expanded to photons from infrared to X-rays/γ-rays for astronomy, material analysis, and plasma physics. In 2019, the participants in LTD-18 increased to 420 by adding biomacromolecules and rare event search [7]. Figure 1.1 illustrates the expansion of particle types on the occasions of LTD-1 in 1987, LTD-11 in 2005, and LTD-18 in 2019. The proceedings of LTD-19 and LTD-20 are available [8, 9]. In addition to the series of the LTD workshops, the workshop on superconductive particle detectors held in 1987 contained many ideas and insights on several types of superconductor detectors [10]. There are also many other conferences contributing to the advances in superconductor detectors. A very brief timeline for epochal events on all the superconductor detector types is mentioned below; the details of historical overviews are described in each Chapter.

Long before LTD-1, following the discovery of superconductivity in mercury in 1911 [11], the thermal detection of infrared radiation using a sharp superconducting transition was proposed by Goetz in 1939 [12]. The successful detector operation was realized in 1942 by Andrews et al., who invented the basic principles of temperature self-regulation for superconductor phase transition (SPT) thermometers weakly connected to a cold stage by adding a parallel shunt resistor [13]. Although it was a bolometer that aimed to measure particle flux (power) rather than single particles (kinetic energy), it was an ancestor of the modern TESD for single particle detection by Irwin in 1995 [14]. Andrews also invented a current-carrying superconductor strip detector that successfully detected MeV α-particles in 1949 [15].

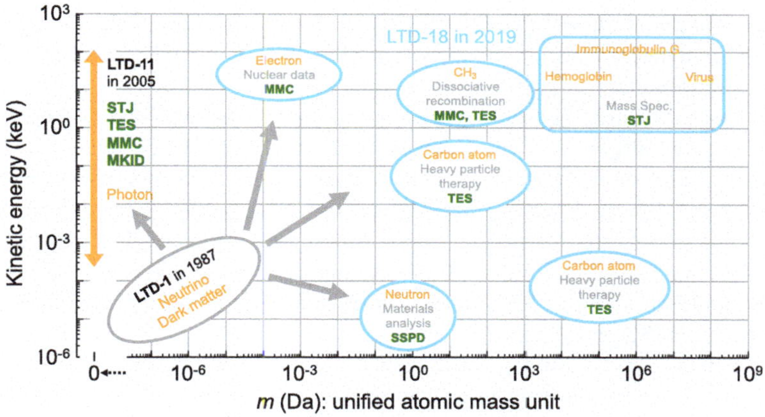

Fig. 1.1 Expansion of LTD community from neutrino and dark matter to photons and nonzero-mass particles. Photons are massless, which is indicated by the dotted arrow to distinguish photons from nonzero-mass particles, which include neutrinos, electrons, neutrons, atoms, molecules, proteins, and viruses. The unit of mass is expressed by Da that is one-twelves of the ^{12}C mass

The superconductor strip detector that had a good thermal contact with a substrate exhibited the highest particle count rate when it was biased in the middle of superconducting transition. It was an ancestor of the modern SSPDs for photons near 1 eV by Gol'tsman et al. in 2001 [16].

Quantum detection of sub-millimeter wave radiation with normal-metal/insulator/superconductor (NIS) or superconductor/insulator/superconductor (SIS) junctions was proposed by Burstein et al. in 1961 [17], just after the discovery of electron tunneling in NIS and SIS junctions by Giaever in 1960 [18]. Both ideas of NIS and SIS aimed to measure particle flux. The SIS junction was an ancestor of the modern STJDs for single particle detection. Two years later, the Josephson effect was theoretically predicted by Josephson in 1962 [19] and experimentally confirmed by Anderson and Rowell in 1963 [20]. The idea of STJDs with a small energy gap for single particles by Wood and White dates back to 1965 [21], although the successful detection of 5.1-MeV α-particles was delayed for four years in 1969 [22, 23]. Kurakado et al. confirmed their result in 1980 [24]. The Josephson effect led to the development of a voltmeter called the superconductor low-inductance undulatory galvanometer (SLUG) or superconductor quantum interference device (SQUID) by Clarke in 1966 [25]. SQUID amplifiers are indispensable to signal readout electronics for superconductor detectors such as TESD and MMCD as well as magnetocardiography (MCG), magnetoencephalography (MEG), mineral exploration, and other magnetic sensor applications [26]. Low-temperature calorimetry was proposed for detecting particles due to rare events such as neutrinoless double-beta decay by Fiorini and Niinikoski in 1984 [27]. Single X-ray photon detection with a calorimeter with a semiconductor thermistor thermometer was proposed and realized by Moseley et al.

and McCammon et al. in 1984 [28, 29]. These papers are important for superconductor thermal detector development; we have been following their guides to improve detector performance.

The range of single photon energy measurement was extended to visible and infrared region with an STJD by Peacock et al. in 1996 [30], and with a TESD by Cabrera et al. in 1998 [31]. Both detectors had an energy resolution of ~ 0.1 eV for ~ 1 eV photons. This extension enabled energy-dispersive spectroscopy (EDS), which was possible exclusively for an X-ray range before, in a range of telecommunication wavelengths or far infrared rays. The extension to massive particles was reported by Twerenbold in 1996: an STJD successfully detected high-mass proteins in mass spectrometry (MS) [Ref. 32].

New detector types invented afterward are MMCD and MKID. The magnetic calorimetry using Er-doped garnets was proposed and demonstrated for 5.5-MeV α-particles by Bühler and Umlauf in 1988 [33, 34]. A metallic magnetic sensor with a fast response time was proposed by Bandler et al. in 1993 [35] and Enss et al. successfully detected single X-ray and γ-ray photons with an MMCD in 2000 [36]. MKID with a large array format for imaging was demonstrated for X-rays by Day et al. in 2003 [37]. The MKID-detectable range was later extended to an optical-to-near-infrared region in 2013 by Mazin et al. [38].

The tremendous improvement has been achieved until now, gaining benefit of modern micro-to-nano fabrication technology, dry cryogenics without liquid cryogen supply, and sophisticated signal processing with semiconductor devices. One of the overviews of all the detector types appeared in a book in 2005 [39].

1.3 Superconductor Materials

Most materials for superconductor detectors are low temperature superconductors (LTSs). A lower T_c is preferable to achieve a better energy resolution due to the resulting lower energy gap or reduced heat capacity. Some examples are Ir, Al, WSi, Nb, Ta, NbN, NbRe: T_c ranges from 0.1 to 16 K. In addition to single layer films, bilayers such as Nb/Al, Mo/Cu, and Ti/Au are utilized for controlling T_c with the proximity effect [40]. Many physical aspects of detector operation using LTSs are understandable based on the standard Bardeen-Cooper-Schrieffer (BCS) theory, which postulates isotropic superconducting energy gaps and *s*-wave symmetry.

In quantum detection, a superconducting energy gap ($2\Delta_g$) is utilized as the scale for measuring particle kinetic energy through the number of quasiparticles created by breaking Cooper-pairs. The origin of the superconducting gaps is totally different from that of the bandgaps in semiconductors; nevertheless, they are analogous from statistical physics point of view. The similarity and dissimilarity between superconductors and semiconductors are described in Chap. 3, Sect. 3.2. The operating temperature shall be well below T_c to reach to a temperature range where the temperature dependence of Δ_g is negligible and to suppress thermal excitation of quasiparticles.

A recommended temperature range is less than $0.1T_c$: it is below ~ 1 K for Nb or below ~ 0.5 K for Nb/Al bilayers.

In thermal detection, a cryogenic temperature range is required to reduce the heat capacity of absorbers. Although thermistor-based thermometers can operate in the cryogenic temperature range, it is difficult to find materials that have high temperature sensitivity. Sharp superconducting transition is suitable for cryogenic thermometry. Superconductor films are biased in the middle of superconducting transition widths. T_c can be adjusted by using the proximity effect in bilayer films of a superconductor and a normal-conductor. For single particle detection, T_c is normally at ~ 0.1 K or below.

The operating temperatures are reachable with adiabatic demagnetization or ^{3}He-^{4}He dilution for a temperature range below 0.1 K, and ^{3}He sorption refrigeration for ~ 0.3 K. The refrigerators are significantly larger than liquid nitrogen cryostats or Peltier coolers for semiconductor detectors but essential for the extraordinarily high performance of superconductor detectors. Huge refrigerator equipment is an obstacle to installing superconductor detectors into analytical instruments. However, the irreplaceable high performance of superconductor detectors can outweigh the obstacle. Moreover, the development of fully automated cryostats that do not require the supply of liquid cryogens such as nitrogen or helium has accelerated the dissemination of easy-to-use superconductor detectors.

Searching superconductors with a higher T_c is a natural trend for the applications of superconductors. Is it possible to use HTSs? Just after LTD-1, an HTS system of LaBaCuO was discovered in 1986 [41]. HTS applications to detectors were already discussed in 1987 [10]. Many HTSs have been discovered until now. HTS Josephson junctions (JJs) were fabricated and applied to SQUIDs soon after the HTS discovery [42, 43] HTS JJs can be explained by the resistively shunted junction (RSJ) model [44] at a liquid nitrogen temperature of 77 K. However, the *d*-wave symmetry of HTSs complicates the detection scheme when we consider quasiparticle excitation. Uniformity of HTS thin-films is another issue. Multi-element composition precludes homogeneous thin film deposition; T_c easily varies with oxygen content in the cuprate HTSs such as $YBa_2Cu_3O_x$ (YBCO) [45]. Moreover, HTS tunnel junctions or SIS junctions with a similar quality to LTSs have not been realized at 77 K due to an extremely short coherence length of less than 1 nm. There was a possibility of an SIS type detector with intrinsic JJs embedded in BiSrCaCuO (BSCCO)[46] crystals. Blocks of one unit cell are bounded by a weak van-der-Waals force; a stuck of SIS alternation appears along the *c*-axis [47]. The intrinsic JJs are utilized to generate and detect THz radiation [48, 49]. THz emission was also observed from YBCO [50]. In addition to the THz range, we considered utilizing the intrinsic Josephson effect as an X-ray detector. However, BSCCO is unsuitable for X-ray detection due to the *d*-wave symmetry and natural staking-faults of different phases of $Bi_2Sr_2CuO_x$ (2201), $Bi_2Sr_2CaCu_2O_x$ (2212), $Bi_2Sr_2Ca_2Cu_3O_x$ (2223), and $Bi_2Sr_2Ca_3Cu_4O_x$ (2234) [51]. A long-range modulation superstructures is another obstacle to utilize the intrinsic Josephson junctions [52]. Therefore, HTS-STJDs have never been realized until now.

There is an exception to HTSs for superconductor detectors. Because the SS type is essentially a particle counting device, not measuring single-particle energy, it is an analogue of Geiger-Müller (GM) counters. The thermal noise is tolerable and thus an operating temperature of $T_c/2$ may be sufficient. One successful HTS-SSPD for telecom-wavelength photons consists of MgB_2, which is a *s*-wave superconductor with multiple energy gaps. The effect of the multiple gap structure is negligibly small for single-photon or single-molecule counting [53, 54]. Another cuprate superconductor LaSrCuO with the *d*-wave symmetry is close to single photon detection [55]. YBCO is attractive in single particle detection at 77 K [56–59]. A dark count experiment with YBCO exhibited a promising result for single photon detection in future [60]. Furthermore, a small 5-μm square SSPD, which was mechanically exfoliated from a bulk BSCCO single crystal at a van-der-Waals gap, detected single photons at temperatures up to 25 K [61]. Concurrently, it was reported that a BSCCO nanobridge that was 250 nm wide and ~ 1-μm long detected single photons up to 20 K [62].

Table 1.1 lists the fundamental properties of the above-mentioned typical superconductors. One of the properties, the Pearl length ($\Lambda = \frac{2\lambda^2}{t_f}$), [63] may be unfamiliar to some readers, where λ is the magnetic penetration depth for bulks and t_f is the film thickness. The Pearl length, which is important for SSPD operation, is the penetration depth for a magnetic field perpendicular to film surfaces.

Table 1.1 Fundamental properties of typical superconductors for detectors

	Nb	bcc Ta	Al	NbN	MgB_2	YBCO	BSCCO-2212
T_c (K)	9.23[a]	4.48[e]	1.2	16.15[h]	39.2[k]	89[n]	90[s]
$\Delta_g(0)$ (meV)	1.52[a]	0.7[e]	0.17	2.46[h]	2, 10[k]	20[p]	42[t]
$\xi(0)$ (nm)	23[b]	–	147[b]	5[h]	39, 35[m]	2.5, 0.8[q]	3.1, 0.4[u]
$\lambda(0)$ (nm)	47[c]	–	52[f]	194[i]	107, 120[m]	92, 635[n]	260, 40000[v]
τ_{ep} (ps)	330[d]	–	395 000[g]	88[j]	2[g]	1.1[r]	0.11[w]
$\Lambda(0)$ (μm)	0.19	–	0.037	15	0.65	21	340

T_c is the critical temperature, Δ_g is the single particle energy gap, ξ is the coherence length, λ is the magnetic penetration depth, τ_{ep} is the electron–phonon scattering time, and Λ is the Pearl length in most cases at 0 K. The τ_{ep} values are at ~ $0.5T_c$ except specified. The τ_{ep} has T^{-n} dependence in the dirty-limit, where n is an integer between 2 and 3. Electron excitation levels also affect τ_{ep}. For anisotropic MgB_2, $YBa_2Cu_3O_x$ (YBCO), $Bi_2Sr_2Ca_{n-1}Cu_nO_{2n+4}$ (BSCCO), where n is an integer between 1 and 4, Δ_g, ξ, and λ values are within the *a-b* plane (left) and along the *c*-axis (right). Since the *c*-axis of MgB_2, YBCO, and BSCCO films tends to be perpendicular to substrate surfaces, $\lambda_{a\text{-}b}$ is applicable for SSPD. For MgB_2, YBCO, and BSCCO, $\lambda_{a\text{-}b}$ and ξ_c values are adopted for calculating Λ for films with a thickness of ξ_c

[a] AIST data: T_c is the middle of transition of a sputtered polycrystalline Nb film with a residual resistance ratio (RRR) of 4.4, and Δ_g was obtained by extrapolating Al thickness dependence to zero thickness on *I-U* curves of $Nb/Al/AlO_x/Al/Nb$ junctions

[b] Polycrystalline film [64]

[c] Polycrystalline cylinder [65]

[d] 30-nm-thick polycrystalline Nb film [66]

[e] Ref [67]

[f] Ref. [68], [g] Ref. [69], [h] Ref. [70]. [i] Ref. [71]

[j] 15-nm-thick polycrystalline NbN film at 8 K[72]

[k] Ref. [73], [m] Ref. [74]

[n] $YBa_2Cu_3O_{6.995}$ single crystal: $\lambda_{ab}(0) = (\lambda_a + \lambda_b)/2 = 92$ nm and $\lambda_c(0) = 635$ nm [75]. For epitaxially grown thin films with defects, $\lambda_{ab}(0)$ is typically in a range between 180 and 200 nm [76, 77]

[p] Ref. [78], [q] Ref. [79], [r] Ref. [80]

[s] Among $Bi_2Sr_2Ca_{n-1}Cu_nO_{2n+4}$ phases, where n is an integer between 1 and 4 in most cases, $Bi_2Sr_2CaCu_2O_8$ (2212) is thermodynamically most stable. The 2212 phase with optimum doping typically has a T_c value of 90 K even in a monolayer thin film [81]

[t] Δ_g at the optimum doping level. It is known that normal-state pseudogap coexists with superconducting gap [82, 83]

[u] Ref. [84], coherence length increases when a hole doping level departs from the optimum level [85]

[v] Ref. [86]

[w] Two time scales for electron–phonon interaction were reported: phonon generation for 0.11 ps and phonon decay for 2 ps [87]

References

1. A.I. Braginski, J. Supercond. Novel Mag. **32**, 23 (2019)
2. H. Rogalla, P. Kes, Physica C **482**, 1 (2012)
3. M. Ohkubo, J. Appl. Phys. **134**, 081101 (2023)
4. N. Booth, D.J. Goldie, Supercond. Sci. Technol. **9**, 493 (1996)
5. D. Twerenbold, Rep. Prog. Phys. **59**, 349 (1996)

6. M. Ohkubo, K. Mitsuda, H. Takahashi (eds.), *Proceedings of the 11th International Workshop on Low Temperature Detectors*, Nucl. Instrum. Methods in Phys. Res. A **559**(2) (2006)
7. M. Faverzani, E. Ferri, A. Giachero, A. Nucciotti, A. Puiu, M. Sisti (eds.), Special Issue: Low Temperature Detectors LTD18. J. Low Temp. Phys. **199**(1–4) (2020)
8. J. Ullom, J. Austermann, K. Koehler, J. Hubmayr (eds.), Special Issue: Low Temperature Detectors LTD19. J. Low Temp. Phys. **209**(3–6) (2022)
9. Y.-H. Kim, W. Cui, H-S. Jo, L. You (eds.), Special Issue: Low Temperature Detectors LTD20. J. Low Temp. Phys. **214**(3–4), **215**(3–4), **216**(1–2) (2024)
10. A. Barone, *Superconductive Particle Detectors* (World Scientific, 1987)
11. H. Kamerlingh-Onnes, Akad. van Wetenschappen, Amsterdam **14**, 818 (1911)
12. A. Goetz, Phys. Rev. **55**, 1270 (1939)
13. D.H. Andrews, W.F. Brucksch, W.T. Ziegler, E.R. Blanchard, Rev. Sci. Instrum. **13**, 281 (1942)
14. K. Irwin, Appl. Phys. Lett. **66**, 1998 (1995)
15. D.H. Andrews, R.D. Fowler, M.C. Williams, Phys. Rev. **76**, 154 (1949)
16. G.N. Gol'tsman, O. Okunev, G. Chulkova, A. Lipatov, A. Semenov, K. Smirnov, B. Voronov, A. Dzardanov, C. Williams, R. Sobolewski, Appl. Phys. Lett. **79**, 705 (2001)
17. E. Burstein, D.N. Langenberg, B.N. Taylor, Phys. Rev. Lett. **6**, 92 (1961)
18. I. Giaever, Phys. Rev. Lett. **5**, 147 (1960); ibid **5**, 464 (1960)
19. B.D. Josephson, Phys. Lett. **1**, 251 (1962)
20. P.W. Anderson, J.M. Rowell, Phys. Rev. Lett. **10**, 230 (1963)
21. G.H. Wood, Master thesis, *An Investigation of the Possibility of Charged Particle Detection Using Tunneling Between Superconductors* (University of British Columbia, 1965). https://doi.org/10.14288/1.0085916
22. G.H. Wood, Doctor thesis, *The Detection of Alpha Particles with Superconducting Tunnel Junctions*(University of British Columbia, 1969). https://doi.org/10.14288/1.0084727
23. C.H. Wood, B.L. White, Appl. Phys. Lett. **15**, 237 (1969)
24. M. Kurakado, H. Mazaki, Phys. Rev. B **22**, 168 (1980)
25. J. Clarke, Phil. Mag. **13**, 115 (1966)
26. R. Stolz, M. Schmelz, V. Zakosarenko, C. Foley, K. Tanabe, X. Xie, R.L. Fagaly, Supercond. Sci. Technol. **34**, 033001 (2021)
27. E. Fiorini, T.O. Niinikoski, Nucl. Instrum. Methods in Phys. Res. **224**, 83 (1984)
28. S.H. Moseley, J.C. Mather, D. McCammon, J. Appl. Phys. **56**, 1257 (1984)
29. D. McCammon, S.H. Moseley, J.C. Mather, R.F. Mushotzky, J. Appl. Phys. **56**, 1263 (1984)
30. A. Peacock, P. Verhoeve, N. Rando, A. van Dordrecht, B.G. Taylor, C. Erd, M.A.C. Perryman, R. Venn, J. Howlett, D.J. Goldie, J. Lumley, M. Wallis, Nature **381**, 135 (1996)
31. B. Cabrera, R.M. Clarke, P. Colling, A.J. Miller, S. Nam, R.W. Romani, Appl. Phys. Lett. **73**, 735 (1998)
32. D. Twerenbold, J-L. Vuilleumier, D. Gerber, A. Tadsen, B. van den Brandt, P. M. Gillevet, Detection of single macromolecules using a cryogenic particle detector coupled to a biopolymer mass spectrometer. Appl. Phys. Lett. 68, 3503 (1996). https://doi.org/10.1063/1.115772
33. M. Bühler, E. Umlauf, Europhys. Lett. **5**, 297 (1988)
34. E. Umlauf, M. Bühler, *Low Temperature Detectors for Neutrinos and Dark Matter*, ed. by N.E. Booth, G.L. Salmon (Editions Frontieres, Gif-sur-Yvette, 1991), p. 229
35. S.R. Bandler, C. Enss, R.E. Lanou, H.J. Maris, T. More, S.F. Porter, G.M. Seidel, J. Low Temp. Phys. **93**, 709 (1993)
36. C. Enss, A. Fleischmann, K. Horst, J. Schoönefeld, J. Sollner, J.S. Adams, Y.H. Huang, Y.H. Kim, G.M. Seidel, J. Low Temp. Phys. **121**, 137 (2000)
37. P.K. Day, H.G. LeDuc, B.A. Mazin, A. Vayonakis, J. Zmuidzinas, Nature **425**, 817 (2003)
38. B.A. Mazin et al., Publ. Astron. Soc. Pac. **125**, 1348 (2013)
39. C. Enss, *Cryogenic particle detectors* (Springer, 2005)
40. W.L. McMillan, Phys. Rev. B **175**, 537 (1968)
41. J.G. Bednorz, K.A. Müller, Zeitschrift für Physik B Condensed Matter **64**, 189 (1986)
42. J. Gao, W.A.M. Aarnink, G.J. Gerritsma, H. Rogalla, Physica C **171**, 126 (1990)

43. R. Gross, P. Chaudhari, M. Kawasaki, M.B. Ketchen, A. Gupta, IEEE Trans. Mag. **27**, 2565 (1991)
44. M. Ohkubo, G. Uehara, J. Beyer, M. Mimura, H. Tanaka, K. Ehara, S. Tanaka, T. Noguchi, E.E. Mitchell, C.P. Foley, R.L. Fagaly, Supercond. Sci. Technol. **35**, 045002 (2022)
45. M. Ohkubo, T. Kachi, T. Hioki, J. Appl. Phys. **58**, 1782 (1990)
46. H. Maeda, Y. Tanaka, M. Fukutomi, T. Asano, Jpn. J. Appl. Phys. **27**, L209 (1988)
47. A.A. Yurgens, Supercond. Sci. Technol. **13**, R85 (2000)
48. H. Minami, M. Tsujimoto, T. Kashiwagi, T. Yamamoto, K. Kadowaki, IEICE Trans. Electron. **E95-C**, 347 (2012)
49. R. Kleiner, H. Wang, J. Appl. Phys. **126**, 171101 (2019)
50. M. Hangyo, S. Tomozawa, Y. Murakami, M. Tonouchi, M. Tani, Z. Wang, K. Sakai, S. Nakashima, Appl. Phys. Lett. **69**, 2122 (1996)
51. M. Ohkubo, J. Geerk, G. Linker, O. Meyer, Appl. Phys. Lett. **67**, 2403 (1995)
52. M. Ohkubo, R. Fromknecht, E. Brecht, J. Geerk, G. Linker, O. Meyer, Phys. Rev. B **54**, 9035 (1996)
53. H. Shibata, H. Takesue, T. Honjo, T. Akazaki, Y. Tokura, Appl. Phys. Lett. **97**, 212504 (2010)
54. N. Zen, H. Shibata, Y. Mawatari, M. Koike, M. Ohkubo, Appl. Phys. Lett. **106**, 222601 (2015)
55. H. Shibata, N. Kirigane, K. Fukao, D. Sakai, S. Karimoto, H. Yamamoto, Supercond. Sci. Technol. **30**, 074001 (2017)
56. N. Zen, Y. Chen, K. Suzuki, M. Ohkubo, S. Miki, Z. Wang, IEEE Trans Appl. Super. **19**, 354 (2009)
57. R. Arpaia et al., Phys. Rev. B **96**, 064525 (2017)
58. P. Amari et al., Supercond. Sci. Technol. **31**, 015019 (2018)
59. X. Xing, K. Balasubramanian, S. Bouscher, O. Zohar, Y. Nitzav, A. Kanigel, A. Hayat, Appl. Phys. Lett. **117**, 032602 (2020)
60. M. Ejrnaes, L. Parlato, R. Arpaia, T. Bauch, F. Lombardi, R. Cristiano, F. Tafuri, G.P. Pepe, Supercond. Sci. Technol. **30**, 12LT02 (2017)
61. I. Charaev et al., Nat. Nano. **18**, 343 (2023)
62. R.L. Merino, P. Seifert, J.D. Retamal, R.K. Mech, T. Taniguchi, K. Watanabe, K. Kadowaki, R.H. Hadfield, D.K. Efetov, 2D Mater. **10**, 021001 (2023)
63. J. Pearl, Appl. Phys. Lett. **5**, 65 (1964)
64. G. Brammertz, A. Poelaert, A.A. Golubov, P. Verhoeve, A. Peacock, H. Rogalla, J. Appl. Phys. **90**, 355 (2001)
65. B.W. Maxfield, W.L. McLean, Phys. Rev. **139**, A1515 (1965)
66. A.L. Leo, G. Grimaldi, R. Citro, A. Nigro, S. Pace, R.P. Huebener, Phys. Rev. B **84**, 014536 (2011)
67. C. Kittel, *Introduction to Solid State Physics*, 6th edn. (Wiley, New York, 1986)
68. F. Behroozi, M.P. Garfunkel, Physica **55**, 649 (1971)
69. L. Parlato, R. Latempa, G. Peluso, G.P. Pepe, R. Cristiano, R. Sobolewski, Supercond. Sci. Technol. **18**, 1244 (2005)
70. E.A. Antonova, D.R. Dzhuraev, G.P. Motulevich, V.A. Sukhov, Sov. Phys. JETP **53**, 1270 (1981)
71. B. Komiyama, Z. Wang, M. Tonouchi, Appl. Phys. Lett. **68**, 562 (1996)
72. L. Zhang, L. You, W. Peng, Z. Wang, Physica C **579**, 1353773 (2020)
73. T. Ekino, T. Takasaki, T. Muranaka, J. Akimitsu, H. Fujii, Phys. Rev. B **67**, 094504 (2003)
74. J.C. Loudon, S. Yazdi, T. Kasama, N.D. Zhigadlo, J. Karpinski, Phys. Rev. B **91**, 054505 (2015)
75. T. Pereg-Barnea et al., Phys. Rev. B **69**, 184513 (2004)
76. A.G. Zaitsev, R. Schneider, G. Linker, F. Ratzel, R. Smithey, P. Schweiss, J. Geerk, R. Schwab, R. Heidinger, Rev. Sci. Inst. **73**, 335 (2002)
77. S.Y. Lee, J.H. Lee, W.I.L. Yang, J.H. Claassen, IEICE Trans. Electron. **E89-C**, 132 (2006)
78. H.L. Edwards, D.J. Derro, A.L. Barr, J.T. Markert, A.L. de Lozanne, Phys. Rev. Lett. **75**, 1387 (1995)
79. H. Jiang, T. Yuan, H. How, A. Widom, C. Vittoria, J. Appl. Phys. **73**, 5865 (1993)

80. M. Lindgre, M. Currie, C. Williams, T.Y. Hsiang, P.M. Fauchet, S.H. Moffat, R.A. Hughes, J.S. Preston, F.A. Hegmann, Appl. Phys. Lett. **74**, 853 (1999)
81. Y. Yu, L. Ma, P. Cai, R. Zhong, C. Ye, J. Shen, G.D. Gu, X.H. Chen, Y. Zhang, Nature **575**, 156 (2019)
82. S. Hüfner, M.A. Hossain, A. Damascelli, G.A. Sawatzky, Rep. Prog. Phys. **71**, 062501 (2008)
83. Y. Kohsaka et al., Nature **454**, 1072 (2008)
84. B. Batlogg, T.T.M. Palstra, L.F. Schneemeyer, R.B. van Dover, R.J. Cava, Physics C **153–155**, 1062 (1988)
85. J. Hwang, Sci. Rep. **11**, 11668 (2022)
86. T. Jacobs, S. Sridhar, Q. Li, G.D. Gu, N. Koshizuka, Phys. Rev. Lett. **75**, 4516 (1995)
87. L. Perfetti, P.A. Loukakos, M. Lisowski, U. Bovensiepen, H. Eisaki, M. Wolf, Phys. Rev. Lett. **99**, 197001 (2007)

Chapter 2
Particle Detection

Abstract Chapter 2 outlines two major particle detection schemes: quantum detection and thermal detection, which apply to both superconductor and non-superconductor detectors such as semiconductor detectors. This comparison prepares the readers for a deeper understanding of detector physics.

Keywords Quantum detection · Thermal detection

2.1 General

Most basic principles of quantum and thermal detection physics are common to both superconductor and non-superconductor detectors. The conventional treatments prepare the readers to understand how superconductivity enables innovations in detector performance. Therefore, we begin with non-superconductor detectors. Afterward, the conventional treatments for semiconductor detectors are extended to superconductor detectors in the subsequent sections.

One of the important detector-performance figures is energy resolution (ΔE), which corresponds to the smallest detectable difference between two adjacent peaks [1]. A ΔE value is in general expressed in full width at half maximum (FWHM) of a peak for the incident events of mono-energetic particle E_0 on a histogram of the particle count against detected energy. A full-energy absorption peak for mono-energetic X-ray photons is a suitable measure of ΔE because all the photon energy transfers to a detector material. The peak shape is also called the response function. Appendix A.1 explains the important probability functions for detectors: Poisson, Gauss, and binomial distributions. High or better resolution means a small ΔE value, which corresponds to a high resolving power defined by $E_0/\Delta E$. Another important measurand is particle flux, radiation power, or wave intensity, not measuring single particles. Although this book focuses on single particle spectroscopy, we also touch on flux measurement below to better understand detector operation.

Figure 2.1 illustrates a comparison between the quantum and thermal detection schemes. The quantum detection in Fig. 2.1a uses an electronic energy gap E_g that

M. Ohkubo, *Superconductor Detectors*, https://doi.org/10.1007/978-3-031-84193-4_2

acts as a threshold for electronic excitation. A typical example of E_g is the bandgap in semiconductors. In semiconductor quantum detectors, the excited charge carriers, whose number is roughly equal to E_0/E_g, are collected by applying an electric field to two electrodes sandwiching a semiconductor. An output current pulse is produced for each particle incidence event as shown in the right panel in Fig. 2.1a. The quantum detection implies that the electronic excitation induced by particle incidence directly produces an output electric pulse. The integration of the current pulse represents the number of excited charge carriers or E_0, as shown in (b). Note that the number of excited charge carriers is equal to the number of electron–hole (e–h) pairs, as explained later.

The thermal detection in Fig. 2.1c relies on temperature rise due to an energy absorbed by a temperature dependent resistor. When a metallic thermometer such as Pt with a good linear temperature dependence in (c) is used, the resistance change is proportional to the power per unit area in (d), where n is the flux of the particles with E_0. Cryogenic thermal detectors often use a neutron transmutation doped (NTD) Ge thermistor, which has a strong negative nonlinear temperature dependence below ~ 100 mK: a typical R–$T^{-0.5}$ curve is depicted in (e). The resistance change leads

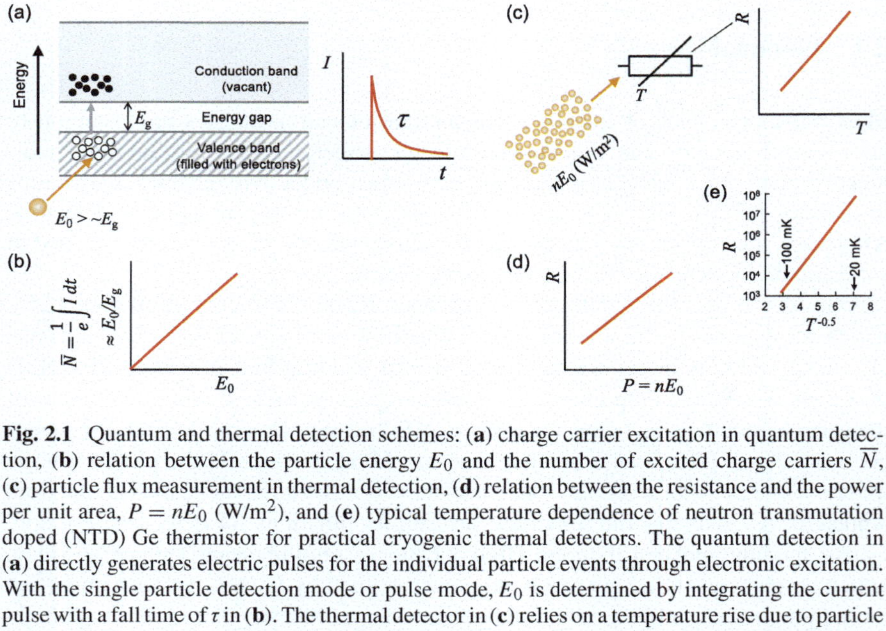

Fig. 2.1 Quantum and thermal detection schemes: (**a**) charge carrier excitation in quantum detection, (**b**) relation between the particle energy E_0 and the number of excited charge carriers $\overline{N}$, (**c**) particle flux measurement in thermal detection, (**d**) relation between the resistance and the power per unit area, $P = nE_0$ (W/m^2), and (**e**) typical temperature dependence of neutron transmutation doped (NTD) Ge thermistor for practical cryogenic thermal detectors. The quantum detection in (**a**) directly generates electric pulses for the individual particle events through electronic excitation. With the single particle detection mode or pulse mode, E_0 is determined by integrating the current pulse with a fall time of τ in (**b**). The thermal detector in (**c**) relies on a temperature rise due to particle flux (radiation power), which is measured with a resistance thermometer. With the flux detection mode or current mode, the particle flux n (m^{-2} s^{-1}) conveys the power per unit area, which induces a resistance change in (**d**). The typical NTD thermistor has a large negative temperature coefficient in a temperature range between 20 and 100 mK, obeying a nonlinear curve of $\log_{10}R - T^{-0.5}$ in (**e**). It should be noted that the quantum detectors can also measure particle flux or power. Similarly, the thermal detectors can measure single particles. The detection mode roughly depends on a relation between τ and $1/n$

to an electric signal through a readout circuit. Figure 2.1c shows a simple classical case, in which the temperature-sensitive resistor combines the thermometer and the absorber.

Historically, quantum detectors were utilized for detecting single particles such as ionizing particles and photons, whereas thermal detectors were often utilized to measure radiation power. These days, both quantum and thermal detectors can operate on either single-particle detection or power/flux measurement, generally depending on a relation between the detector response time τ and the inverse of the particle flux $1/n$. In the field of analytical instrumentation, the single particle detection and the power/flux measurement are designated as "pulse mode" and "current mode," respectively. Especially for quantum detection, the detection of particles with $E_0 > E_g$ is called "direct detection." On the other hand, the detection of particles with $E_0 < E_g$ is called "coherent detection" or "heterodyne detection," in which diodes or superconductor-insulator-superconductor (SIS) mixers are utilized for detecting waves rather than single particles.

2.2 Introduction to Quantum Detection

In the quantum detection, charge carriers such as electrons, holes, or ions are excited by particle incidence into a solid or gaseous absorber. The charge carriers are e–h pairs in semiconductors, while they are called quasiparticle pairs in superconductors. The term "pair" emphasizes the simultaneous generation of electron-like and hole-like excitations, without implying any binding between them. The difference between the e–h pairs and the quasiparticle pairs is explained in Chap. 3, Sect. 3.2.3. Figure 2.1a shows an undoped intrinsic semiconductor that has the energy gap or the bandgap E_g. The average number $\overline{N}$ of e–h pairs created by a particle with an energy of E_0 is roughly proportional to E_0/E_g. The precise discussion on carrier generation in semiconductors appears in Chap. 3, Sect. 3.1. When all the charge carriers are collected by using a semiconductor p–n junction or a surface barrier, it is possible to estimate E_0 by calculating $\overline{N} \times E_g$ for each particle incidence event. When we take a Poisson distribution with an average of $\overline{N}(\approx E_0/E_g)$, the standard deviation σ of the number of the charge carriers is given by $\sigma \approx \sqrt{\frac{E_0}{E_g}}$. The standard deviation in a unit of eV is given by $\sigma \approx E_g\sqrt{\frac{E_0}{E_g}} = \sqrt{E_0 E_g}$. This means that a smaller E_g leads to a smaller standard deviation. Semiconductors cannot detect photons in a range lower than E_g, which means that the semiconductors are transparent. This is a rough sketch of quantum detection. In reality, however, $\overline{N}$ is considerably smaller than E_0/E_g because only a part of E_0 is consumed for creating charge carriers. This means that an average energy of $E_0/\overline{N}$ consumed for creating one e–h pair is considerably larger than E_g, which induces statistical fluctuation in N. The details are described in Chap. 3, Sect. 3.1.

The first quantum detector, which utilized gas ionization, dates back to 1896, when it was developed by Thomson and Rutherford [2] shortly after the discovery of X-rays by Röntgen in 1895. Note that Thomson also initiated chemical analysis using positive ions; it is an ancestor of mass spectrometry (MS) [3]. The gaseous detectors are known as ionization chambers or proportional counters. When an X-ray photon or an α-particle passes through an absorber gas, a photoelectron emitted by an X-ray absorption event (or electrons by the α-particle), which is ionization, loses the initial kinetic energy through inelastic collisions with gaseous atoms. A cascade of ionization occurs along the trajectory of the electron, creating electron–ion pairs. The electron–ion pairs in the gas are collected on a wire electrode to which a high voltage is applied. A current pulse is produced for each particle incidence event. Historically, the average energy consumed for creating one electron–ion pair was denoted by W for gaseous ionization detectors. The similar average energy consumed for creating one e–h pair in semiconductors is expressed by ε. The W value of air is 34 eV, which relates to the ionization potential that is equivalent to the energy gap in Fig. 2.1a. The gaseous detectors are often applied for detecting α-particles with a kinetic energy range of MeV or high energy cosmic rays. Due to the large W value, energy resolution of proportional counters is poor compared with modern quantum detectors. Nonetheless, the gaseous detectors, which are still widely used for monitoring nuclear radiation, can measure particle energy. An extreme case of the gaseous detectors is Geiger-Müller (GM) counters, which produce a constant pulse height due to avalanche ionization under a high voltage across two electrodes. The GM counters have no energy resolution, but they have a high sensitivity to nuclear radiation. Similarly, semiconductor avalanche photodiodes (APDs) are known as fast and highly sensitive photon detectors. Furthermore, a type of superconductor detectors, the superconductor strip particle detector (SSPD), operates in an avalanche mode driven by Joule self-heating; it is categorized as a thermal detector type in this book.

Modern quantum particle detectors have a solid absorber. They are called "solid state detectors." The first semiconductor p–n junction detector with Ge having an E_g value of 0.67 eV was invented in 1949 by McKay [4]. Semiconductor detector development has two directions: smaller-energy-gap materials for a higher energy resolution or for a wider coverage toward low energy photons such as infrared and far-infrared rays; and larger-energy-gap materials for room temperature operation to eliminate cryogenic cooling or thermal excitation. The small-gap materials need to be cooled to suppress thermal excitation. One example of the first direction is a compound semiconductor, HgCdTe, with a tunable bandgap between 0 and 1.6 eV. The bandgap varies with composition from HgTe to CdTe. The former has no gap due to the overlap between the conduction and valance bands, whereas the latter has an E_g value of 1.6 eV. HgCdTe with a small ε value is sensitive to infrared photons. However, imperfect crystallinity mainly originating from multi-elements precludes precise measurement of photon energy due to carrier trapping in defects. For the second direction, materials for room temperature operation include wide-bandgap semiconductors such as SiC, ZnO, and GaN with E_g values above 3 eV. Furthermore, by adding Zn into CdTe, $Cd_{1-x}Zn_xTe$ (CZT) with $x = 0.1$–0.3 has a bandgap range of 1.60–1.66 eV. They are promising for detecting X-rays and γ-rays at room temperature without cooling, unlike high-resolution Si detectors with E_g = 1.1 eV. Note that the latest Si drift detectors (SDDs) operate at a moderate energy resolution at room temperature. For further higher temperature operation, insulators can be utilized. Diamond has an E_g value of 5.47 eV, which is still sufficient to produce ~ 200 e–h pairs for a-few-keV X-rays.

The development of superconductor quantum detectors follows the first direction using an extremely small gap in a meV range: superconductors have gaps that are ~ 1000 times smaller than those of semiconductors. Another striking advantage of superconductors is that no single crystal is required because superconductivity is a macroscopic phenomenon that is insensitive to local defects. For superconductors maintaining a superconducting state, the carriers are called quasiparticles that exhibit both electron and hole characters. We presume that they are either quasi-electrons or quasiholes to explain detector operation statistically. The quasiparticles directly produce an electric output pulse, similar to e–h pairs in semiconductors.

2.3 Introduction to Thermal Detection

In the thermal detection, a thermometer measures a temperature rise due to radiation or particle incidence. Figure 2.1c shows the thermal detection with a resistive thermometer. The flux n (s^{-1} m^{-2}) of particles with an energy of E_0 is equivalent to the power per unit area, $P = nE_0$ (W/m^2). Within a linear R–T region, R is proportional to P as shown in Fig. 2.1d. Cryogenic thermal detectors use a non-linear thermistor, which has a large negative temperature coefficient. The large coefficient is indispensable for measuring single-particle energy. The typical cryogenic thermistor is the NTD Ge shown in Fig. 2.1e. The temperature rise due to radiation from a base temperature is mainly determined by the heat capacity C of the absorber. The relaxation or recovery time to the base temperature depends on a heat dissipation into surrounding. Thermodynamical energy fluctuation in the absorber determines the energy resolution limit of the thermal detectors for single particle detection. The standard deviation of the energy fluctuation is proportional to $T \times C^{1/2}$. This means that energy resolution becomes better by lowering operating temperature. The precise discussion on the thermodynamical fluctuation appears in Chap. 4, Sect. 4.1.

The first thermal detector dates back to 1800 when infrared radiation in the sunlight was discovered with a mercury thermometer by Herschel [5]. The mercury served as both radiation absorber and thermometer using thermal expansion; modern thermal detectors normally consist of discrete two components of an absorber and a thermometer. The absorber with a heat capacity of C is connected to a heat bath with a thermal conductance G. The thermometer with a sensitivity of dR/dT is attached to the absorber. The mercury thermometer is replaced by a thermistor, thermopile, or semiconductor p–n junction. Instead of reading the mercury thermometer rise, temperature-dependence of resistance generates electric signals [6].

The development of superconductor thermal detectors benefits from cryogenic operation that leads to a reduced C value. In addition, by using a sharp superconducting transition, superconductor thermometers can have a dR/dT value that is significantly higher than that of thermistors. This combination yields an ultimate energy resolution. The details are explained in Chap. 4, Sect. 4.2.

When we adopt an appropriate design, calorimeters outperform quantum detectors regarding energy resolution. The entire kinetic energy of a particle is finally converted to the thermal energy, so that the energy resolution is determined by thermal fluctuation of a canonical ensemble, which is proportional to $TC^{0.5}$. In contrast, quantum detectors with E_g lose a part of E_0 because excitations in an energy range below E_g do not contribute to electronic carrier generation, which leads to fluctuation proportional to $(E_0\varepsilon)^{0.5}$. On the other hand, the electronic excitations in quantum detection are collected quickly or relax to a steady state faster than thermal recovery, which results in a high count-rate. Furthermore, the energy gap ensures a higher operating temperature and a higher tolerance to base temperature fluctuation.

The superconductor strip (SS) type or SSPD is classified into thermal detection in this book. The initial interaction with photons or nonzero-mass particles can be treated with Cooper pair breaking or vortex motion within a framework of quantum detection. However, output pulse generation is due to Joule self-heating with electrothermal feedback (ETF) deriving from shunt circuitry, as described in Chap. 4, Sect. 4.2.5. Superconductor strips are biased near superconducting-normal (S–N) transition by applying a current close to critical current I_c at a temperature well below T_c. This SSPD operation is not the same as superconductor films used for bolometers or calorimeters known as the transition edge sensor (TES) type that is biased in the middle of S–N transition. Nevertheless, the SS type that operates close to S–N transition can be categorized as the TES type. The basic principles of parallel shunt resistance and ETF are common to both SS and TES types; therefore, a seamless explanation of detector physics is possible.

References

1. IEC Electropedia "energy resolution": https://www.electropedia.org/iev/iev.nsf/display?openform&ievref=395-03-106
2. J.J. Thomson, E. Rutherford, Lond. Edinburgh Dublin Philos. Mag. J. Sci. **42**, 392 (1896)
3. J.J. Thomson, *Rays of Positive Electricity and Their Application to Chemical Analysis* (Green and Co., Longmans, 1913)
4. K.G. McKay, Phys. Rev. **76**, 1537 (1949)
5. W. Herschel, Philos. Trans. **90**, 284 (1800)
6. R.L. Richards, J. Appl. Phys. **76**, 1 (1994)

Chapter 3
Quantum Detection

Abstract Chapter 3 for quantum detection describes the electronic excitation in quantum detectors, the fundamental statistical physics on energy resolution, and the operation of practical devices such as STJD and MKID to readout the quasiparticle excitation. A historical overview highlights the works of pioneers. Similarity and dissimilarity between semiconductors and superconductors bring a deep understanding of superconductor quantum detectors. Unexpectedly, even in semiconductor detectors, there are unsolved issues on the average energy ε consumed for generating one electron–hole (e–h) pair and the Fano factor. Additionally, superconductor quantum detectors encounter a problem of spatial inhomogeneity of detector output with respect to particle absorption locations. Physics of superconductor quantum detectors is lacking in some fundamental data, which represents an important area for future research. Nevertheless, the superconductor quantum detectors achieve an energy resolution of 2–4 eV in a soft X-ray range below 1 keV and a high count rate of ~ 200 kcps, which enables high throughput imaging analysis of advanced materials.

Keywords Quantum detection · Energy gap · Semiconductor detector · Superconductor detector · Average energy · Fano factor · Electron–hole pair · Quasiparticle · Phonon · Josephson vortex · Proximity effect · Spatial inhomogeneity

The original version of the chapter has been revised. A correction to this chapter can be found at https://doi.org/10.1007/978-3-031-84193-4_7

M. Ohkubo, *Superconductor Detectors*, https://doi.org/10.1007/978-3-031-84193-4_3

3.1 Energy Resolution of Quantum Detectors

3.1.1 Carrier Number Fluctuation

It is imperative for us to begin with semiconductor detectors, partly because our claim is that superconductor detectors outperform semiconductor detectors. Furthermore, reconsidering and even modifying the established fundamental bases are crucial to understanding and renewing the detector operation of both semiconductor and superconductor detectors comprehensively. Section 3.1 primarily describes semiconductors with references to superconductors, followed by Sect. 3.2, which provides the details of superconductor quantum detectors.

Particle impact on semiconductors creates numerous e–h pairs by exciting electrons in the valence band to the conduction band beyond the bandgap E_g, leaving holes behind in the valence band, as shown in Fig. 3.1. The electrons and holes are collected by a high electric field applied to the depletion layer with a very high resistance of R_d, which is practically infinity. Under the high electric field, bound states called excitons are not formed so that all the electrons and holes are collected separately. In a condition of $R_d \gg R_b \gg R_{out}$ and R_L, a current flow from the bias circuit to the semiconductor detector is negligible. Because the electrons and the holes circulate the closed circuit of the detector and the charge-sensitive preamplifier, the collected charge Q_0 is not equal to the sum of the number of electrons and the number of holes but the number of e–h pairs, keeping charge neutrality. This is also important for superconductor quantum detectors. When STJDs are also read out by a similar charge-sensitive preamplifier, charge carrier integration should end up a similar result to e–h pairs in semiconductors to keep charge neutrality. Therefore, Q_0 in STJDs equals the number of quasiparticle pairs.

Output current pulses typically have a decay time of ~ 10 ns in semiconductor detectors. The total charge Q_0 ($= \int I(t)\mathrm{d}t$) accumulates in the feedback capacitor C_f, converting a current pulse to a voltage step of $U_{out} = -Q_0/C_f$ by using the inverting input of the operational amplifier. The readout circuit in Fig. 3.1 forms a charge-sensitive preamplifier with resistive feedback. The output voltage decays at a time constant of $C_f R_f$; for example 200 μs, to accept subsequent events. The time constant is extremely longer than the current pulse decay; therefore, the output voltage can be virtually treated as a step function, the height of which is proportional to E_0. Technically, it is also possible to skip R_f and add a periodic reset circuit, which is called the gated integrator. Plotting the number of photons within a small bin as a function of U_{out} constructs a pulse height spectrum.

After subtracting electronic noise due to the charge-sensitive preamplifier and subsequent readout circuits, the fluctuation of detector output for monochromatic E_0 particles represents the uncertainty around the average number $\overline{N}$ of e–h pairs. We consider a case of X-ray detection. The details of X-ray absorption and subsequent processes are explained in Appendix B. Briefly, X-ray absorption by an inner electron shell of an atom leads to cleavage of a chemical bond in covalent crystals such as Si and Ge and results in ionization of the atom, emitting a secondary electron. The

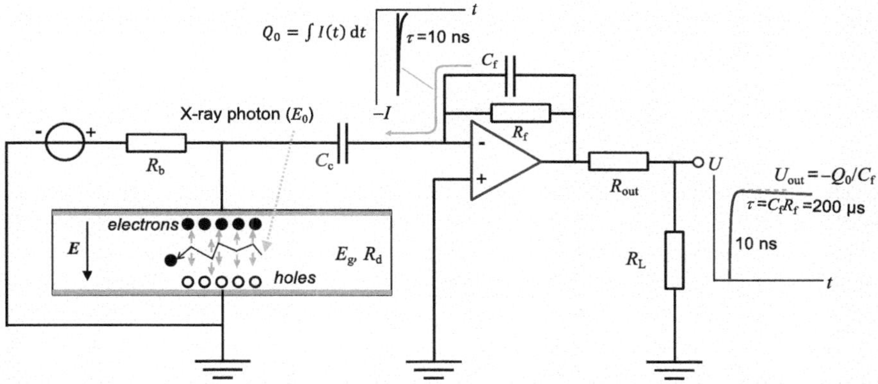

Fig. 3.1 Schematic illustration of combination of semiconductor quantum detector and charge sensitive preamplifier. R_b is the bias resistance, E_g is the bandgap, R_d is the resistance of the depletion layer in a reverse bias condition, C_c is the coupling capacitance, C_f is the feedback capacitance, R_f is the feedback resistance, R_{out} is the output resistance, and R_L is the input resistance of a pulse shaping circuit or an analog–digital converter (ADC). At liquid nitrogen temperature, R_d can be regarded as infinity in Si. Normally, a condition of $R_d \gg R_b \gg R_{out}$ or R_L holds. Under a high electric field E, all the electrons and holes are separately collected without forming excitons. The C_f value is adjusted to keep output voltage within 10 V in general at an expected Q_0 value and a photon count rate. The typical values are 100 MΩ for R_b, 2 pF for C_f, and 100 MΩ for R_f; the feedback circuit with R_f restores the output voltage to zero at a time constant of 200 μs. R_f can be replaced by a periodic reset circuit, which is called the gated integrator. A similar charge-sensitive amplifier is utilized for STJDs

energies of the secondary electron are in a range of keV region. The secondary electron is either photoelectron from the shell absorbing the X-ray photon or Auger electron from an upper electron shell. In addition, the hole in an inner shell causes shake-off electrons [1, 2]. The energetic electron ionizes another atom or collides with other electrons, encountering energy down-conversion. The inelastic collisions create a number of e–h pairs before the electron stops in a semiconductor crystal. When no escape of secondary electrons occurs from a detector surface, the semiconductor absorbs the total energy of the particle E_0.

It is arguable whether an exact probability distribution function of the number N of e–h pairs is analytically obtainable; we cannot trace physical processes inside the detector material accurately. However, one plausible statistical model for the e–h pair generation is based on binomial probability distribution expressed by

$$P(x) = \frac{n!}{(n-x)!x!} p^x (1-p)^{n-x}, \tag{3.1}$$

where n is the number of the trials to create e–h pairs, each trial has a success probability p, and x is the number of the success times. The x value is equivalent to the number N of e–h pairs. Other important distribution functions such as Poisson and Gauss for radiation detectors appear in Appendix A.1. The energy resolution associated with the fluctuation of N under the binomial distribution is given by

$$\Delta E = 2.355\sqrt{\left(\frac{1}{p} - 1\right)E_0 E_g}. \tag{3.2}$$

The mathematical derivation of Eq. (3.2) in FWHM is explained in Appendix A.2. Note that Eq. (3.2) is different from the formula including the Fano factor F in most books. When we adopt the binomial distribution or other analytical functions for the carrier generation, the Fano factor disappears from the ΔE formula. Equation (3.2) expresses the theoretical limit when charge carrier generation follows the binomial function. The Fano factor F is defined as the variance ratio of an exact distribution to a Poisson distribution [3], both of which have the same average $\overline{N}$ $(= pn)$. In the binomial distribution, F is given by

$$F = 1 - p = 1 - E_g/\varepsilon. \tag{3.3}$$

A general formula of F that is valid for any probability distribution functions appears in Eq. (A.7) in Appendix A.2. Substituting Eq. (3.3) for Eq. (3.2), it is transformed to the well-known formula of

$$\Delta E = 2.355\sqrt{FE_0\varepsilon}. \tag{3.4}$$

For detectors obeying the Poisson statistics, $F = 1.0$ and $\Delta E = 2.355\sqrt{E_0\varepsilon}$. Detectors with high energy resolution have a F value considerably less than 1.0. Equation (3.4) is often referred to as the statistical and theoretical limit of the energy resolution of quantum detectors. However, it is difficult to identify the exact probability distribution function, so that the F value is theoretically indeterminable. Experimentally, F is slightly smaller than $1 - E_g/\varepsilon$ in modern Si detectors. This means that the exact distribution function is narrower than the binomial distribution. In contrast to the correlation in Eq. (3.3), F and ε values are often determined independently from experimental results or Monte Carlo simulations. This signifies that the analytical treatment of quantum detection is incomplete even at present.

The F and ε values are the primary parameters of interest in single-particle quantum detectors. It is worth surveying the current situation of these two parameters in semiconductor detectors, occasionally comparing with superconductor detectors. There are a vast number of experimental and theoretical papers on semiconductor detectors. It is believed that F and ε of semiconductors have been well established experimentally and theoretically. However, we will realize soon in the following two sections that it is still controversial even in Si. In experiments, full-energy absorption peak positions are insensitive to peak-broadening factors such as electronic noise and photoelectron escape tailing, so that the variation of the ε values obtained from the peak positions is small. On the other hand, the peak-broadening factors directly affect F. Therefore, in general, the variation of experimental F values is larger than that of ε in the literature. Monte Carlo simulation results are similar to experiments.

Furthermore, disagreement between experiments and calculations on semiconductor detectors is large in an energy range below ~ 1 keV.

3.1.2 Fano Factor

Radiation phenomena are often treated with the above-mentioned Poisson statistics. A typical one is radioactive decay. The probability of observing x radioactive decay events within a certain time interval is expressed by the following equation:

$$P(x) = \frac{(\overline{x})^2 e^{-\overline{x}}}{x!}, \tag{3.5}$$

where $\overline{x}$ is the average number of the decay events in an entire observation time that must be significantly shorter than the half-life of the radioisotope. The $\overline{x}$ value is equal to np appearing in Eq. (3.1). This Poisson distribution is the limit of the binomial distribution when n approaches infinity while keeping a constant np, which means $n \to \infty$ and $p \to 0$. It is called the Poisson limit theorem. The variance of the Poisson distribution is equal to the average $\overline{x}$.

The Fano factor was originally devised to treat a difference between experimental ΔE values and the prediction of $\Delta E = 2.355\sqrt{E_0 W}$ based on the Poisson distribution in gaseous ionization detectors [3]. The experimental ΔE values of the gaseous detectors were substantially smaller or better than that of the Poisson distribution. If an exact analytical probability distribution function for carrier generation is obtainable, we need not use F like Eq. (3.2). The Fano factor is the variance ratio of the exact probability distribution function to the corresponding Poisson distribution with the same $\overline{x}$ or $\overline{N}$. However, when it is unknown, Eq. (3.4) is convenient with adjustable parameter F, which indicates the deviation degree of the unknown distribution from the Poisson distribution.

The modern quantum detectors have distribution functions that are far narrower than the Poisson statistics and slightly narrower than the binomial statistics. Therefore, instead of analytical approaches, the Fano factor can be empirically defined by replacing σ of an unknown distribution function in Eq. (A.6) with the experimental standard deviation of a full-energy absorption peak [4]. Monte Carlo simulations resemble experimental results. To be more practical, firstly, the ε value can be obtained by dividing E_0 by the number of detected charge carriers. Secondly, empirical F is obtained from experimental ε values and experimental ΔE values by using Eq. (3.4), providing that the peak shape is Gaussian. In this empirical definition, any combinations of F and ε are possible without any constraints. This is a convenient way because any full-absorption peaks in pulse height spectra can be explained by Eq. (3.4). However, this has no physical and statistical background. The correlation between F and ε should exist according to the definitions of F and ε.

The experimental F values for high-quality semiconductors are significantly smaller than 1.0; for example, an early-stage value of 0.059 [5] and recent values of

~ 0.12 in Si [6]. These extremely small F values are explained by many theoretical models on e–h pair generation statistics; for example, statistical partitioning models including “crazy carpentry model” [7] in 1965, and “phenomenological models of electron cascades” [8] in 2006, and “bathtub-whiskey-glass model” [9] in 2008. A low energy region, in which only a few charge carriers are created, is treated with the Conway-Maxwell-Poisson distribution, which is a member of the weighted Poisson distribution family covering both cases of F 1 and F 1 in 2018 [10]. It provides freedom on a $\overline{N}-F$ parameter space. The basic statistical physics for quantum detection even in semiconductors is still controversial ever since the Fano’s famous paper published 78 years ago in 1947 [3]. Although the recent variation of F values in experiments and calculations is smaller than those observed in the early years of research, it is concluded that the Fano factor remains far less well-established even in Si [6].

3.1.3 *Relation Between ε and E_g*

3.1.3.1 Comparison Between Experiments and Theories

The experimental ε values in Si and Ge detectors became smaller in the 1960s as the quality of crystals was improved [11]. The high crystalline quality of modern Si detectors practically eliminates both trapping loss and e–h recombination almost completely. There are countless reports on the ε value measurement in Si. Here, we list typical values: $\varepsilon = 3.650(9)$ eV with $F = 0.128(1)$ in a commercial charge coupled device (CCD) camera at 185 K with a readout noise of 5e for X-rays of 5895 eV that is a weighted average of the Mn K-L_3 and K-L_2 lines from ^{55}Fe X-ray sources [12], and $\varepsilon = 3.752(2)$ eV with $F = 0.1190(25)$ in a Skipper-CCD at 123 K with a sub-electron noise of 0.2e achieved by non-destructive multiple sampling for 5895-eV X-rays [13]. The numbers in parentheses denote the standard deviation values. These two data are based on the extraordinary low readout-noise; however, the differences of the ε and F values are significant. The bandgap E_g varies with temperature: 1.15 eV at 185 K and 1.16 eV at 123 K according to the Varshni’s empirical equation [14] or the O’Donnell’s semi-theoretical one [15]. The reported difference in the ε values is larger than that expected from the temperature dependence of E_g. This suggests that experimental fluctuation is as large as 3%, presumably depending on materials, experimental setups, or analysis protocols. A recent experimental trend is that ε values fluctuate around 3.75 eV at ~ 100 K ($E_g = 1.16$ eV) and 3.66 eV at 300 K ($E_g = 1.12$ eV) for different radiation sources in an energy range above 1 keV [6]. These values almost agree with Monte Carlo simulations; for example, $\varepsilon = 3.645$ eV and $F = 0.117$ at 300 K [16]. Although the experimental uncertainty in ε and F for Si is as large as 3–7%, it is smaller than other semiconductor materials. Consequently, Si detectors are frequently utilized as a reference to determine F and ε values of other materials such as compound semiconductors. However, it is important to recognize that Si, as the reference material, possesses this level of uncertainty.

For covering a wide E_g range up to ~ 10 eV, we need to survey compound semiconductors and insulators in addition to Si and Ge. Imperfect carrier collection is common in compound detectors; therefore, correction in carrier mobility, carrier lifetime, and electric field strength is necessary for precisely estimating ε and F values [17]. Although we should pay attention to the correction in the literature, the experimental data on a variety of semiconductor materials undoubtedly indicate that the relation between ε and E_g is not a simple relation like $\varepsilon = E_g/p$. Furthermore, the reported $\varepsilon - E_g$ relations are largely scattered as described below.

A well-known empirical formula obtained in 1960s–1980s from the data on Si, Ge, InSb, GaP, SiC, C, etc. for α-particles between 5 and 30 MeV is

$$\varepsilon(\text{eV}) = aE_g + b = 2.73E_g + 0.55.[18\text{–}21] \tag{3.6}$$

This relation was seemingly explained by both analytical treatment and Monte Carlo simulation by Klein [18] and Alig et al. [19–21]. The explanation has been the basis of semiconductor detector physics for a long time. The green straight line tagged by "Alig$^+$" in Fig. 3.2 plots Eq. (3.6); other lines will be discussed later. It is often stated that Eq. (3.6) is independent of particle types and energy. In fact, an experiment on α-particles in a range between 5.344 and 8.785 MeV resulted in $\varepsilon =$ 3.70 eV for Si at 290 K ($E_g = 1.13$ eV) [22]; this ε value is not largely different from an experimental value of 3.752 eV at 123 K ($E_g = 1.16$ eV) for the 5895-eV X-rays [13]. Furthermore, it was reported that the difference in ε values was within 1% for a variety of particle types: electrons (114.86 keV–1.0481 MeV), γ-rays (121.97 keV), and α-particles (5.344–8.785 MeV) [22]. These results are reasonable because the incidence of all the particle types end up with the energy loss processes of hot electrons and holes after ionization. However, strictly speaking, the a and b factors should be dependent on particle types, particle energies, and detector materials, considering that initial interactions of particles with absorber elements are diverse. Intriguingly, the experimental ε data on Si was explained by many combinations of a and b values, which is shown by the different branches in Fig. 3.2.

Equation (3.6) with $a = 2.73$ is consistent with the classical Klein model of $\varepsilon = 2.8E_g + b$. The origin of $a = 2.8$ is attributed to an average kinetic energy of electrons and holes below a threshold energy E_{th}, as will be discussed in the subsequent sections. The b factor attributed to optical phonon emission from electrons and holes above E_{th} [18], although no exact b value is predicted. Equation (3.6) is often used to estimate ε and then F, as in Ref. [23]. However, the combination of $a = 2.73$ and $b = 0.55$ in Eq. (3.6) is now open to debate both experimentally and theoretically.

Even during the early stages, the branch deviated from Eq. (3.6) was reported for HgI_2, AgBr, and AgCl, which is depicted by the green line tagged with "Alig$^-$" ($\varepsilon = 2.73E_g - 1.5$) in Fig. 3.2 [19]. The similar branch was also observed for other materials such as PbI_2, TlBr, SiC, C, and AlN [24]. This "Alig$^-$ " branch is physically unrealistic due to the negative b value, which challenges the validity of $b = 0.55$ eV in Eq. (3.6). In addition, it is surprising that the branch line for the compound semiconductors is lower than that for the pure Ge and Si, because the crystalline quality of the compound semiconductors is significantly lower than that of

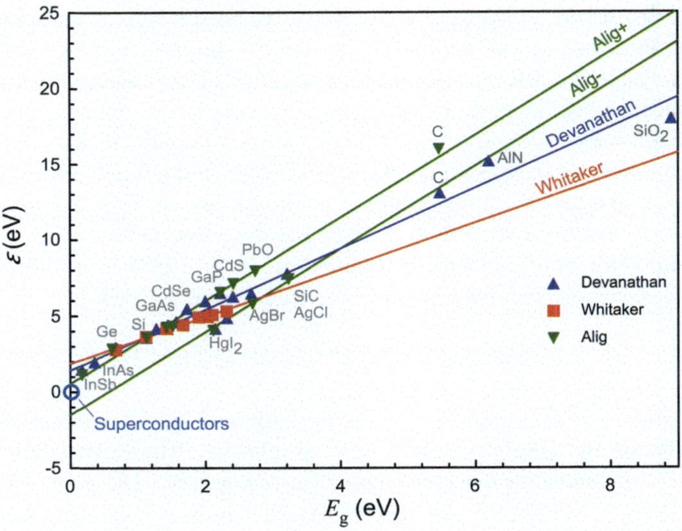

Fig. 3.2 Empirical relations between ε and E_g in semiconductors and insulators. The lines express the different branches tagged by Alig$^+$ ($\varepsilon = 2.73E_g + 0.55$) [19], Alig$^-$ ($\varepsilon = 2.73E_g - 1.5$) [19], Devanathan ($\varepsilon = 2.0E_g + 1.43$) [27], and Whitaker ($\varepsilon = 1.54E_g + 1.89$) [25]. The Alig$^+$ branch is consistent with the Klein relation [18] that is based on a free particle approximation with the momentum conservation law and the energy conservation law. The most recent Whitaker branch has a reasonably-small proportional coefficient of $a = 1.44$ and a large residual constant of $b = 1.89$ eV. Superconducting energy gaps and ε values are located on the origin at this scale

the single-element semiconductors. Poor crystalline quality leads to imperfect charge collection; therefore, the experimental ε values for the compound semiconductors should be located higher than the line for Si and Ge.

Currently, the experimental $\varepsilon - E_g$ relation is a hot topic [25, 26]. A variety of experimental branches with different a and b values in addition to the two branches have been reported recurrently. The branches intersect at Si because Si is a reference material. Focusing on recent experimental data, two relations were proposed. When new data are added to some of the green "Alig$^+$" data [19], the relation is altered to

$$\varepsilon = 2.0\,E_g + 1.43, \text{[27]} \tag{3.7}$$

which is indicated by the dark blue branch tagged by "Devanathan." Another branch called the Bertuccio–Maiocchi–Barnett (BMB) relationship [28] was reported by adding the data of $Al_{0.6}Ga_{0.4}As$ to those of Ge, Si, GaAs, AlGaAs, InGaP, and AlInP:

$$\varepsilon = 1.54\,E_g + 1.89, \text{[26]} \tag{3.8}$$

which is the "Whitaker" red branch with the smallest a and the largest b values. The recent experimental trend is that the proportional coefficient a is decreasing, while the residual constant b is increasing, keeping the same value for Si. The relation of Eq. (3.8) solves the problem that the Ge data point considerably deviates upward from the "Alig$^+$" branch.

The b value of 0.55 eV in Eq. (3.6) is the smallest among the different branches. For 5895-eV X-rays from a ^{55}Fe source, the number $\overline{N}$ of the e–h pairs created in Si is approximately 1600; average energy transferred to the phonons $b \times \overline{N}$ is 880 eV. The phonon escape or phonon loss occupies 15% of the absorbed energy E_0; it is the lowest loss among the realistic three branches. Most Monte Carlo simulations did not report the a and b values separately but resultant ε values only. One exception is Ref. [6], which yields $a = 1.70$ and $b = 1.74$ for considering optical phonon emission at 63 meV. These values are close to Eq. (3.8) rather than the Alig$^+$ relation in Eq. (3.6), although both relations predict $\varepsilon = \sim 3.6$ eV for Si. The validity of the latest relation is discussed in Sect. 3.1.4 in this Chapter.

3.1.3.2 Mobile Carrier Generation Model for Semiconductors

The details of initial cascade processes of photoelectrons, Auger electrons, characteristic X-rays and shake-off electrons are described in Ref. [29]. These initial high-energy processes are occasionally neglected in calculations [20]. Additionally, in an excitation energy range above a few 10 eV, one of the electron energy loss processes is plasmon emission at a few 10 eV. Plasmons rapidly decay and generate e–h pairs because plasmon energies are higher than E_g [6, 20]; they efficiently contribute to generating e–h pairs. Therefore, it is often reported that ε is insensitive to the plasmon excitation.

The models for the carrier generation related to a and the phonon loss related to b were devised and diagramed by Shockley [30], Klein [18], and Alig [20] to explain the experimental data in Fig. 3.2. Figure 3.3a shows a modified diagram for non-doped insulating Si. We omit the initial ionization cascade processes and the plasmon excitation, focusing on the processes of hot electrons and holes only. The electrons and holes are treated as equal free particles with the same mass and the same physical property [20, 21]. Such electrons and holes are called free carriers collectively. Figure 3.3b for Nb in a superconducting state is shown here to visually reveal the difference between semiconductors and superconductors according to full-analytical treatment [18, 20, 31]; the details are described in Sect. 3.2.3.2. In this section, we focus on Si to better understand superconductors.

Figure 3.3a for Si illustrates competitive processes of e–h pair generation and longitudinal optical (LO) phonon emission in the course of relaxation processes below 6.3 eV after initial ionization cascades induced by X-ray photon absorption. The kinetic energy E and the threshold energy E'_{th} are relative to the bottom of the conduction band (band edge) for the sake of simple mathematical operation for semiconductors, while the energy E_k is relative to E_F. Theoretical studies often assume single energy values for phonon emission: the LO phonons at 63 meV (Ω_{LO})

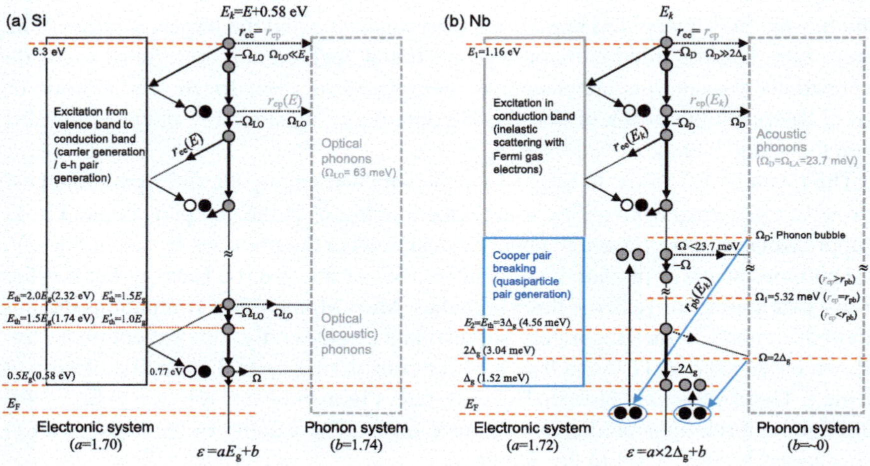

Fig. 3.3 Comparison in conceptually simplified models for free carrier generation between semiconductors and superconductors : (**a**) e–h pairs in Si with a bandgap of 1.16 eV [18, 20] and (**b**) Bogoliubov quasiparticle pairs in Nb with a Δ_g value of 1.52 meV [31]. The closed and open circles denote electrons and holes, which are collectively called free carriers both in (**a**), (**b**). The gray circles symbolize either electron or hole at all energy levels in (**a**), whereas they are either electron or hole above the Debye energy Ω_D and Bogoliubov quasiparticles below Ω_D in (**b**). The horizontal red dashed lines indicate the major threshold energy levels (see the text in detail), following the Klein-Alig model [18, 20] for Si and the Kozorezov model for Nb [31]. The dotted red line at 1.74 eV in (**a**) is a new suggestion to explain the large difference between the theory and the recent experimental and Monte Carlo data (see the text at the end of Sect. 3.1.4). The energy E_k and E_{th} in (**a**), (**b**) are relative to E_F, while the energy E and E'_{th} in (**a**) are conventionally relative to the bottom of the conduction band (the band edge) for the sake of simple mathematical expression for semiconductors. The range of the vertical energy axis is 6.3 eV in (**a**), at which the e–h pair generation rate r_{ee} (E) equals to the phonon emission rate r_{ep} (E). In (**b**), at 1.16 eV ($E_1 = 49 \times \Omega_D$), the electron–electron scattering rate r_{ee} (E_k) equals to the phonon emission rate r_{ep} (E_k). The arrow lengths schematically represent the energy dependence of the scattering rates except the blue arrows for Cooper pair breaking in (**b**). The Fermi level E_F in Si is located at the middle of the bandgap, whereas E_F in Nb is at 5.32 eV relative to the bottom of the conduction band. It is postulated that in (**b**) E_0 is transformed to the total energy of the Ω_D phonons in Nb, which is called the phonon bubble that breaks Cooper pairs [31]. The a and b values in the $\varepsilon - E_g$ or $\varepsilon - \Delta_g$ relation at the bottom of (**a**), (**b**) originate from the Monte Carlo simulations for Si [6] and Nb [63]. It is presumed that the recombination between electrons and holes in Si or between quasielectrons and quasiholes in Nb is negligible. The explanation of Nb in (**b**) appears in Sect. 3.2.3.2

in Si and the longitudinal acoustic (LA) phonons at 23.7 meV ($\Omega_D \simeq \Omega_{LA}$) in Nb. The rate ratio r_{ep} $(E)/r_{ee}$ (E) of phonon emission (dotted arrows) to carrier generation (solid arrows) for semiconductors is given by

$$\frac{r_{ep}(E)}{r_{ee}(E)} = A\frac{105(E - \hbar\omega_0)^{\frac{1}{2}}}{2\pi\left(E - E_g\right)^{\frac{7}{2}}}, \tag{3.9}$$

where $\hbar\omega_0$ is the phonon energy that is often fixed at 63 meV for the LO phonon in Si, and A is the adjustable parameter, which is usually assumed to be 5.2 eV^2 [19, 20]. The exact phonon DOS profile of Si has a wide spectrum as shown in Fig. 3.4a [32]; nevertheless, the postulation of the phonon emission at 63 meV is modestly acceptable because electron–phonon interaction strength increases with phonon energy. Meanwhile, recent studies indicate that the contribution of acoustic phonons is nonnegligible, which will be discussed in Sect. 3.1.3.4.

In Fig. 3.3a, according to Eq. (3.9), an electron at 10 eV has a rate ratio r_{ep}/r_{ee} of 0.13 in Si, indicating that phonon emission is significantly smaller than the e–h pair generation but meaningfully nonnegligible. The r_{ep} $(E)/r_{ee}$ (E) rapidly increases as E decreases, approaching E_g. Therefore, the b factor discussed below is governed by the phonon emission loss in an energy range below ~ 10 eV. The first threshold energy level is 6.3 eV, at which the rate of e–h pair generation becomes equal to the rate of phonon emission ($r_{ep} = r_{ee}$). The left-pointing solid arrows denote the carrier generation process creating the e–h pairs symbolized by a pair of closed (electron) and open(hole) circles. The right-pointing dotted arrows denote the phonon emission, assuming the LO phonons at 63 meV as often adopted in theoretical treatments. The lengths of the arrows in Fig. 3.3a roughly represent Eq. (3.9); r_{ee} decreases as the energy of carriers E decreases below 6.3 eV, while r_{ep} is rather constant. The gray circles symbolize either electron or hole at all energy levels. The dual transition

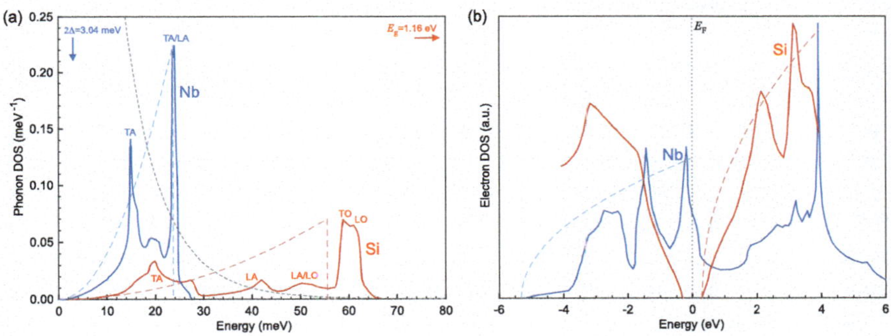

Fig. 3.4 Comparison in DOS profiles between Si (red) and Nb (blue). In (**a**) for phonon DOS, the solid red and blue curves depict the DOS profiles calculated with DFTP at 100 K for Si and measured at 293 K for Nb, respectively [32, 38]. The dashed blue and red curves are based on the Debye approximation with Ω_D values of 55.6 meV for Si and 23.7 meV for Nb. The short-dashed black curve depicts the Bose–Einstein distribution in thermal equilibrium at 100 K in arbitrary units. Theoretical calculations for carrier generation often adopt the fixed phonon energies at 63 meV for Si and 24 meV for Nb, omitting the dispersion. In (**b**) for electron DOS, the energy is relative to E_F that is situated at the middle of the gap in Si and at 5.32 eV from the bottom of the conduction band in Nb. The solid red and blue curves depict the electron DOS profiles calculated with density functional perturbation theory (DFPT) for Si and calculated with density functional theory (DFT) for Nb [40, 41]. The blue and red dashed lines for electron DOS profiles in (**b**) are based on the free particle approximation, which yeilds the $\sqrt{E}$ dependence in Eq. (3.10). The regions below and above E_F correspond to the valence band and the conduction band of Si, respectively. The red profile below E_F for Si is not a hole DOS shown in Fig. 3.8 but the electron DOS

arrows of the primary gray circle located on the 6.3 eV level show the possibilities of both carrier generation toward the left side and the phonon emission toward the right side. The same dual arrows are applied to the gray circle on 2.32 eV. Other gray circles have either carrier generation or phonon emission for the sake of simplicity. After carrier generation, the three carriers, which are an e–h pair and the scattered primary carrier, are expediently located on the same energy level, although exact energy levels depend on collision condition. When a primary carrier emits one LO phonon, the energy of the carrier decreases by 63 meV as symbolized by $-\Omega_{LO}$. Cascade processes occurs such that the excited particles induce either carrier generation or phonon emission until the energy of the primary carrier decreases below the threshold energy $E_{th} = 2.0E_g$ (2.32 eV) relative to E_F, which corresponds to E'_{th}=1.5E_g relative to the band edge [21]. We assume that particles below E_{th} cannot generate e–h pairs. The carriers below E_{th} thermalize with the lattice, emitting phonons. Based on the Klein-Alig model [18, 20], the relaxation below E_{th} determines the a factor of the ε–E_g relation, while the phonon emission from carriers above E_{th} leads to energy loss that determines the b factor. Any phonons in Si generate no e–h pairs in $\Omega_D \ll E_g$. Note that the dotted level at $E_{th} = 1.5E_g$ (1.74 eV) is a suggestion to explain the difference between the average excitation energy model (Klein-Alig model) and recent experimental and theoretical values at the end of Sect. 3.1.4.

3.1.3.3 Proportional Factor "a" in ε–E_g Relation

The second threshold energy level in Si is located at 2.32 eV, at which a carrier can generate the final e–h pair. The physical basis for the experimental a value of 2.73 in Fig. 3.2 is analytically explained by an average energy of excited carriers [18, 19]. It is assumed that an electron with E_{th} (E'_{th}) has an ability to create one e–h pair, but not another one. When all carriers relax below E_{th}, which occurs within a few 10 fs, no more e–h pairs are generated. Therefore, the number of the e–h pairs and the fluctuation at this stage govern ε and F. The carriers below E_{th} thermalize with lattice temperature by emitting phonons and finally reach a stable distribution profile. For phonon–phonon interaction, the decay of optical phonons to acoustic phonons has a time scale of 1 ps [33]. The decay of short-wavelength acoustic photons to long-wavelength phonons occurs on a time scale of several ns [34], which is a typical operation duration of semiconductor detectors after absorbing an X-ray photon.

We assume that the density of the states (DOS) for electrons and holes is proportional to $\sqrt{E}$ within the free particle approximation in semiconductors and furthermore all the available states below E_{th} are equally occupied. The sum of the average kinetic energies of the electrons and the holes occupying the states between 0 and E'_{th} relative to the band edge is calculated by

$$E_{\text{e-h}} = \frac{2\int_0^{E_{th}'} E\sqrt{E}\mathrm{d}E}{\int_0^{E_{th}'} \sqrt{E}\mathrm{d}E}. \tag{3.10}$$

Simple mathematics results in $E_{\text{e-h}} = 6/5\ E'_{\text{th}}$, so that the average energy necessary to generate one e–h pair is given by $\varepsilon = E_{\text{g}} + 6/5\ E'_{\text{th}}$. A primary hot carrier creates e–h pairs along the trajectory while losing a small part of its kinetic energy whenever collision occurs. The relation between E'_{th} and E_{g} is obtained by assuming the free-particle approximation [18, 19]. The momentum of the primary carrier after the collision must be always equal to or larger than those of the generated electron and hole. We consider the case that the kinetic energy of the primary carrier is reduced to E'_{th} and the final e–h pair is generated. The momentum conservation law in the free-particle collinear collision predicts that the three product particles should have the same momentum of $P/3$, where P is the momentum of the primary carrier before the final collision. Therefore, the kinetic energy of the final three particles is $E'_{\text{th}}/9$ each. Then, the energy conservation law before and after the collision is expressed by $E'_{\text{th}} = 3\left(E'_{\text{th}}/9\right) + E_{\text{g}}$. As a result, we obtain the relation of $E'_{\text{th}} = 1.5E_{\text{g}}$ (1.74 eV relative to the band edge in Si or 2.32 eV from E_{F}), which is depicted by the dashed red line above the dotted red line in Fig. 3.3a. Note that the level of 1.74 eV from the band edge is different from the dotted red line marked "$E_{\text{th}} = 1.5\ E_{\text{g}}$ (1.74 eV)." This relation yields $\varepsilon = 2.8E_{\text{g}}$. We call 2.8 the Klein factor. The three product carriers lay on the level of $0.67E_{\text{g}}$ (= 0.77 eV) relative to E_{F} after the final collision, which is slightly higher than the band edge. The relation $E'_{\text{th}} = 1.5E_{\text{g}}$ derived in the 1960s [18, 35] is frequently used for Monte Carlo simulations to halt further tracking of collision cascades even these days [2, 36]. The Klein factor of 2.8 is close to 2.73 that is one of the experimental results in Fig. 3.2, neglecting the residual constant b. A similar analytical approach is applied to superconductors later in Sect. 3.2.3.3. The level of $E_{\text{th}} = 2.0E_{\text{g}}$ will be modified to $E_{\text{th}} = 1.5E_{\text{g}}$ such that the final states lay on 0.58 eV on the band edge. This new threshold level can better explain experimental a values as discussed in Sect. 3.1.4.

3.1.3.4 Residual Constant "*b*" in ε–E_{g} Relation

The residual constant b for semiconductors within the Klein model is attributed to the energy loss through the optical phonon emission of the carriers above E_{th}. Early papers often stated that the dominant phonon branch is longitudinal-optic (LO) [18, 37]. Before discussing the b factor, it is worth explaining phonon branches. Semiconductors such as Si and Ge have primitive cells with two atoms, which leads to the separation into three acoustic branches and three optical branches: one LA, two transverse acoustic (TA1 and TA2), one LO, and two transverse optical (TO1 and TO2) branches. For the transverse modes, the notations of TA or TO are usually sufficient, omitting 1 and 2. The acoustic and optical branches have completely different dispersion curves, depending on directions and magnitudes of a wave vector $\boldsymbol{K}$. In Si, the LO/TO phonon branches form phonon states near ~ 60 meV, whereas the acoustic phonons have a broad distribution between 0 and ~ 50 meV. It should be noted that no optical phonon branches exist in simple metals such as bcc Nb and fcc Al, which have primitive cells with one atom. Therefore, superconductivity in simple metals stems exclusively from interaction between electrons and acoustic phonons.

Figure 3.4 exhibits differences in phonon DOS between Si and Nb in (a) together with electron DOS in (b). Figure 3.4a shows the phonon DOS profile calculated by density functional perturbation theory (DFPT) for Si [32] at 100 K and the phonon DOS measured with neutron scattering for Nb [38] at 293 K with the notations of major contributions from the different phonon branches. Experimental data are in good agreement with calculations: for example, a recent density functional theory (DFT) calculation for Nb [39]. The dashed lines depict the Debye approximation curves with Debye energies (Ω_D) of 55.6 meV for Si and 23.7 meV for Nb. The E_g value of 1.16 eV at 100 K with the red arrow is beyond the *x*-axis scale, which indicates that phonons in Si cannot create e–h pairs. On the other hand, the vertical blue arrow at a $2\Delta_g$ value of 3.04 meV for Nb indicates that phonons in Nb can break Cooper pairs.

Figure 3.4b depicts the calculated electron DOS profiles for Si [40] and Nb [41]. The energy scale is relative to E_F, which corresponds to the middle of the bandgap in Si and 5.32 eV relative to the bottom of the conduction band in Nb. The DFTP calculation in (b) underestimates E_g of Si [40]. The dashed electron DOS curves with $\sqrt{E_k}$ dependence are based on the free particle approximation. The electron DOS profiles in Fig. 3.4b and the Fermi level positions visualize the clear difference in electronic properties between Si and Nb. Electron–phonon coupling strength, which depends on phonon DOS rather than electron DOS [42, 43], primarily governs electric conductivity and superconductivity. In thermal equilibrium, phonon occupation is obtained by multiplying the phonon DOS and the Bose–Einstein distribution; the electron occupation follows the Fermi–Dirac distribution. The hot electron relaxation dynamics or electron–phonon coupling in semiconductors is reviewed in Ref. [44]. Electron–phonon interaction in a nonequilibrium state was studied for a simple fcc metal Al [45]. The contribution of the different phonon branches to coupling strengths was retrieved individually; the order of the electron–phonon coupling strengths is LA, TA1, and TA2 in simple metals [45].

Most theoretical calculations on Si at the early stages assumed a single LO phonon energy of 62–63 meV for the phonon emission, as mentioned before; still they yielded a reasonable theoretical prediction of $\varepsilon = {\sim}3.65$ eV [6, 20]. This value agrees with experimental data at room temperature. Strictly speaking, the phonon escape in semiconductor detectors depends on electron-phonon coupling strengths of the different phonon branches. Recent computational power improvement enables us to consider all phonon branches and their dispersions instead of the single phonon energy. Intriguingly, electron-phonon interaction also plays an important role for power dissipation or carrier mobility in nanoscale Si devices and very large scale integrated (VLSI) circuits [46, 47]. Recent Monte Carlo simulations and DFT calculations predict that the scattering rate with acoustic phonon emission is larger than that with optical phonon emission in Si and GaAs [47, 48]. This is in contrast to the common belief that hot electrons in semiconductors lose their energies mostly via optical phonon emission. One recent DFT report argues that the interaction with optical phonons predominantly contributes to hot electron energy loss when both phonon emission and absorption by electrons over an energy range up to 8 eV are considered [49]. However, the contribution of acoustic phonon emission in that report is significant at ~ 20 % of the total energy loss rate. At this moment, therefore, it is concluded that both acoustic and optical phonons should be considered to predict a and b value more precisely. The contribution of acoustic phonons causes an increase of phonon emission loss or an increase of the b factor. Therefore, the Monte Carlo simulation result [6] with $a = 1.70$ and $b = 1.74$ for Si can be modified such that they approach the recent experimental result of $a = 1.54$ and $b = 1.89$ in Eq. (3.8) [26] when we add the contribution of the acoustic phonons.

Semiconductor detectors still have unsolved issues. It is uncertain whether the above-mentioned a and b factors for Si are applicable to a variety of other semiconductor materials. Because each material has own phonon dispersion and DOS profiles, it is natural that each material has own b value related to phonon loss, which can partly explain the different branches in Fig. 3.2. The empirical b value of 0.55 eV in Eq. (3.6) may be applicable exclusively to a limited number of materials. We need to elucidate whether a universal linear ε–E_g relation exists across various types of semiconductors regardless of direct or indirect transitions. Finally, Si is the most popular semiconductor material for detectors; nevertheless, more studies for unsolved/open questions are necessary unexpectedly.

3.1.4 Lessons from Statistical Physics of Quantum Detection

In preceeding sections, we surveyed the current status of semiconductor detectors, which promotes better understanding of superconductor quantum detectors in Sect. 3.2. Supposing that Poisson statistics is valid for semiconductors, which corresponds to $F = 1$, and that ε equals to E_g, Eq. (3.4) predicts that Si with $E_g = 1.16$ eV has an energy resolution of 196 eV for 6-keV X-rays. For a superconductor Nb with $2\Delta_g = 3.04$ meV, the expected ΔE value is 10 eV that is 20 times better than that of Si, following $\sqrt{E_g/2\Delta_g} \approx 20$. This kind of estimation occasionally appears in the literature and reflect part of the advantage of superconductors over semiconductors. However, it contains a serious error on statistical physics. That is the assumption that ε and F are independent and thus a relation of $\varepsilon = E_g$ is compatible with $F = 1$.

According to the statistical definition of F, a very small F value close to zero, which is far end of the above-mentioned Poisson statistics, implies strong dependency among carrier generation trials along energetic-electron trajectories. In other word, carriers are generated efficiently. This holds true regardless of any probability distribution functions governing carrier generation processes. In $F \ll 1.0$, ε/E_g should be close to 1.0. Therefore, the assumption of $\varepsilon = E_g$ is incompatible with $F = 1$. Under the condition of $\varepsilon = E_g$, all the incidence-particle energy E_0 is consumed for generating e–h pairs. When all the carriers are experimentally collected, there is no chance that the number of carriers fluctuates, so that F and ΔE become zero. Consequently, there must be a correlational constraint between ε and F as the binomial distribution satisfies Eq. (3.3). The binomial distribution cannot incorporate all the microscopic processes of the e–h pair generation precisely; however, it explains experimental ΔE values naturally better than the Poisson distribution. The success probability p

in the binominal distribution represents E_g/ε; $(1-p)$ equals the variance ratio of the binominal distribution to the Poisson distribution, which is the Fano factor F. The p value is always less than 1.0, which stems from the energy loss processes without contributing to the e–h pair generation.

The carrier generation model in Fig. 3.3a for Si contains two types of energy loss processes: the phonon emission of carriers with energies below E'_{th}, which corresponds to the average excitation energy and determines the a factor, and the phonon emission of carriers with energies above E'_{th}, which determines the b factor. In the previous section, we understood that the a and b values in Eq. (3.6) are controversial. Even though theories and experiments on Si detectors commonly yield $\varepsilon = 3.65$–3.75 eV [29, 50], a variety of combinations of a and b result in the same ε value for Si. In the following part of this section, we present an experimental method for determining ε and F, shedding light on a common problem. Afterward, an alternative new model is proposed to better explain the ε–E_g relations of different semiconductor materials.

Among different radiation sources to determine ε and ΔE experimentally, the photoelectric effect of X-rays conveys all of the photon energy E_0 to an absorber through the emission of energetic photoelectrons, Auger electrons, and shake-off electrons [1, 2, 51], when neither electron escape nor characteristic X-ray photon escape from the absorber occurs. Therefore, monochromatized X-ray photons are the best to experimentally determine ΔE, F, and ε in Eq. (3.4), and a and b in Eq. (3.6). Even when escape peaks appear, full-energy absorption peaks in pulse height spectra can be utilized to evaluate $\overline{N}$ and then ε $(= E_0/\overline{N})$, where $\overline{N}$ is the average number of e–h pairs. We can obtain material dependent E_g values by observing the edges of optical absorption or reflection; those values in the literature are dependable. The $\overline{N}$ is determined from the position or detected charge of a full-energy peak, provided that carrier collection loss is negligible. There is another method using photocurrent [52]. Figure 3.5 illustrates a supposed pulse height spectrum for monochromatized X-rays, comparing two detectors with the different peak positions and different ΔE values. The x-axis is divided into narrow bins with the same energy width, which are called channels. The product of the channel number and the bin width in eV gives the detected photon energy. The number of absorbed X-ray photons in Fig. 3.5 is 1000.

Si detectors are frequently utilized as a reference for charge output. Although the a and b values of Si are controversial, the number of charges detected experimentally or the e–h pair number is accurate. Even when a Si detector is not calibrated precisely, we can use general values between 3.62 and 3.75 eV in a temperature range between 5 and 300 K with a relative standard deviation of 1.1% by assuming the rectangular distribution with an average value of $\varepsilon_{Si} = 3.70$ eV. By comparing the full-energy peak position of the device under test (DUT) with that of the Si detector, the ε_{DUT} is simply calculated by $\varepsilon_{DUT}/\varepsilon_{Si} = \overline{N}_{Si}/\overline{N}_{DUT}$, where $\overline{N}_{Si}$ and $\overline{N}_{DUT}$ are the average carrier numbers. The ratio of $PH_{Si}/PH_{DUT} = 800/600$ can replace $\overline{N}_{Si}/\overline{N}_{DUT}$ by employing the same readout circuit for the Si and DUT detectors. The ε_{DUT} value is calculated by $3.70 \times 800/600 = 4.93$ eV. Alternatively, it is possible to directly measure ε by utilizing a well-calibrated readout circuit [22].

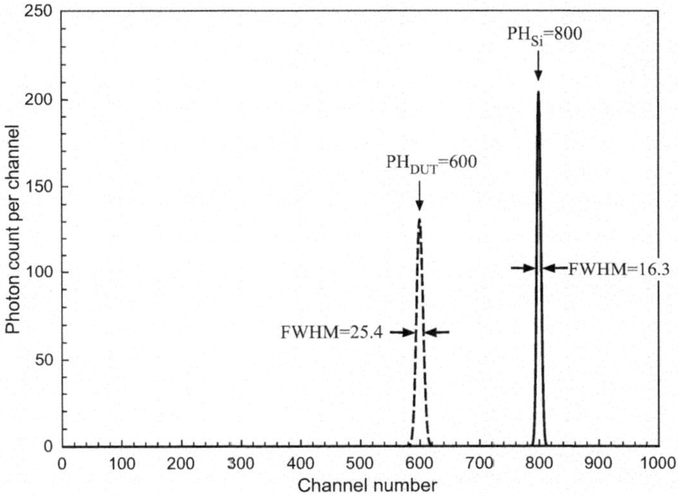

Fig. 3.5 Experimental determination method for average energy ε consumed for generating one e–h pair and Fano factor F of semiconductor detectors. It is supposed that a Si detector and a detector under test (DUT) absorb 1000 monochromatic X-ray photons at 5895 eV from a ^{55}Fe source. The Si detector as a reference is assumed to generate a full-energy peak with a FWHM value of 120 eV, which corresponds to $\varepsilon_{Si} = 3.70$ eV and $F_{Si} = 0.12$, in this supposed X-ray pulse height spectrum with a channel bin width of 7.369 eV. The ε value of the DUT detector is calculated by $\varepsilon_{DUT} = \varepsilon_{Si} \frac{PH_{Si}}{PH_{DUT}} = 4.93$ eV. The F value of the DUT detector is obtained by using Eq. (3.4): $F_{DUT} = 0.22$. At present, there is no similar method for superconductor detectors to measure the number of generated quasiparticles due to many uncertainties of quasiparticle dynamics and imperfect quasiparticle collection

The ΔE values in FWHM are determined by using Gaussian fit to the full-energy peaks, as shown in Fig. A.1 in Appendix A. Finally, one can calculate F_{DUT} by using Eq. (3.4): $F_{DUT} = 0.22$. The F value determination is affected by noise sources besides fundamental carrier-number fluctuation; therefore, appropriate subtraction is necessary using the propagation law. A similar method may be valid for superconductor detectors; however, it is impractical at this moment due to many uncertainties in quasiparticle generation and collection processes in STJDs, as explained in Sect. 3.2.4.

The above-mentioned experimental procedure determines ε and F with no correlation constraint between the two parameters. The classical Alig$^+$ relation has a large ε_{Si}/E_g ratio of 3.2, while empirical F_{Si} values are as small as ~ 0.12 in the literature. This combination of the large ε_{Si} and the small F_{Si} is inconsistent with the statistical definitions of ε and F. We presumably overlook important basics. If e–h pair generation follows the binomial distribution, there is the apparent constraint between ε and F according to Eq. (3.3). However, the experimental F values of Si detectors is considerably smaller than the F value of the binomial distribution. One possibility is that the exact distribution function is narrower than the binomial distribution. Another possibility is that phonon emission by hot carriers with energies larger than

E'_{th} should be excluded from e–h pair generation trials. There is naturally no chance for e–h pair generation by phonons in semiconductors in $\hbar\omega_0 \ll E'_{th}$. Therefore, the energy related to the b factor can be subtracted from E_0 before calculating ε and F.

Let us compare calculations based on the Alig$^+$ branch and the Whitaker branch in Fig. 3.2 to bridge the difference between the experimental results and the binomial statistics. For Si with $E_g = 1.16$ eV at 100 K, the Alig$^+$ branch predicts a ε_{Si} value of 3.72 eV, which gives $\overline{N} = 1580$ for the 5895-eV X-rays. The total energy $b \times \overline{N}$ for the phonon loss due to the carriers above E'_{th} is 870 eV at $b = 0.55$ eV in the Alig$^+$ branch, which occupies 15% of the absorbed energy E_0. This energy should be excluded from the binomial statistics because no phonons contribute to e–h generation trails above E'_{th}. On the other hand, we can regard the phonon emission by particles below E'_{th} as the carrier generation trials because it occurs after e–h pair generation. In this context, the total energy of 870 eV should be subtracted from E_0; the ε value becomes 3.18 eV instead of 3.72 eV. Assuming that the binomial distribution is valid for Si, the value of 3.18 eV gives $p = 0.365$ and $F = 0.635$ that is still considerably larger than an experimental value of $F = \sim 0.12$ [6]. Furthermore, the corresponding ΔE is calculated to be 257 eV for the 5895-eV X-rays, which is also considerably larger than an experimental value of ~ 120 eV [53]. In contrast, when we use the Whitaker branch with $a = 1.54$ and $b = 1.89$ in Eq. (3.8), it gives a $\overline{N}$ value of 1600, $\varepsilon = 1.79$ eV, $p = 0.648$, $F = 0.352$, and $\Delta E = 144$ eV with a phonon loss energy of 3030 eV that should be subtracted from 5895 eV. The ΔE value is close to experimental values of ~ 120 eV. This new procedure based on the binomial distribution can explain experimental data better than the Alig$^+$ branch.

The Whitaker branch is supported by a Monte Carlo simulation that reports the a and b values in Si:

$$\varepsilon = 1.70\,E_g + 1.74. \text{ [6]} \tag{3.11}$$

The recent sophisticated Monte Carlo and ab initio calculations indicate that contribution of acoustic phonons to electron–phonon coupling cannot be ignored [46–49]. Therefore, the phonon loss should be enhanced beyond 3030 eV or the b factor further increases from 1.74, which approaches the Whitaker branch further.

It is worth noting here that there are two types of semiconductors: direct bandgap and indirect bandgap. Si and Ge are indirect semiconductors; the bottom of the conduction band and the top of the valence band have different momentums on $\boldsymbol{k}$ space. The electron transition from the valence band to the conduction band requires either phonon absorption or emission. On the other hand, GaAs, CdSe, and CdS are direct semiconductors. The $Al_xGa_{1-x}As$ alloys can be either direct or indirect, depending on x. The direct semiconductors are advantageous over indirect ones for photon detection in an energy range near bandgaps or ~ 1 eV. In an indirect semiconductor Si, photon absorption near E_g requires either phonon absorption or emission, although direct transition occurs even in Si in a photon energy range above ~ 3 eV. Therefore, in an X-ray energy range, connecting the data of both direct and indirect semiconductors in Fig. 3.2 is meaningful. When the photon energy is close to E_g, the transition types must influence the ε–E_g relation.

The $\varepsilon - E_g$ relation in semiconductor detectors is a long-standing controversy for 57 years after the Klein's paper in 1968 [18]. Currently, there is no conclusion on the different branches in Fig. 3.2. Further systematic experimental and theoretical studies are still required even for commonly-used semiconductors. Nevertheless, the basic statistical physics established for semiconductor detectors provides guidance in developing quantum detectors with a higher energy resolution. Surveying semiconductors, HgCdTe is a well-known material with tunable E_g; the energy gap can be tuned between ~ 0 and 1.6 eV by adjusting composition. Therefore, in principle, HgCdTe detectors with a small gap can have a high energy resolution, outperforming Si and Ge detectors. However, it is difficult to obtain a high-quality crystal of the ternary compound semiconductor unlike single element materials such as Si and Ge. High-quality single crystals without both lattice defects and impurities are mandatory to avoid imperfect charge collection. In fact, the history of semiconductor detector development indicates that better crystal quality led to lower ε values.

If there is a polycrystalline or amorphous material with a low and uniform energy gap and negligible carrier loss, it is a possibility of overcoming the performance limit of single-crystalline semiconductor detectors. Superconductors meet the above criterion; they possess a uniform energy gap insensitive to local defects and impurities. Owens and Peacock plotted ΔE values expected for different superconductors as a function of $2\Delta_g$, extrapolating the data of the semiconductors toward a lower energy gap region [24]. The Fano factors are assumed to be 0.14 for semiconductors and 0.22 for superconductors. When we plot superconductors on Fig. 3.2, all superconductors are located on the origin; $2\Delta_g$ is in a meV range in contrast to ~ 1 eV in Si. It is empirically evident that a smaller Δ_g value is advantageous for a higher energy resolution. Ta junctions with a Δ_g value of 0.7 meV usually outperform Nb junctions with a Δ_g value of 1.52 meV. However, it is not at the stage where we can compare ε values of different superconductor materials quantitatively. It should be emphasized that another advantage of superconductors is that the b factor is close to zero because the Debye energy is higher than $2\Delta_g$.

The Klein model predicts $a = 2.8$ instead of 1.5–2.0 in the recent experimental and Monte Carlo data in Eqs. (3.7), (3.8) and (3.11). Are there any rooms to refine the Klein's analytical treatment? The semiconductor detectors operate in a high electric field. The external force affects electrons and holes, so that the momentum conservation law is broken. When we adopt the energy conservation law only, eliminating the momentum conservation law and assuming zero kinetic energy for the three product carriers after collision, the threshold level for the final e–h pair generation lay on the dotted red line marked "$E_{th} = 1.5E_g$ (1.74 eV) $E'_{th} = 1.0E_g$" in Fig. 3.3a. We obtain $a = 2.2$ from Eq. (3.10), which approaches the recent data of $a = 1.5$–2.0. In addition, when we assume an occupation profile biased toward lower energy states on the $\sqrt{E}$-dependent DOS in contrast to equal occupation below E'_{th}, the average energy is further lowered. An analytical a value further approaches the recent data. Consequently, the modified average excitation-energy treatment explains the basis of semiconductor detectors. Application of this average-energy treatment to superconductors is described in the next section. This is an alternative approach to explain the energy loss due to phonon emission below $2\Delta_g$; the details appear in Sect. 3.2.3.3.

3.2 Superconductor Quantum Detectors

3.2.1 A Historical Overview

A prediction of 20 times better energy resolution was made by Wood and White in 1965 in the Wood's master thesis [54]. They acquired the idea that superconductor tunnel junctions (STJs) were promising for reading out quasiparticles excited by energetic single particles such as α-particles in nuclear physics. University of British Columbia had a Van de Graff accelerator. Because no high electric field exists in superconductors, a tunnel junction with an insulating barrier with a thickness of ~ 1 nm is required to readout the increase of quasiparticles excited by particle impact. After a trial with Pb-based junctions, they successfully observed output pulses with $Sn/SnO_2/Sn$ tunnel junctions for 5.1-MeV α-particles. Although no pulse height spectra were acquired in 1969 [55], the predicted energy resolution for a Nb-based tunnel-junction quantum detector was about 2 eV for 525-eV photons corresponding to the O K-$L_{2,3}$ (Kα) line.

In the 1960s, they faced practical problems of superconductor materials, fabrication, cryogenics, and readout electronics. In his master thesis [54], he concluded that "The experimental and theoretical study reported in this thesis has shown the plausibility of successfully developing a superconducting charged particle detector. There is, however, a considerable difference between a device working in theory and a device working in practice." In addition, he wrote that "The most formidable problem was degradation of $Al/AlO_x/Pb$ junctions showing junction resistance increased with time after fabrication." Similarly, Pb-based Josephson junctions were employed in the Josephson computer project initiated by IBM in 1964 [56]. The Pb-based junctions were unfortunately unstable against heat cycles between room temperature and 4.2 K.

Another problem of the Wood's junctions was the operating temperature of 1.2 K, at which thermally excited quasiparticles masked any increase in tunnel current induced by particle incidence. The temperature of 1.2 K was the lowest they were able to attain by pumping liquid ^{4}He. It is known today that temperatures of ~ 0.3 K or below are required for proper operation, suppressing thermal excitation. In addition, the α-particle energy of 5.1 MeV was sufficiently high to induce local superconducting-to-normal (S-N) transition. Even for absorption of X-rays in a few keV range, a matter of concern is possible formation of a normal core or a reduced Δ_g core [31, 57, 58]. When they are formed, no quasiparticle generation occurs due to energy gap collapse, or the number of quasiparticles decreases. At present it is known that STJDs perform well in a low energy region below ~ 2 keV.

After Wood and White, Kurakado and Mazaki utilized a Sn-based junction for detecting α-particle in 1980 [59]. The Sn-based junction successfully acquired pulse height spectra, which was the first spectroscopic measurement, although they were not able to observe a distinct full-energy peak for the 5.1-MeV α-particles. A Monte Carlo simulation on ε and F values was performed for Sn by Kurakado in 1982 [60]; the result was $\varepsilon = 1.68\Delta_g$ ($= 0.969$ meV) and $F = 0.195$ for single quasiparticles when Sn absorbed 5.75 eV ($= 10^4 \times \Delta_g$). This result is consistent with subsequent Monte Carlo simulations [61–63] and a full-analytical calculation [31] described in Sect. 3.2.3.2. The ΔE values expected by using the Monte Carlo simulation on Sn and Eq. (3.4) are 2.5 eV for the 5895-eV X-rays from a ^{55}Fe source, 0.74 eV for 525 eV (O K-L_3), and 0.64 eV for 392 eV (N K-L_3), assuming that all the quasiparticles are collected. These values provide the lower limit of the energy resolution of Sn-STJDs. They are far better than the performance of Si: 121 eV for 5895 eV, 36 eV for 525 eV, and 31 eV for 392 eV according to the Monte Carlo simulation values of $\varepsilon = 3.752$ eV and $F = 0.119$ [13]. It is difficult to reach the lower limits experimentally with STJDs; nevertheless, experimental ΔE values exceed the Si limits in the literature. For example, AIST reported 3.6 eV at 400 eV with a Nb/Al-STJD, and Lawrence Livermore National Laboratory (LLNL) reported 1.5 eV at 250 eV and 3.5 eV at 1 keV with a Ta/Al-STJD [64].

In addition to the high energy resolution due to the meV gaps in superconductors, another advantage is a range extension of single-photon energy measurement to lower-energy photons. The lower detectable limit of Si detectors is ~ 50 eV, whereas Nb- and Ta-based STJDs record a relative energy resolution of $\Delta E/E_0 \approx 10\%$ at ~ 1 eV in a range from ultraviolet to near-infrared light (200 nm–2 μm) [65–67]. Similar performance was also realized by thermal detection with an TESD [68]. Superconductor detectors are uniquely capable of enabling energy measurement of single photons near visible light without utilizing diffraction gratings.

In addition to photons, Twerenbold et al. utilized a Sn-based junction to detect 20-keV lysozyme protein ions (14 305 Da) in mass spectrometry (MS) in 1996 [69, 70]. Hilton et al. improved energy resolution for biomolecules by using a normal-insulator-superconductor (NIS) junction in 1998, enabling charge state discrimination for bovine serum albumin ions (BSA, 66 430 Da) [71]. Wenzel et al. realized detection of macromolecules such as immunoglobulin M (IgM, 1 MDa) and von Willebrand Factor (1.5 MDa) with charge state discrimination by using a 16-pixel Nb/Al-STJD in 2005 [72]. Shiki et al. detected doubly-charge diatomic ions such as $^{14}N_2{}^{2+}$ eliminating *m/z* overlap with $^{14}N^+$ in 2008 [73]; the doubly-charged ions are presumably responsible for atmosphere escape from planets. Ohkubo et al. realized direct MS of neutral molecules in 2011 [74]; neutral radical fragments after dissociation can be a source for prebiotic organic molecules in the interstellar space.

Concerning detector fabrication, the Sn-based junctions as well as Pb-based junctions were unfortunately unstable against thermal cycling, as mentioned before. Stable tunnel junctions were obtained in the course of Josephson computer projects worldwide [75], which required an LSI-level integration of Josephson junctions. Numerous efforts were devoted to produce high-quality stable tunnel junctions with a high V_m value, which is the product of junction critical current I_c and resistance R_d in a subgap region below $2\Delta_g/e$ on *I*–*U* curves. Here, "*I*–*U*" is used instead of "*I*–*V*" for ensuring the consistency with units of ampere A for *I* and volt V for *U* to avoid V for *V*, following the IEC/ISO standard. Junctions with soft metals such as Sn and In had a better tunneling barrier than those with refractory metals such as Nb. However, the soft-metal junctions are unstable against thermal cycling. Contrarily, the refractory-metal junctions were stable during thermal cycling; the tunneling barrier of natural oxides such as NbO_x was unsatisfactorily leaky [76]. In 1972, the invention of AlO_x tunnel barrier layer, which was formed by surface oxidization of Al, resulted in the successful development of high-quality refractory Nb-based tunnel junctions [77]. In this first Nb/Al junction paper, interestingly, the thickness of the Al layer covering a Nb electrode was as thick as 80 nm. Therefore, an *I–U* characteristic curve at 4.2 K exhibited a low subgap resistance or a low V_m due to a low Δ_g from the thick Al layer, which led to thermally excited quasiparticles. Following studies using thinner Al layers of 3–6 nm exhibited ideal *I–U* curves with high V_m values of 35–88 mV at 4.2 K [78, 79], which were considerably larger than 12 mV of Nb/NbO_x/Nb junctions [80].

It is intriguing to notice that the first Nb/Al junction with such a thick Al layer is ideal for detectors. Thick Al layers are disadvantageous to logic circuits at 4.2 K, whereas they are adopted for modern tunnel junction detectors to increase output charge. The sandwich layer structure of STJDs is the same as that of Josephson tunnel junctions. However, the Josephson effect is suppressed by a small magnetic field in parallel to a junction surface. STJDs measure a change of quasiparticle tunnel current, which is called the Giaever mode. In addition, unlike Josephson-junction applications in SQUIDs and logic circuits, which often involve sub-μm junctions, junctions used for particle detection typically have much larger areas, around 100 μm or more, to achieve a practical sensitive area. Sixty years later after Wood and White, we appreciate the advances in peripheral technology to solve all the Wood's difficulties. STJDs have been installed in analytical instruments in synchrotron radiation (SR) facilities and analytical laboratories, outperforming semiconductor detectors. They are in routine use in analyzing advanced functional and structural materials.

Another quantum detector type has a superconductor strip resonator configuration to sense a surface impedance change in a microwave frequency range, based on Cooper pair breaking. It was invented by Day et al. in 2003; microwave kinetic inductance detector (MKID) enables a large-scale array format for imaging [81]. Quasiparticle excitation due to radiation varies a quality (Q) factor. A change in either RF transmission or phase shift produces output pulses. Each pixel has its own resonance frequency; it is possible to readout all pixels through one excitation line and one readout line. A MKID camera with 20 440 pixels for a wavelength band of 800–1400 nm (0.89–1.55 eV) was commissioned as a visiting instrument at the Subaru telescope in 2018 [82].

3.2.2 Superconducting Energy Gap

The nature of the energy gap in superconductors is completely different from that in semiconductors or insulators, although the statistical treatment methods for carrier generation are similar. The superconducting gaps serve as a scale for converting particle kinetic energy into the number of charge carriers, similar to bandgaps in semiconductors, as explained in Sect. 3.1.3.2. In addition, the existence of energy gaps in the electronic states ensures stable detector operation even when temperature drift or fluctuation occurs.

In semiconductors or insulators, energy gaps or bandgaps correspond to a forbidden band between the top of a valence band, which is formed by the electrons on the outmost electron shell, and the bottom of a conduction band, in which the electrons are freely mobile. The bandgaps stem from the electron-lattice interaction, creating covalent bonds or chemical bonds. The Fermi level E_F is located at the middle of the bandgap in non-doped intrinsic semiconductors and insulators. The semiconductors are insulating at 0 K because all electrons are tied with crystal lattice. Near room or elevated temperatures, some electrons in the valence band are excited to the conduction band according to the Fermi–Dirac distribution function, which results in mobile electrons in the conduction band and mobile holes in the valence band. Semiconductor detectors for spectroscopy must be normally cooled at a liquid nitrogen boiling point of 77 K to avoid thermal excitation. For room temperature operation, wide-bandgap semiconductors are beneficial to suppress thermally excited carriers. In exchange, the energy resolution of wide-gap semiconductors such as 4H-SiC with a bandgap of 3.23 eV is worse than that of Si with an energy gap of 1.16 eV. Nevertheless, operation at room temperature is advantageous, eliminating cooling apparatus that is often much bigger than detectors. The latest Si drift detectors (SDDs) operate at a moderate energy resolution even at room temperature [83].

In superconductors, the Fermi level is located in the conduction band because they are metals in a normal-conducting state in $T > T_c$. In a superconducting state well below T_c, two electrons with opposite wave vectors and opposite spin directions on the $\boldsymbol{k}$ space $(\boldsymbol{k}_\uparrow, -\boldsymbol{k}_\downarrow)$ interact attractively via the acoustic phonons in simple metal superconductors: the phonon dispersion is shown in Fig. 3.4a. The pair condensation of electrons produces the Cooper pairs that are bosons. The Cooper pairs can occupy the same state and thus supercurrent can flow without resistance or excitation. The number of the Cooper pairs is dependent on T. Due to this condensation, the gap Δ_g, which is proportional to the order parameter Δ in the phenomenological theory, appears at the Fermi level. In this book, we use Δ_g and Δ equivalently in some cases for the sake of simple notation. They are used distinctively when it is necessary. In electromagnetic absorption experiments, an absorption edge appears at a wavelength corresponding to a superconducting energy gap $2\Delta_g$ in a meV range, which corresponds to a THz region. The factor 2 comes from the fact that two quasiparticles are excited by breaking a Cooper pair. The two quasiparticles are approximated by a pair of quasielectron and quasihole in detector operation, similar to an e–h pair in semiconductors. On the other hand, the half gap Δ_g affects single quasiparticles, so that Δ_g governs electronic heat capacity instead of $2\Delta_g$.

The BCS theory predicts temperature dependence of Δ, which is also equivalent to the binding energy of Cooper pairs [84]. The Δ value is zero at T_c and maxim at 0 K, which is given by

$$\Delta(0) = 1.76 k_B T_c. \tag{3.12}$$

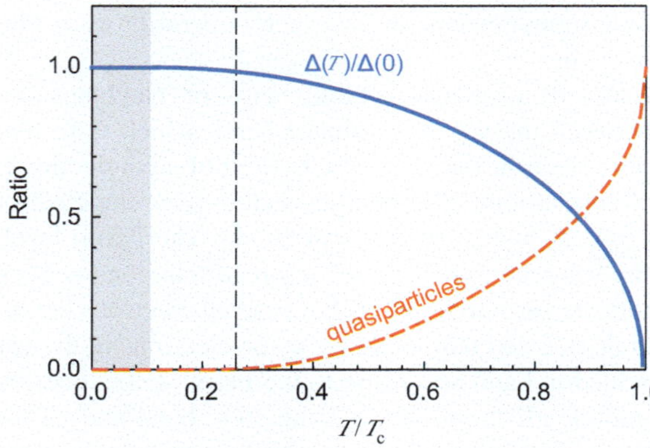

Fig. 3.6 Schematic temperature dependence of Δ, depicted by blue solid line, and the ratio of thermally excited quasiparticles to electrons forming Cooper pairs at 0 K, depicted by red dashed line, based on the BCS theory. The gray shadow region shows a recommended operation range for quantum detectors ($< T_c/10$) to suppress thermal excitation of quasiparticles. The variation of $\Delta(T)$ below $T_c/10$ is negligible; no precise temperature control is necessary. This is important for linear operation of detectors. The vertical dashed black line shows an effective temperature when a 200-nm-thick Nb film absorbs a 6-keV X-ray photon, provided that quasiparticles spread in a diameter of 10 μm. The position of the vertical dashed line at $T/T_c = 0.26$ indicates that quasiparticle generation is nonlinear at 6 keV because Δ is reduced

Equation (3.12) indicates that the gap energy is an order of the thermal energy at T_c. Furthermore, a lower T_c reduces Δ, making a superconductor with a lower T_c advantageous for achieving higher energy resolution according to Eq. (3.2) or (3.4). Figure 3.6 shows the temperature dependence of $\Delta(T)$. Near T_c, $\Delta(T)$ is approximated by

$$\Delta(T) \approx 1.74\Delta(0)\sqrt{1 - T/T_c}. \tag{3.13}$$

The BCS theory explains the properties of many elemental superconductors. However, some elements significantly deviate from the prediction of the BCS theory. A history of the advances in experimental and theoretical studies on elemental superconductors and simple alloys can be found in Ref. [85]. The special issue in 2015 describes the overview of all conventional and unconventional superconductors [86]. Although the prediction is imprecise for some superconductors, the BCS theory based on the electron-phonon interaction is the essential basis of superconductivity.

The $\Delta(T)$ values are almost constant until a substantial number of quasiparticles are excited by particle incidence, similar to thermal excitation at high temperatures. For detecting particles, the number of thermally excited quasiparticles should be negligible to precisely measure the number of the excited quasiparticles. Therefore, superconductor quantum detectors must be cooled below ~ $T_c/10$ in the shadowed region in Fig. 3.6. For example, Nb/Al-based detectors are normally operatable at temperatures below ~ 0.5 K. This operating temperature is extremely lower than that for semiconductor detectors. Nevertheless, the advances in liquid-cryogen free refrigeration resulted in the development of practical analytical instruments equipped with superconductor detectors, freeing users from supplying liquid nitrogen and helium.

Figure 3.6 indicates an upper limit of particle energy for linear response. The linearity is preferable for spectroscopy, which requires a constant Δ. The vertical dashed black line indicates an effective temperature T/T_c=0.26, which corresponds to the density of quasiparticles generated by a 6-keV photon, assuming a quasiparticle diffusion diameter of 10 μm inside a 200-nm thick Nb film. The Δ reduction is nonnegligible at the vertical black dashed line. When a larger amount of quasiparticles are generated along energetic electron tracks after X-ray absorption, Δ decreases significantly. It even induces transition to a normal state with zero Δ [31, 57, 58]. Incidence of MeV α-particles in the experiment by Wood and White conveys three orders of magnitude higher energy than that of X-rays to a superconductor [55]. Such high energy results in a strong nonlinear response, although STJDs still generate output pulses. Examples of nonlinear response will be shown later in Fig. 3.17 and Fig. 3.18 in this Chapter.

Which kind of superconductor films can form STJDs, holding a constant Δ value throughout a necessary sensitive area? Fig. 3.7 shows the surface morphology of a polycrystalline Nb film with a thickness of 250 nm, which is employed for Nb/Al junction detectors. It was deposited on a Si wafer by DC magnetron sputtering. An atomic force microscope (AFM) visualizes the elongated columnar nanoplates standing on the wafer surface vertically; each column is single-crystalline. In general, the Nb(110) plane is in parallel to a substrate surface. The deposition of an Al layer on the Nb surface for forming an AlO_X tunnel barrier exhibits a preferential orientation of Al(111) columnar structure [87]. There is no epitaxial relation between the Nb crystals and the Si substrate. This phenomenon is known as preferential growth. The Nb column tops have an area of ~ 100 × ~ 10 nm^2; the RMS surface roughness is 0.9 nm in Fig. 3.7. There are two distinct limits in superconductivity: the clean limit in $\xi \ll \ell$ and the dirty limit in $\xi \gg \ell$, where ξ is the coherence length corresponding to the size of Cooper pairs, and ℓ is the electron mean free path. In general, the polycrystalline Nb films with small grains are treated within the dirty limit: $\xi = 35$ nm and $\ell = 3.9$ nm [88]. While ξ is comparable with the grain size, ℓ is shorter than the grains. Even in the dirty limit, there is no meaningful change in T_c and Δ. Therefore, the Nb polycrystalline film can be regarded as a uniform material with a constant energy gap without spatial variation. Almost no quasiparticle trapping is anticipated. This is the reason why no single-crystalline superconductors are necessary for superconductor quantum detectors, although a long mean free path in single crystalline films contributes to a better performance for infrared–visible light [65]. It is noteworthy that superconductor X-ray detectors with the Nb polycrystalline absorber experimentally outperform conventional single-crystalline Si detectors by a factor of ~ 10 in a soft X-ray range. This is the strength of the macroscopic quantum effect of superconductivity.

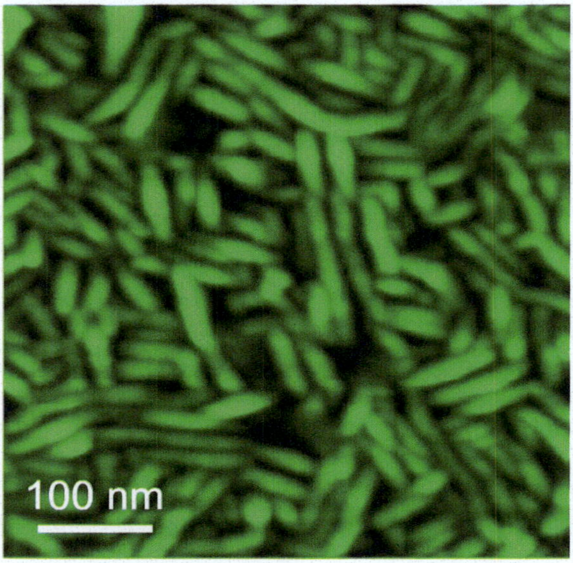

Fig. 3.7 Polycrystalline feature of superconductor Nb absorber film normally used for superconductor tunnel junction detectors (STJDs). The atomic force microscope (AFM) image visualizes the Nb single-crystalline columnar nanoplates standing vertically on a Si substrate. This particular Nb film deposited with DC sputtering has a thickness of 250 nm and a root-mean-square (RMS) surface roughness of 0.9 nm. It is remarkable that detectors with this polycrystalline superconductor film outperform single-crystalline Si detectors by a factor of ~ 10 regarding energy resolution as shown in Fig. 3.16. AFM analysis of each layer for Nb/Al/AlO$_x$/Al/Nb tunnel junctions appears in Ref. [89]. Engineering challenge for obtaining a low leakage current of ~ nA is to realize stress-free smooth films by optimizing deposition conditions

3.2.3 Carrier Excitation

3.2.3.1 Electronic Density-of-State (DOS) Profiles

There are two types of charge excitations in metals, semiconductors, insulators, and superconductors: electron-like excitation (quasielectron) and hole-like excitation (quasihole). In metals and semiconductors, excitations have either pure electron character or pure hole character. In superconductors, on the other hand, excitations are a superposition of both electron and hole characters, collectively referred to as quasiparticles. Energetic particles of photons, electrons, protons, atoms, and even proteins create quasiparticles by breaking Cooper pairs. Let us consider the case of X-ray photons, as other particles except proteins follow the same process as X-rays except for the initial interaction with atoms in a detector material. Basically, the excitation due to an X-ray photon can be treated as electron energy loss processes after the emission of a photoelectron or an Auger electron from an atom [29]. The highly excited electrons (hot electrons) relax to the order of eV equivalent to E_g

of semiconductors at ~ 0.1 ps after photon absorption, through inelastic scattering or ionization, as diagramed in Fig. 3.3a. They relax to the order of meV equivalent to Δ of superconductors in an elapsed time range of 1 ns–1 μs through electron–electron or electron–phonon scattering [90], as shown in Fig. 3.3b. The electron and hole DOS profiles in these excitation energy ranges below ~ 10 eV are depicted in Fig. 3.8, which is based on the free particle approximation. In this section, we describe the charge carrier excitation in semiconductors first as guidance in excitation in superconductors.

With the free particle approximation, as formulated in many books, the DOS profiles of electrons and holes in semiconductors are given by

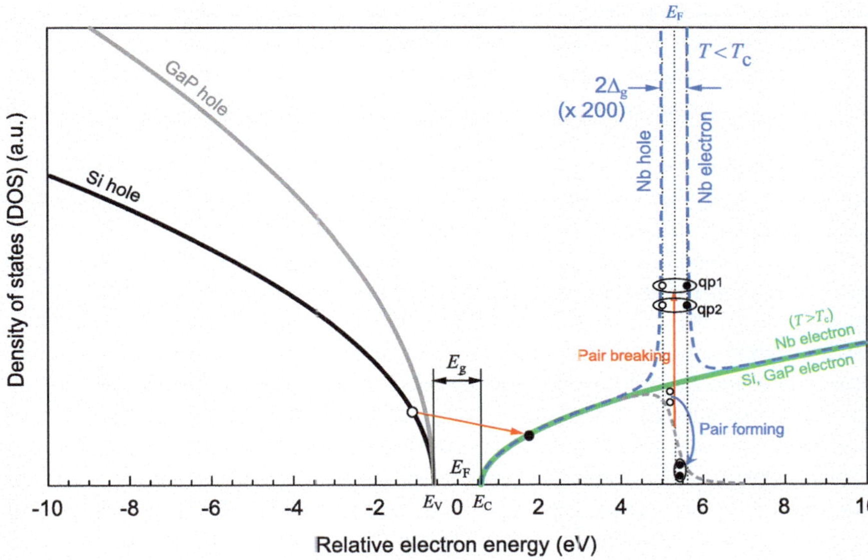

Fig. 3.8 Density of the stale (DOS) profiles for quasielectrons and quasiholes in normal-conducting Nb, superconducting Nb, Si, and GaP. The green solid line depicts the electron DOS curves of normal-conducting Nb, Si, and GaP. The back and gray solid lines depict the hole DOS curves of Si and GaP. The superconducting energy gap $2\Delta_g$ is magnified by a factor of 200. It is assumed that the normal-conducting Nb, Si, and GaP have the same energy gap (E_C–E_V) and the same DOS curve for the sake of explanation. Note that the hole DOS in this figure is different from the electron DOS in the valence band of Si in Fig. 3.4b. Energy absorption at several eV excites an electron in the valence band to the conduction band, which occurs in all materials. Examples of the initial excitation path is denoted by the horizontal red arrow. The electrons in Nb form the conduction band with an E_F value of 5.32 eV relative to the bottom of the conduction band. Conventionally, Cooper pairs are located on E_F based on the semiconductor-model description for superconductors. The curved blue arrow denotes Cooper pair formation, leaving holes below E_F; the vertical red arrow denotes Cooper pair breaking, producing two Bogoliubov quasiparticles. Reproduced from Ohkubo [144], with the permission of AIP Publishing

$$N_{\mathrm{e}}(E_k) = \frac{4\pi}{h^3}(2m_{\mathrm{e}}^*)^{\frac{3}{2}}\sqrt{E_k - E_{\mathrm{c}}}(E_k > E_{\mathrm{c}}) \text{ and} \tag{3.14}$$

$$N_{\mathrm{h}}(E_k) = \frac{4\pi}{h^3}(2m_{\mathrm{h}}^*)^{\frac{3}{2}}\sqrt{E_{\mathrm{v}} - E_k}(E_k < E_{\mathrm{v}}), \tag{3.15}$$

where E_k is the energy relative to E_{F}, E_{C} is the bottom of the conduction band relative to E_{F}, E_{V} is the top of valence band relative to E_{F}, h is the Plank constant, $m_{\mathrm{e}}{}^*$ is the effective mass of electrons, and $m_{\mathrm{h}}{}^*$ is the effective mass of holes. When we use the energy E relative to the bottom of the conduction band for electrons or relative to the top of the valence band for holes, both E_k–E_{c} and E_{v}–E_k are replaced simply by E, as seen in Eq. (3.10): it is convenient for mathematical operation. However, superconductors require the notation with E_k. Therefore, we use the E_k notation here to compare semiconductors and superconductors.

When single effective mass cannot be defined, one can use the density-of-state effective mass expressed by $m_{\mathrm{e}}^* = \left(m_{\mathrm{et}}^{*2} \cdot m_{\mathrm{el}}^*\right)^{\frac{1}{3}}$, where m_{et}^* is the transverse mass and m_{el}^* is the longitudinal mass of electrons; and $m_{\mathrm{h}}^* = \left(m_{\mathrm{hl}}^{*\frac{3}{2}} + m_{\mathrm{hh}}^{*\frac{3}{2}}\right)^{\frac{2}{3}}$, where m_{hl}^* is the light mass and m_{hh}^* is the heavy mass of holes. For Si, one can obtain $m_{\mathrm{e}}^* = 0.33m_0$ and $m_{\mathrm{h}}^* = 0.56m_0$, where m_0 is the free electron mass.

Figure 3.8 illustrates the simplified DOS profiles of Si, GaP, and Nb. The green line for electrons for all the materials, the gray line for holes in GaN, and the black line for holes in Si. For the sake of simplicity, we assume that the electron DOS curves of Nb in a normal-conducting state, Si, and GaP are identical, and moreover the gaps between E_{v} and E_{c} for all the materials are also equal. The $\sqrt{E}$ dependence of DOS curves near E_{F} is simplified from the exact DOS profiles in Fig. 3.4b. In semiconductors at 0 K, all the states above E_{c} are vacant and all the states below E_{v} are occupied by electrons. The Fermi level of intrinsic semiconductors is located at the middle of the energy gap, which is different from the highest energy level occupied by electrons in metals at 0 K. This difference prompts frequent questions about E_{F} between E_{V} and E_{C} in semiconductors. The answer is that the E_{F} position is defined as an energy level where the Fermi distribution function at an elevated temperature becomes 0.5; it is at the middle of the gap in intrinsic semiconductors and insulators.

In superconductors within the framework of the BCS theory, E_k is expressed by

$$E_k = \sqrt{\Delta_{\mathrm{g}}^2 + \xi_k{}^2}, \tag{3.16}$$

where E_k is the quasiparticle excitation energy and ξ_k is the normal-state single-particle kinetic energy of the state $\boldsymbol{k}$ relative to E_{F}. By equating the numbers of normal-conducting states and superconducting states with a relation of $N_{\mathrm{n}}\ (\xi_k)\mathrm{d}\xi = N_{\mathrm{s}}\ (E_k)\mathrm{d}E_k$, we can obtain the DOS profiles for quasielectrons and quasiholes in superconductors:

$$\frac{N_s(E_k)}{N_n(\xi_k)} = \frac{d\xi_k}{dE_k} = \frac{|E_k|}{\sqrt{E_k^2 - \Delta_g^2}} \ (|E_k| < \Delta_g)$$
$$= 0 \ (|E_k| < \Delta_g). \tag{3.17}$$

The N_n (ξ_k) values are almost constant near Δ_g in a condition of $\Delta_g/E_F{\approx}0.001$; the DOS profiles for quasielectrons and quasiholes can be treated symmetrically for excitation within ~ $10 \times \Delta_g$. At 0 K, the states are occupied by electrons up to E_F in a normal-conducting Nb, which is a step function that may be realized by applying a magnetic field to superconducting Nb. In a superconducting state at 0 K, on the other hand, the electron distribution is like a Fermi function at an elevated temperature, forming Cooper pairs. In $T < T_c$, spontaneously, the superconducting energy gap $2\Delta_g$ appears at E_F in the conduction band according to Eq. (3.17). The DOS curve shapes of quasiparticles in superconductors are totally different from those in semiconductors. Semiconductors show a square root dependence with zero DOS at the gap edges at E_c for electrons and E_v for holes according to Eqs. (3.14) and (3.15). On the other hand, BCS superconductors follow Eq. (3.17) such that DOS is divergent at $\pm \Delta_g$ relative to E_F. The exact quasiparticle DOS profiles of some elements are considerably different from the BCS one, as in Pb [91]; the extension of the BCS theory to the Eliashberg theory may be necessary [85]. Note that $2\Delta_g$ in Fig. 3.8 at 0 K is magnified by a factor of 200 to visualize the gap at the energy scale of Fig. 3.8. It visualizes that DOS values of quasielectrons and quasiholes are asymmetric at high energy excitation like ~ $100 \times \Delta_g$. With this semiconductor-model description, we can recognize the dissimilarities between the bandgap in semiconductors and the superconducting gap. Nevertheless, semiconductors and superconductors are similar on the statistical detector physics using Eq. (3.2) or (3.4), when we consider the fluctuation of the number of free charge carriers produced by particle incidence.

The hole DOS in semiconductors is higher than the electron DOS due to a heavier effective mass as shown in Fig. 3.8. Even in this asymmetry, normally, electrons and holes are equivalently treated in theoretical studies: hole properties are not known exactly [21]. It should be noted that holes in Si are mobile, while holes in Nb are located on inner shells so that they are almost immobile. These holes are different from mobile quasiholes in a superconducting state. For quasiparticles in BCS superconductors, it is expected that both m_e^* and m_h^* are not considerably larger than m_0 [92, 93], except for heavy Fermion superconductors. Quasielectrons and quasiholes in superconductors are also treated equally in theoretical studies.

Cooper pairs are located near E_F in Fig. 3.8; they are formed by electrons within ~ $k_B T_c$ (0.8 meV in Nb) from E_F. Although the Cooper pair breaking process by energy absorption is often described in textbooks, the forming process catches less attention. Both forming and breaking processes are essential for quantum detection or quasiparticle generation. The downward curved blue arrow depicts the stochastic Cooper pair forming process, in which two electrons are excited above E_F, leaving two holes behind. These holes in the conduction band are different from holes in the valence band or in inner shells. The holes play a pivotal role in the character of quasiparticles when Cooper pairs are broken. Although this Cooper pair forming process requires an excitation energy, energy reduction or condensation energy due to the acoustic phonon mediated electron–electron interaction surpasses the excitation energy. The Fermi-function-like dashed gray curve in Fig. 3.8 depicts the electron occupation function in the BCS ground state at 0 K. It is different from the step-like Fermi function at 0 K in normal-conducting metals. When superconductivity is broken beyond a coherence length, for example, heating-up, the two electrons forming the Cooper pair return to the former two electrons, not generating quasiparticles, by tracking the blue arrow backward.

The upward red arrow in Fig. 3.8 depicts the Cooper pair breaking process while keeping the coherency of superconductivity, which is the case of superconductor quantum detectors. Two quasiparticles are excited simultaneously and thus the superconducting energy gap is $2\Delta_g$. Precisely, we should consider the excitation on the momentum $\boldsymbol{k}$ space. A photon or a phonon with an energy above $2\Delta_g$ generates two quasiparticles by breaking a Cooper pair, obeying the laws of momentum and energy conservation. When a Cooper pair $(\boldsymbol{k}_\uparrow, -\boldsymbol{k}_\downarrow)$ is broken, a pair of excitations appears at unoccupied states $(\boldsymbol{k}_{1\uparrow}, -\boldsymbol{k}_{1\downarrow})$ and $(\boldsymbol{k}_{2\uparrow}, -\boldsymbol{k}_{2\downarrow})$. The character of the quasiparticles changes between electron-like and hole-like, depending on how momentum and energy are transferred to the two excitations [94]. When the quasiparticle excitation level is high enough beyond $2\times\Delta_g$, the quasiparticles are close to either pure electron or pure hole [95, 96]. On ensemble average, the quasiparticles have the 50%-electron and 50%-hole character, keeping charge neutrality in neutral perturbation like photons and phonons. Even when the number of electron-like quasiparticles is unequal to that of hole-like quasiparticles, which is called charge imbalance (see Sect. 3.2.3.4), a change of the number of superconducting electrons holds the charge neutrality [95, 96]. These quasiparticles in superconductors are called the Bogoliubov quasiparticles or Bogoliubons. The original paper was published in 1958 [97]; it is still an active research area [98], influencing many fields beyond superconductivity [99]. The Bogoliubov quasiparticles are ascribed as the superposition of electron amplitude u_k and hole one v_k, which is particle-hole mixture. The character of Bogoliubov quasiparticles changes continuously between hole-like excitation and electron-like excitation within the single-particle picture.

One quasiparticle can have an intermediate charge in principle. A Fano factor analysis of shot noise in a SIS-type Josephson tunnel junction indicated a sign of fractional charge tunneling with a value less than $1e$ in a bias range close to $2\Delta_g/e$ [100]. It was concluded that each tunneling event denoted a superposition of electron and hole, which meant that the tunneling charge of a quasiparticle was a fractional value less than $1e$. However, another shot-noise measurement of a mesoscopic SIS junction exhibited a sign of only $1e$ tunneling, observing neither fractional charge nor $2e$ [101]. Therefore, experimental validation whether each quasiparticle has a fractional charge remains as an unsolved subject for future studies. Alternatively, we can consider that each tunneling has a charge of $1e$ because an electron cannot be divided [94]; the sign of fractional charge tunneling in the shot noise is attributed to an ensemble average. For superconductor quantum detectors, we follow an approximation that the absorption of an energy quantum above $2\Delta_g$ results in the excitation of one quasielectron and one quasihole by breaking a Cooper pair, provided that the number of quasiparticles is small enough to hold Δ_g constant.

3.2.3.2 Mobile Carrier Generation Model for Superconductors

The comparative explanation of semiconductors and superconductors is effective in understanding the mobile carrier generation in superconductors. Figure 3.3 in Sect. 3.1.3.2 illustrates the conceptually simplified models for mobile carrier generation, comparing semiconductors and superconductors. Figure 3.3b is based on the full-analytical treatment for superconductors [31]. We consider the quasiparticles to be free mobile carriers of quasielectron-quasihole pairs in superconductors like e–h pairs in semiconductors. Note that the term "pair" emphasizes the simultaneous generation of electron-like and hole-like excitations, without implying any binding between them. In superconductors, two quasiparticles are simultaneously excited when a Cooper pair is broken. To be exact, the quasiparticles in superconductors are Bogoliubov quasiparticles in Fig. 3.8. In Fig. 3.3b, electron-like quasiparticles and hole-like quasiparticles are equally treated in the same way as electrons and holes in the theories on semiconductors. It is worth recalling that, before reaching the energy scale in Fig. 3.3, the absorption of an E_0 photon in a keV range leads to excitation of an electron in an inner shell of an atom in both Nb and Si, producing a photoelectron with an energy of a few keV, an Auger electron with an energy between a few 100 to a few keV, or shake-off electrons with 2–20 eV. The total energy E_0 is converted to energetic electrons. The energy of these energetic electrons decreases through secondary ionization, electron–electron scattering, and phonon emission. The secondary ionization corresponds to a process creating a hole in an inner shell or an outer shell, emitting a photoelectron or other electrons in both Si and Nb.

In an energy range below ~ 100 eV for Si, no more secondary ionization of atoms is induced, and electron excitation begins to occur from the valence band to the conduction band, reducing the energy of primary electrons further. On the other hand, in an energy range below ~ 200 eV for Nb, no more secondary ionization of atoms is induced; electronic excitation in the conduction band begins to occur, adding inelastic scattering with Fermi-gas free-electrons. The valence electrons in Nb are free, released from the outermost shells and forming the conduction band. This is in contrast to Si, in which the valence electrons are bounded around atoms, forming chemical bonds. Another electron energy loss process is plasmon excitation at 15–20 eV. Nevertheless, the plasmon loss can be neglected in Nb in a superconducting state because all the plasmons break Cooper pairs, which is not a loss process.

When the energy of electrons or holes in Si is reduced to a range below a few10 eV within a sub-picosecond after the photon absorption, e–h pair generation is a central process of the energy reduction of hot electrons. At 6.3 eV, the rate of e–h pair generation r_{ee} becomes equal to the phonon emission rate r_{ep}. The left-pointing solid arrows and the right-pointing dotted arrows in Fig. 3.3a denote r_{ee} and r_{ep}, respectively. The r_{ee} decreases as the electron or hole energy decreases, whereas the r_{ep} is rather constant in a range just below 6.3 eV. It is assumed that no generation of e–h pairs occurs below the threshold energy of $E_{\mathrm{th}} = 2.32$ eV relative to E_{F}, which corresponds to $E'_{\mathrm{th}} = 1.5E_{\mathrm{g}} = 1.74$eV relative to the bottom of the conduction band. The number of the e–h pairs generated up to this stage correspond to the total charge of an output pulse. Further relaxation, which is thermalization with equilibrium phonons, does not contribute to e–h pair generation; it represents the *a* factor for Si in Eq. (3.6). For Nb, on the other hand, the carrier generation continues even in an energy range well below 1 eV. The lengths of the left-pointing solid arrows for r_{ee} and the right-pointing dotted arrows for r_{ep} in Fig. 3.3 represent the energy dependence of the rates. All the threshold levels for Nb follow the notation for energy down-conversion processes of $E_1 \to \Omega_{\mathrm{D}} \to \Omega_1 \to E_2 \to \Delta$ in Refs. [31, 102].

At E_1 (= 1.16 eV) in Nb, the rate r_{ee} of inelastic electron scattering with Fermi-gas free-electrons becomes equal to the phonon emission rate r_{ep}. The Debye energy Ω_{D} is equal to the energy of LA phonons (23.7 meV) in Nb. In the relaxation from E_1 to the level Ω_{D}, the r_{ee} quadratically decreases as the quasiparticle energy decreases, whereas the r_{ep} is rather constant between E_1 and Ω_{D} in the same way as Si. The r_{ep} cubically decreases below Ω_{D}. The electron–phonon interaction predominantly leads to the LA phonon emission at 23.7 meV in Nb, because the interaction strength is strongest for the highest phonon energy.

It is postulated that the total photon energy E_0 is eventually transferred to the Ω_{D} phonons during the relaxation from E_1 to Ω_{D}, forming so-called the phonon bubble [31]. The phonon bubble is often the starting point of Monte Carlo simulations [60]. The fixed phonon energy is analogous to the Si case, in which the energy of emitted phonons is fixed at 63 meV for the LO phonon branch. The clear difference between Si and Nb is that the LO phonon emission in Si is the energy loss process related to the *b* factor, whereas the Debye phonons in Nb are the source for generating quasiparticle pairs. Two initial phonon-bubble dispersion spectra based on the Debye model and the neutron scattering experiment resulted in no meaningful change in the Mote Carlo simulation: $\varepsilon = 1.747\Delta_{\mathrm{g}}$ and $F = 0.22$ for Nb [61]. More realistic initial condition of one photoelectron with an energy of 56 eV with subsequent electron scatting cascades led to a combination of $\varepsilon = 1.72\Delta_{\mathrm{g}}$ and $F = 0.214$ for Nb [63], which is slightly different from the above result of the phonon bubble. The conclusion of the full-analytical treatment is that Nb, Ta, Al, Zn, Mo, and Hf have a universal relation of $\varepsilon \approx 1.65\Delta_{\mathrm{g}}$, whereas Tl, Hg, Sn, In, and Pb should have a material-dependent relation [31]. For Nb and Ta, a variation in $\varepsilon/\Delta_{\mathrm{g}}$ is between 1.65 and 1.75, which has a relative standard deviation of 1.6% by assuming the rectangular distribution with an average value of 1.70. The value of 1.6% is slightly larger than a relative standard deviation of 1.1% for Si; nevertheless, the $\varepsilon/\Delta_{\mathrm{g}}$ variation in superconductors is adequate for detector physics, considering that there are no reports on the ε–Δ_{g} relation in superconductors and both experiments and theories are unmatured compared with those of semiconductors.

In an energy range below Ω_{D}, Fig. 3.3b illustrates the relaxation and generation processes of Bogoliubov quasiparticles, provided that the coherency of superconductivity is maintained and Δ_{g} is constant. Quasiparticle generation in Nb enters into the pivotal stage below Ω_{D}. Below Ω_{D}, the r_{ee} is negligible, whereas the electronic relaxation rate r_{ep}, which is the phonon emission rate, is predominant. In addition, near Ω_D, the rate r_{pb} of phonon Cooper pair breaking, which is the energy relaxation mechanism of the phonons, is lower than the r_{ep}. Therefore, in a range between Ω_D and Ω_1, the relaxation of the electron and phonon systems is governed by the change of phonon spectral distribution At Ω_1 (= 5.32 meV), r_{ep} becomes equal to r_{pb}. In a region below Ω_1, r_{pb} is higher than r_{ep} because r_{ep} has a cubic relation to E_k and r_{pb} is proportional to E_k. Therefore, the relaxation of the system is governed by the change of quasiparticle spectral distribution below Ω_1. The energy transfer occurs from the phonon system stored between Ω_D and Ω_1 to the electronic system. Figure 1 of Ref. [31] shows the various rates as a function of E_k.

The E_2 level is the final important threshold, at which a quasiparticle emits the final phonon that has an ability to break a Cooper pair; no additional Cooper pair breaking occurs. The E_2 value is assumed to be equal to $3\Delta_g$, considering the energy conservation law. Below E_2, the quasiparticle spectral distribution further relaxes toward Δ_g, emitting phonon lower than $2\Delta_g$ (= 3.04 meV in Nb). This energy level corresponds to the dotted red line "$E_{th} = 1.5E_g$ (1.74 eV) $E'_{th} = 1.0E_g$", which is the newly suggested level for the final e–h pair generation in Si. When there is no phonon escape to surroundings, the quasiparticle phonon-emission with $\Omega < 2\Delta_g$ is only the energy loss process, which amounts ~ 40% of E_0 [102]. This loss corresponds to the a factor, leading to a relation of $\varepsilon \approx 1.7\Delta_g$ [31]. During this relaxation stage, no more quasiparticles are generated, so that the number of quasiparticle pairs $\overline{N}_{qp}$ is fixed when the energy of quasiparticles is reduced to $3\Delta_g$. The transition time from E_1 to E_2 ranges from ~ 10 ps in Nb to ~ 2 μs in Hf [31]. The output signal of STJDs lasts for a quasiparticle lifetime, which is normally 1–10 μs in Nb/Al STJDs, depending on the Al thickness.

The full-analytical solution for the transition from E_1 to E_2 yields $\varepsilon = E_0/2\overline{N}_{qp} = 1.65\Delta_g$ for Nb and Ta [31]. This value reasonably agrees with those by the Monte Carlo simulations: 1.72 for Nb and 1.78 for Ta [63]. More precisely, analytical solutions depend on material parameters, especially the magnitude order of Ω_1 and E_2 [31]. The group of Tl, Hg, In, and Pb with $E_2 \geq \Omega_1$ should have no electronic relaxation stage from Ω_1 to E_2; the analytical estimation results in $\varepsilon = {\sim}4\Delta$. Although the analytical treatment effectively provides the physical basis for quasiparticle generation and relaxation processes, there is the difficulty in tracing all material properties and relaxation processes precisely, especially when the overlap of the threshold energy levels occurs [31].

Comparison in the energy down-conversion processes between superconductors and semiconductors provides a ε–Δ_g relation for superconductors. Summarizing the model for semiconductors before discussing on superconductors, the a factor is assigned to the average excitation energy in a range between the bottom of the conduction band and E_{th}; it also represents the phonon emission loss in a range below E_{th}. The further relaxation from the nonequilibrium carrier distribution to a thermal equilibrium causes no change in the number of e–h pairs. When we assume that electrons equally occupy all the states below E_{th} at a constant fraction, the average excitation energy represents the a factor. The calculation result by Klein and Alig is $a = 2.8$ for $E_{th} = 2.0E_g$. On the other hand, we propose $a = 2.2$ for $E_{th} = 1.5E_g$ considering only the energy conservation because a high electric field breaks the momentum conservation. The b factor stems from the phonon emission loss of the electrons in a range above E_{th}. The b factor cannot be precisely deduced from the Klein-Alig model. The Monte Carlo simulation predicts $a = 1.70$ and $b = 1.74$ [6]. The phonon-loss related to the b does not contribute to e–h pair generation trials; it can be subtracted from E_0. In this new model, the binomial distribution is a suitable approximation of the exact probability distribution function for carrier generation in semiconductors. The a value approaches the recent experimental and Monte Carlo data of 1.54–2.0.

For superconductors, it is a reasonable postulation that E_0 is converted to the total energy of the Ω_D phonons [31, 60]. The total phonon energy is then transferred to the electronic system, generating quasiparticle pairs by breaking Cooper pairs because $2\Delta_g$ is lower than Ω_D. Superconductor quantum detectors have a sensitivity to phonons. The only loss process is the phonon down-conversion into energies below $2\Delta_g$, which predominantly occurs for quasiparticles below $3\Delta_g$. Even in a range above $3\Delta_g$, the probability of emitting phonons with energies below $2\Delta_g$ is not absolutely zero, but it is normally negligible. Therefore, the ε–Δ_g relation has a form of $\varepsilon = a\Delta_g$; the b value should be close to zero. In the next section, we try to apply the average excitation energy approach to superconductors, provided that quasiparticles occupy the states below $3\Delta_g$ equally in the same way as the Klein-Alig approach for semiconductors. The average excitation energy model gives $a = 1.81$ for single quasiparticles following the BCS DOS profile. This value is close to 1.65 by the analytical calculation for Nb and Ta [31] and 1.68–1.78 by the Monte Carlo simulations for Sn, Nb, Ta, Al, and Hf [63]. Further relaxation occurs with phonon emission below $2\Delta_g$, not generating quasiparticles. Practically, the most striking energy loss process in actual STJDs is the escape of phonons with an energy above $2\Delta_g$ to surroundings [102].

3.2.3.3 Average Excitation Energy Model for the ε–Δ Relation in BCS Superconductors

We apply the Klein-Alig model of the average excitation energy approach for semiconductors to the evaluation of the a factor for superconductors. Although it is a very rough analytical evaluation, it provides an alternative physical basis to Monte-Carlo simulations. The average kinetic energy $E_{e\text{-}h}$ of quasiparticle pair is calculated by

$$E_{e\text{-}h} = 2\left(\frac{\int_{\Delta_g}^{E_{th}} \frac{E_k^2}{\sqrt{E_k^2-\Delta_g^2}}\mathrm{d}E_k}{\int_{\Delta_g}^{E_{th}} \frac{E_k}{\sqrt{E_k^2-\Delta_g^2}}\mathrm{d}E_k} - \Delta_g\right), \tag{3.18}$$

provided that all the states between Δ_g and E_{th} are equally occupied by quasiparticles. The scheme of this average energy is consistent with an experimental result that the dependences of charge output on bias voltage and photon energy indicate the existence of quasiparticles staying at the high energy levels above the edge of Δ_g

[103]. The subtraction of Δ_g in the right side in Eq. 3.18 is to cancel the lower limit of Δ_g for the definite integral, and to obtain an average kinetic energy relative the edge of Δ_g. It is assumed that quasiholes have the same character as quasielectrons, which introduces a factor of 2. One primary quasiparticle, the kinetic energy of which is reduced to E_{th} ($= E_2$), produces one scattered quasiparticle and two new quasiparticles generated by the $2\Delta_g$ phonon that is emitted by the primary quasiparticle. This collision is the final Cooper pair breaking. We follow only the energy conservation law because the collision with phonon emission is inelastic. The E_{th} value is simply equal to $3\Delta_g$. A quasiparticle at E_{th} emits a 2Δ phonon, generating one scattered quasiparticle at Δ_g and the final quasiparticle pair at Δ_g, as depicted in Fig. 3.3b [31]. Further relaxation of quasiparticles generates no quasiparticle pairs.

The solution of Eq. (3.18) is not as easy as Eq. (3.10), but one can obtain $E_{e\text{-}h} = 1.62\Delta_g$. Therefore, the average energy consumed for generating one quasiparticle pair is given by $\varepsilon = 2\Delta_g + E_{e\text{-}h} = 3.62\Delta_g$, When we adjust it to the case of $E_g = 2\Delta_g$ in superconductors, ε per quasiparticle pair is given by

$$\varepsilon = E_0/\overline{N}_{qp} = 1.81 \times 2\Delta_g. \tag{3.19}$$

The factor 1.81 is considerably smaller than the Klein factor of $a = 2.8$ for one e–h pair in semiconductors. This arises from the DOS curve difference between semiconductors and superconductors in Fig. 3.8. When we convert it to ε per single quasiparticle, the average energy consumed for generating one quasiparticle is $1.81\Delta_g$, which is close to ~ $1.65\Delta_g$ by the analytical solution or to $1.68\Delta_g$–$1.78\Delta_g$ by the Monte Carlo simulations [60–63]. Quasiparticles are always created as pairs by breaking Cooper pairs; it is natural that "the average energy consumed for creating one quasiparticle pair is $3.62\Delta_g$ or $1.81 \times 2\Delta_g$," similar to ε per one e–h pair in semiconductors. Note that the output charge of superconductor quantum detectors should be equal to the number of quasiparticle pairs (quasielectron-quasihole pairs) when all the quasiparticles are collected at a time. The b factor for superconductors should be close to zero. Further relaxation from the equal quasiparticle occupation between Δ_g and $3\Delta_g$ induces the emission of phonons with $< 2\Delta_g$. Currently, we have no systematic experimental data on superconductor detectors with different Δ_g values. In contrast to semiconductor detectors, STJDs are inadequate for elucidating the ε–Δ_g relation experimentally because numerous undetermined factors arise from the complexity of quasiparticle dynamics and tunneling processes. Therefore, we cannot perform the experimental validation of the model on superconductors at this stage. Further studies are required to overcome this limitation, which remains an unsolved aspect of the model.

In contrast to the simple BCS DOS profile, proximity bilayers used for actual STJDs should have a non-smooth DOS profile [104]. Therefore, analytical calculation on proximity bilayers is unfeasible, not as easy as obtaining the solution of Eq. (3.18). Experimentally, for both single superconductor and proximity cases, MKIDs for an infrared-optical range or superconductor RF filters are suitable to confirm the ε–Δ_g relation with $a = 1.68$–1.78 and $b = $ ~0. When we consider all the cases in the preceding sections, for typical superconductors such as Nb and Ta used for detectors, ε per quasiparticle pair is approximately given by

$$\varepsilon \approx 1.7 \times 2\Delta_g. \tag{3.20}$$

3.2.3.4 Charge Imbalance

Returning to the Bogoliubov quasiparticles, the quasiparticles exhibit characters varying between electron-like and hole-like properties as described in Sect. 3.2.3.1. The charge of quasiparticles can vary between –1 and + 1 continuously, depending on E_k or ξ_k in Eq. (3.16). We can practically consider that a quasiparticle in $E_k > 2\Delta_g$ has a charge of either –1 or + 1. On the other hand, a quasiparticle in $\Delta_g < E_k < 2\Delta_g$ has an intermediate charge value between –1 and + 1 [94–96]. At the relaxation stages above Ω_1 in Fig. 3.3b, all the quasiparticles practically have either pure electron character or pure hole character. Based on Eq. (3.18) for the average excitation energy model, the quasiparticles that are finally generated after the relaxation equally occupy all the states between Δ_g and $3\Delta_g$ of the BCS DOS curve. The percentage of the quasiparticles with intermediate charges in $\Delta_g < E_k < 2\Delta_g$ is 60%. Further relaxation results in a situation that all the quasiparticles have an intermediate charge if the sates below $2\Delta_g$ can accommodate all the quasiparticles. These intermediate quasiparticles generate an output pulse in STJDs.

On the E_k–ξ_k dispersion curve derived from Eq. (3.16), there are two quasiparticle-disequilibrium modes: energy mode disequilibrium and charge imbalance (or branch imbalance) disequilibrium [94, 95]. The energy mode disequilibrium appears with neutral perturbation like photons or phonons while the electron-like quasiparticle population is equal to the hole-like quasiparticle population. It is a deviation from a thermal equilibrium quasiparticle distribution. On the other hand, the charge imbalance disequilibrium occurs when the electron-like quasiparticle population becomes unequal to the hole-like quasiparticle population, as in a case of electron injection. The charge imbalance was observed in a location where electrons are injected into superconducting thin films [105, 106] or near a phase-slip center in current-carrying superconductor wires [107]. These are the clear cases of observing the charge imbalance in superconductors. Even when the charge imbalance appears, overall neutrality is maintained due to a change in the number of superconducting electrons in the condensed Cooper pair sea [94, 95].

According to the above-mentioned common belief, it is believed that photon absorption in superconductors should not lead to charge imbalance, but to energy mode disequilibrium. It is true for small perturbation in a photon energy range below E_1 in Fig. 3.3b. In addition, it is expected that the impact of keV molecules, which stop on surfaces and produce Ω_D phonons, also induces no charge imbalance. The nonequilibrium superconductivity for low energy excitations near $\pm\,\Delta_g$ was recently reviewed in Ref. [108]. Meanwhile, what happens with absorption of high energy photons like X-rays? It is a photon perturbation, but absorption of an X-ray photon is transformed to the energy loss of a photoelectron at keV, Auger electrons at 100 eV to keV, and shake-off electrons at 2–20 eV. These charged particles can induce charge imbalance locally or transiently. A limited number of studies on charge imbalance in superconductor X-ray detectors were performed as summarized below.

Gulian and Vechten pointed out that the asymmetric electron DOS profiles can lead to charge imbalance for 10-keV X-rays. They called it the geometrical inequivalence of the Fermi sphere [109, 110]. At highly excited states far beyond $10\Delta_g$, the quasihole DOS is significantly lower than the quasielectron DOS, because $N_n(\xi_k)$ in Eq. (3.17) should follow Eq. (3.14) or (3.15) in contrast to the postulation that $N_n(\xi_k)$ is constant within the conventional BCS treatment. They claimed that the number of available states for holes at $-\xi_k$ should be lower than that for electrons at ξ_k. The asymmetric DOS is visualized in Fig. 3.8. In this context, the hole DOS reaches zero at E_c, which corresponds to the center of the Fermi sphere. When the excitation energy is in a few eV range, the $N_n(\xi_k)$ values for quasiholes are considerably lower than those for quasielectrons. They pointed out that the charge imbalance can remain after the excitation energy relaxes to an order of meV near $2\Delta_g$, so that the number of quasielectrons is usually higher than that of quasiholes at the final stage of the relaxation [109]. In addition to the geometrical inequivalence in bulk superconductors, it was also theoretically predicted that, even for absorption of photons near $2\Delta_g$, the charge imbalance generated a voltage across superconducting-normal interface [111]. This effect is enhanced by magnetic impurities. Furthermore, biexponential decay signals were observed in a temporal range of < 4 ns in a Nb film excited by laser pulse of 665 nm (1.86 eV) [112]. This behavior resembles a fast electron signal followed by a slow hole signal in semiconductors, although the origin of the biexponential decay in the superconductor is still unclear.

Considering the relaxation of the charge imbalance, at temperatures near T_c in a superconducting state, inelastic electron–phonon scattering is usually a dominant relaxation process. Each branch of quasiparticles deviated from thermal equilibrium distribution reaches a thermal equilibrium state separately. Through the interaction between quasiparticles and phonons, the populations of quasielectrons and quasiholes become equal. The relaxation time values are 300 ps for Sn and 10 ps for Pb [113]. On the other hand, in $T \ll T_c$, which is the case for superconductor quantum detectors, the inelastic electron–phonon scattering is ineffective because the electron–phonon coupling strength is proportional to T^5 [114]. Effective balancing mechanisms at very low temperatures are nonmagnetic impurity elastic scattering and magnetic pair breaking [115–118]. However, these mechanisms cannot cancel the imbalance completely in contrast to the phonon mediated branch mixing. As a result, nonnegligible charge imbalance may remain in superconductor quantum detectors at a temperature below $0.1T_c$.

Experimentally, we observe output current pulses with a single exponential decay of a few μs for 5.9 keV X-rays, as shown in Fig. 3.14b in Sect. 3.2.4.2. Even when charge balancing is slow at a few ns in $T \ll T_c$, the charge imbalance disappears in a temporal scale of the quasiparticle effective lifetime τ_{eff}. The measured τ_{eff} for Nb/Al junctions ranges from ~ 1 μs to a few 10 μs, depending on temperature, particle energy, and junction layer structure [119, 120]. The τ_{eff} is elongated from the intrinsic lifetime by Cooper pair breaking due to quasiparticle recombination phonons with energies slightly above $2\Delta_g$ [121], so-called the phonon trapping effect. Consequently, it seems that the charge imbalance vanishes. Nevertheless, the charge imbalance in superconductor detectors is intriguing detector physics. Further theoretical and experimental studies are required to fully understand unsolved aspects of the charge imbalance and the effect on detector performance.

3.2.3.5 Summary of Carrier Excitation in Superconductors and Signal Readout

We have learned what occurs when energetic particles such as photons and ions impinge superconductors. The essential property of quantum detection is an energy gap in electronic excitation spectra. Precise data of ε and F are indispensable for modeling superconductor quantum detectors; however, we realized both experimental and theoretical difficulties in determining the ε–Δ_g relation. Nevertheless, the previous sections described three approaches: the full-analytical treatment with $\varepsilon = E_0/\overline{N}_{qp} = 1.65 \times 2\Delta_g$ for Nb and Ta [31]; the Monte Carlo simulations with $\varepsilon = 1.68$–$1.78 \times 2\Delta_g$ for Sn, Nb, Ta, Al, and Hf [63]; and the newly proposed average excitation-energy model with $\varepsilon = 1.81 \times 2\Delta_g$ for BCS superconductors. The variation of 1.65–1.81 is acceptable at this stage, considering that even for semiconductors the ε–E_g relation expressed by $\varepsilon = aE_g + b$ is controversial recently. A problem is that a variety of combinations of the a and b factors yield a single ε value, although experimental ε values on Si are reliable. Therefore, the three approaches for superconductors are consistent each other. We expect that the b factor in superconductors is practically zero in contrast to the large b values reported for semiconductors: $2\Delta_g$ is roughly 1000 times smaller than E_g in semiconductors. In contrast, there is a report that phonon loss into energies less than $2\Delta_g$ is substantial [158], which yields a nonzero b value. An experimental method to precisely determine ε–Δ_g of superconductors is required, which is an unsolved aspect. Meanwhile, we postulate that the b factor of superconductor quantum detectors is zero.

Excitations beyond an energy gap simultaneously generate pairs of electron-like excitation and hole-like excitation in superconductors. Therefore, the statistical fluctuation in the number of the quasiparticle pairs determines the theoretical limit of energy resolution. The total output-charge extracted by a readout external circuit is not equal to the number of the quasiparticles, but the number of the quasielectron-quasihole pairs when all the carriers are collected without loss, similar to semiconductors in Sect. 3.1.1.

Unlike semiconductor detectors, it is impossible to collect all carriers by applying a high electric field to superconductors. In addition, quasielectrons and quasiholes occasionally recombine into Cooper pairs, which causes the quasiparticle number reduction through escape or energy down-conversion of Cooper pair recombination phonons. The quasiparticle recombination in superconductors resembles the exciton recombination in semiconductors: the excitons are not formed in semiconductor detectors under a high electric field. These disadvantages in superconductors result in a degradation of energy resolution; however, this is compensated by the statistical advantage provided by the extremely large number of quasiparticle pairs generated due to a meV-scale superconducting energy gap. Without applying a high electric field to superconductors, there are two readout schemes: superconductor tunnel junction detectors (STJDs) using a subgap tunnel current increase and microwave kinetic inductance detectors (MKIDs) using a RF surface impedance change, described in the subsequent Sects. 3.2.4 and 3.2.5.

Output charge from quantum detectors should be converted to a voltage pulse. Although superconductor circuits will realize this conversion in future, it is currently performed with semiconductor-based amplifier circuits. They are normally charge-sensitive semiconductor preamplifiers, which are similar to Fig. 3.1, even though, carriers in superconductors are totally dissimilar to those in semiconductors. At the final stage of the relaxation in superconductors after a time scale of 10 ps–2 μs, Cooper pair breaking is completed as quasiparticles located near Δ_g. The quasiparticles are in an energy-mode disequilibrium state through a temporal charge imbalance disequilibrium state. Providing that the quasiparticle density is held below a level inducing a meaningful Δ_g reduction, as illustrated by the gray shadow region in Fig. 3.6, all the quasiparticles are in intermediate charge states between electron and hole characters. The electron-like or hole-like quasiparticles in superconductors must be converted to either pure electrons or pure holes in semiconductor readout circuits. The conversion can occur at superconductor/normal-conductor interfaces. Phenomena at superconductor/normal-conductor or superconductor/semiconductor interfaces have been investigated intensively; this topic is left to the literature. We expect that the number of quasiparticle pairs in superconductors is equivalent to the number of e–h pairs in the semiconductor readout circuits for superconductor detectors.

3.2.4 Superconductor Tunnel Junction (STJ) Type

3.2.4.1 Structure of STJ Type

Superconductor tunnel junction detectors (STJDs) have a layer structure of superconductor-insulator-superconductor (SIS), which is identical to the structure of Josephson tunnel junctions. Figure 3.9 shows an SEM top-view image and a TEM cross-sectional image of a Nb/Al-based STJD. In most cases, the top and bottom superconductor electrodes have a proximity bilayer structure of Nb/Al. Normally, the electrodes have a Nb layer with a thickness of 200–300 nm and an Al layer with a thickness of 30–70 nm. The thickness of the AlO_x tunnel barrier is approximately 1 nm. The SiO_2 layer covers the junction edge to isolate the top and bottom electrodes. An example of the fabrication processes is found in Ref. [122]. The details of the STJ type are described below.

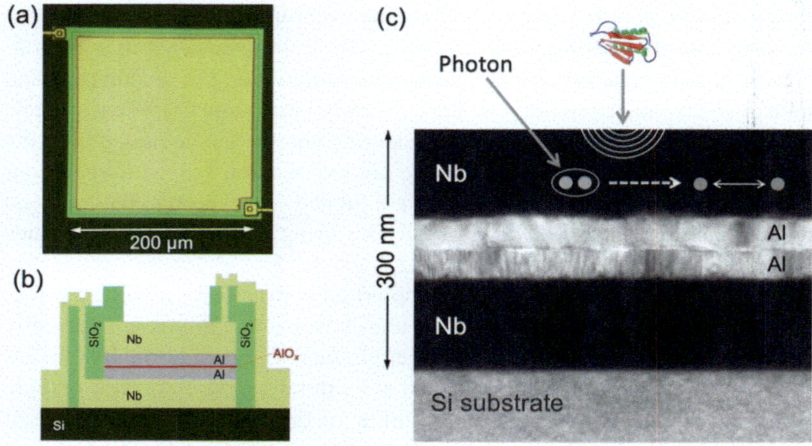

Fig. 3.9 Typical structure of superconductor tunnel junction detectors (STJDs): (**a**) top-view optical micrograph, (**b**) cross-sectional structure along the diagonal direction passing though the contact holes for the top and bottom electrodes, and (**c**) TEM cross-sectional view near the center of the Nb/Al/AlO_x/Al/Nb multilayer junction. The thickness of the AlO_x tunnel barrier is approximately 1 nm, forming a superconductor-insulator-superconductor (SIS) sandwich structure for quantum tunneling. In (**b**), the film thicknesses of the SiO_2 layer and the Nb wiring leads are not scaled. Both Al and Nb layers deposited by DC sputtering have polycrystalline structure. The top Nb layer predominantly acts as the absorber for soft-X-rays and molecules. Relatively thick Al layers (30–70 nm) are effective in achieving a low leakage current and increasing output charge. Typical surface morphology of the Nb films is shown in Fig. 3.7. In (**c**), photon and molecule incidence results in Cooper pair breaking, generating two quasiparticles. Reproduced with permission from Ohkubo [144]. Copyright 2023 AIP Publishing LLC

Analytical science requires a large sensitive area or a large solid angle to acquire signals such as characteristic X-rays emitted from analytes. Therefore, a junction size is more than 100 times larger than that of sub-μm Josephson junctions for the digital and quantum bit applications. The junction areas for analytical science are normally between 100 and 200 μm^2 to retain a proper junction capacitance for semiconductor readout circuits. The SIS structure forms a capacitor [123]. Another technical consideration is that a possibility of pinholes in 1-nm thick tunnel barrier layers increases with the junction size. Nevertheless, the large junctions are still considerably smaller than semiconductor detectors; the latest high-resolution SDDs [124] have a size of a few tens of square mm. The smallness of superconductor detectors is always a deficiency. Therefore, a straightforward way to achieve a practical sensitive area is to arrange many junction pixels two-dimensionally.

Another difference from the digital and quantum bit applications is that the surface of the top superconductor electrode is exposed to air. The junctions for integrated circuits are normally passivated by an insulation layer, which is unsuitable for detection of particles. The bare superconductor surface is necessary for covering an X-ray energy range below ~ 2 keV and enabling detection of large molecules. For example, the O K-$L_{2,3}$ ($K\alpha_{1,2}$) X-rays at 525 keV is attenuated to 60% for a 500-nm thick SiO_2

layer. We use the IUPAC notation for characteristic X-rays with the conventional Siegbahn notation in parentheses; the details appear in Appendix B. It is even worse for large biomolecules such as proteins that completely stop on a detector material surface even when they have a kinetic energy of a few tens of keV. Therefore, the detector surface must be a bare superconductor without any passivation layer or contamination.

We design the contact holes to connect the Nb electrodes and the wiring leads as small as possible to prevent the quasiparticle out-diffusion from the superconductor electrodes to the wiring leads. Furthermore, the small contact holes are indispensable for avoiding undesirable absorption events in the contact hole areas. It is possible to make a 1-μm square contact hole readily with common photolithography as shown in Fig. 3.9a. The ratio of the event number on the contact hole to that for other area is ~ 2×10^{-4}, which is negligibly small.

Figure 3.9b shows the schematic cross-sectional structure of one STJ pixel. The Nb/Al bilayer electrode structure crucially affects leak current and detector performance. The tunnel junctions normally have the Al layers with a thickness of several 10 nm on both sides of the AlO_x tunnel barrier layer. The Al layers have a columnar structure like the polycrystalline Nb layers in Fig. 3.7. The TEM image of the Al layers in Fig. 3.9c reflects the orientation of the Al single-crystalline grains: the Nb layers are totally black due to the large difference in the TEM contrast between Al and Nb. The considerably thick Al layers originate from energy-gap engineering proposed by Booth in 1987 [125] and Kraus in 1989 [126], expecting quasiparticle trapping in the Al layers. This will be discussed in Sect. 3.2.4.4 along with an alternative model based on gap reduction due to the proximity effect [127] in thin film devices.

More detailed surface analysis was performed in the course of fabrication processes for a 5-layer structure of Nb/Al/AlO_x/Al/Nb on a Si substrate [89]. Surface roughness naturally increases as sputtering deposition proceeds from the bottom Nb layer to the top Nb layer. In addition, the TEM cross-sectional view shows that the top Nb/Al electrode has a larger grain size than the bottom Nb/Al electrode. The relationship between polycrystalline features of Nb and Al films and electric properties was investigated [128]. A rough surface of the bottom Nb layer enhances Al diffusion into Nb-grain gaps, which induces a nonuniform tunnel barrier with pinholes and micro-shorts. The Josephson computer development required intensive structural investigation on Josephson tunnel junctions. Recently, structural investigations of tunnel junction layer structures have revived due to the rapid advancement of superconductor qubits for quantum computing. The details of this topic are left to the literature.

High atomic-number (Z) superconductors such as Nb or Ta serve as an X-ray absorber with a meaningful absorption efficiency. Proximity bilayers such as Nb/Al or Ta/Al maintain a T_c value higher than 1.2 K of bulk Al, which is favorable for raising operating temperature. The bottom electrode can be fabricated from a single crystalline superconductor film [65, 129], which is advantageous to optical photon detection. The amorphous AlO_x tunnel barrier layer formed by oxidizing the Al surface has a thickness of about 1 nm, which is a few atomic layers. At this thickness, the tunnel barrier can be inhomogeneous across an sensitive area of 100×100 – 200×200 μm^2. It was reported that less than 10% of the junction area dominantly contributed to tunnel current due to a Gaussian variation of AlO_x thickness without pinholes [130]. The AlO_x barrier layers have an oxygen deficiency near Al/AlO_x interfaces [131]. Relation between electric properties and completeness of AlO_x layers will be discussed later in Fig. 3.14. One of the latest microfabrication technology for array STJDs appears in Ref. [122], in which the junction and wiring layers have a three-dimensional (3D) layer structure, separated vertically by using a planarization process with chemical mechanical polishing (CMP).

3.2.4.2 Electric Properties of Josephson Junctions for Detectors

Current–Voltage (I–U) Characteristic Curves

Current–voltage (I-U) characteristic curves of a Nb/Al junction is shown in Fig. 3.10. The 50-μm square junction has 12-μm wide wiring leads such that full I–U curves beyond the junction I_c is measurable without exceeding the wiring lead I_c. The thickness of Al layers on both sides of the AlO_x barrier is 5 nm; the resulting proximity effect leads to a small hysteretic feature near $2\Delta_g/e$. The I–U curves are described by the proximity effect model specific to the Nb/Al/AlO_x/Al/Nb layer structure [132–134]. In Fig. 3.10a, the DC Josephson current observed at zero voltage in a zero magnetic field precludes observing the subgap region less than $2\Delta_g/e$. For detector operation, the DC Josephson current should be suppressed by applying a magnetic field of several mT in parallel with a junction surface and normally along the diagonal direction. This operation of SIS junctions is called the Giaever mode [135, 136]. Figure 3.10b shows the temperature dependence of I–U curves in the subgap region-below $2\Delta_g/e$. The subgap current quasi-exponentially decreases with decreasing temperature, reflecting the quasi-exponential reduction of thermally excited quasiparticles. The slight difference in Δ_g values between the top and bottom electrodes leads to the kinks near $\pm$ 0.3 mV. The gap disparity is due to the differing Nb layer thickness in the top and bottom electrodes.

The suppression of the DC Josephson effect and the Fiske resonance strongly depends on magnetic field directions and junction shapes. The suppression in a low magnetic field has required various Josephson junction shapes — diamond, sine, and quartic curves, as well as normal functions — proposed so far [137–140]. The diamond shape is better than a simple regular square by applying a magnetic field along the longer diagonal direction. The diamond shape is effective in reducing the necessary strength of the magnetic field [138], although a weak side-lobe Fraunhofer pattern remains. The normal-function shape further improves the suppression and realizes the complete side-lobe suppression in principle [139]. Figure 3.11a shows the normal-function shape and the magnetic field direction. Figure 3.11b plots the magnetic field dependence of DC Josephson currents. The magnetic flux is expressed by $\Phi = Bd\sqrt{A}$, where d is the sum of magnetic penetration depth and the barrier thickness and A is the junction area. A small magnetic field of $2\Phi_0/d\sqrt{A}$ is

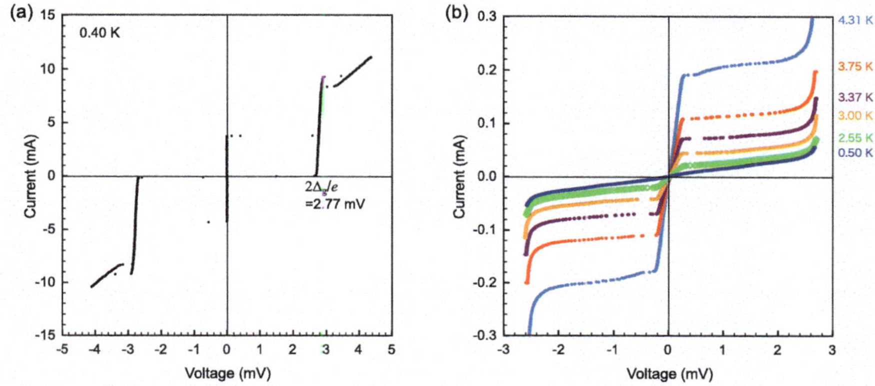

Fig. 3.10 Current–voltage (I–U) characteristic curves of 50-μm square Nb/Al junction: (**a**) I–U curve at 0.40 K in the absence of a magnetic field and (**b**) I–U curves at different temperatures after suppressing the DC Josephson current with a magnetic field of ~ 10 mT along the junction-diagonal direction. The 50-μm square junction has a layer structure of Nb(300)/Al(5)/AlO_x/Al(5)/Nb(100), in which the numbers in parentheses are thicknesses in nm, with a normal state resistance R_n of 0.394 Ω and a critical current density J_c of 160 A/cm^2. The $2\Delta_g$ value of 2.77 meV is slightly reduced from 3.06 meV of bulk Nb due to the proximity effect with the Al layers. The gap values of the top and bottom superconductor electrodes are slightly different at Δ_1(~ 0 K) = 1.52 meV and Δ_2(~ 0 K) = 1.25 meV, which originates from the Nb layer thickness disparity. The gap difference induces the small kinks at bias voltages $\pm |\Delta_1 - \Delta_2|/e$ (± 0.3 mV) in (**b**). At the Al layer thickness of 5 nm, the incomplete Al-coverage [128] of Nb surfaces causes a large leak current, shown by a linear current increase with increasing voltage at 0.50 K in (**b**)

sufficient for suppressing the DC Josephson current completely, where Φ_0 is the flux quantum (2.067 834 × 10^{-15} Wb). For a junction with $A = 10^4$ μm^2 and $d = 50$ nm, it is as low as 0.83 mT, which is approximately equivalent to one flux quantum for a cross-section of 100 μm × 50 nm. When a necessary magnetic field strength decreases, a coil integrated underneath a junction layer is sufficient instead of conventional external large magnet coils [141]. Furthermore, for an annular junction shape, one Josephson vortex trapping is enough to suppress the Josephson effect without an external coil [142]. Another advanced method is to use a magnetic tunnel barrier, which requires no external magnetic field to suppress the DC Josephson effect[143]. Nevertheless, practical array STJDs currently installed in analytical instruments have a simple square shape and an external magnet coil. The regular square shape is favorable to filling a space with as many STJ pixels as possible; a magnetic field is applied along the diagonal direction, imitating the diamond shape. A relatively high magnetic field of a few 10 mT and a high bias point near Δ_g/e are effective in operating many pixels simultaneously and stably. These conditions are not optimum for the best ΔE; nevertheless, we can achieve a ΔE value of 6 eV for the soft X-rays at 400 eV, including electronic noise after integrating all the 100 Nb/Al-STJ pixels [144]. The normal-function-shaped STJDs are promising for compact instrumentation without an external magnetic coil in future.

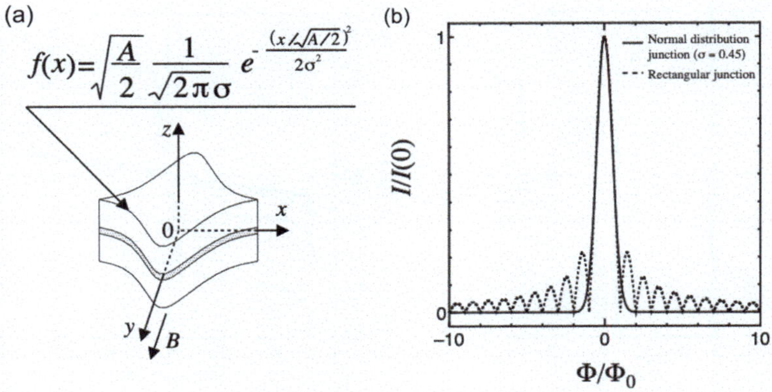

Fig. 3.11 Efficient suppression of DC Josephson current: (**a**) normal-function (normal distribution) shape expressed by the equation and the direction of applied magnetic field B, and (**b**) comparison in Fraunhofer patterns between the simple regular square shape and the normal-function shape [139]. The magnetic field is expressed by $\Phi = Bd\sqrt{A}$, where d is the sum of the magnetic penetration depth λ and the barrier thickness (~ 1 nm) and A is the junction area. For a 100-μm square Nb junction with $\lambda = 39$ nm and $A = 100\ \mu\text{m}^2$, $10\Phi_0$ in (**b**) corresponds to 2.6 mT. Reproduced with permission from Kikuchi et al. [139]. Copyright 2000 IOP

Semiconductor-Model Description for Superconductor Tunnel Junctions

The semiconductor-model description for the SIS structure in Fig. 3.12 explains the I–U curves in Fig. 3.10. In a thermal equilibrium state, thermally excited quasielectrons and quasiholes exhibit no charge imbalance because of the branch mixing through phonons, so that they contribute to the tunnel current equally. As explained earlier in Sect. 3.2.3.1 and Sect. 3.2.3.4, the semiconductor-model description implicitly postulates that quasiparticles have either pure electron character or pure hole character. The tunnel current drastically increases at the bias voltages $U_b = \pm|\Delta_1 + \Delta_2|/e$. The slope in bias regions higher or lower than $\pm\,|\Delta_1 + \Delta_2|$ shows a linear increase with increasing bias voltage, representing a normal state resistance R_n that is a transparency of the AlO_x tunnel barrier. In the bias regions $|U_b|>|\Delta_1 + \Delta_2|/e$, the electrons filling the states lower than $-\Delta_1$ or $-\Delta_2$ tunnel to the vacant states higher than $+\Delta_1$ or $+\Delta_2$. The R_n is independent of temperature. In $|U_b|<|\Delta_1 + \Delta_2|/e$, on the other hand, the subgap current increases with increasing temperature due to the thermally excited quasiparticles, reducing sub-bap resistance. In Fig. 3.12, three quasiparticle tunnel paths at an elevated temperature are denoted by the arrows 1, 2, and 3. The path 1, which is forward tunneling, transfers one electron from the left electrode to the right one. The path 2 transfers one hole from right to left; one electron moves from left to right. The path 3 that breaks and creates a Cooper pair is equivalent to the path 2 for quasihole tunneling and called back tunneling or the Gray effect [145]. The path 3 excites one quasiparticle in the left electrode, while one quasiparticle disappears from the right electrode, still transferring one electron from left to right. Therefore, in the bias condition in Fig. 3.12. All particle absorption events either in the left (top)

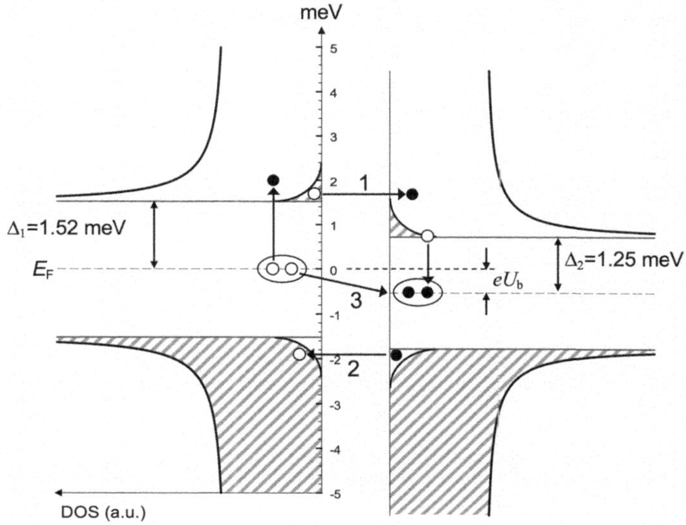

Fig. 3.12 Semiconductor-model description for quasiparticle tunneling processes in superconductor tunnel junction biased at eU_{b} at elevated temperature. The BCS-DOS curves in Fig. 3.8 are rotated 90° such that tunneling occurs horizontally through the ~ 1-nm thick AlO_x insulation barrier in the middle. The Δ_{g} values of 1.52 and 1.25 meV represent the particular case in Fig. 3.10. The proximity effect of the Al layers is omitted: the DOS curves follow Eq. (3.17). No charge imbalance between quasielectrons and quasiholes is expected in a thermal equilibrium state. Three quasiparticle tunnel paths are denoted by the arrows: the path 1 represents electron tunneling from left to right, and the path 2 represents hole tunneling from right to left, and the path 3, which breaks and creates a Cooper pair, is equivalent to the path 2. The path 3 is called back-tunneling or the Gray effect [145]. All the paths of 1, 2, 3 transfer an electron from left to right. Reproduced with permission from Ohkubo [144]. Copyright 2023 AIP Publishing LLC

and right (bottom) electrodes generate the same current flow direction from right to left.

In $T < 0.1T_{\mathrm{c}}$, an X-ray absorption event creates nonequilibrium quasiparticles in either left or right electrode of the STJ in Fig. 3.12. The electron-charge transfer due to a bias voltage of U_{b} still always occurs from left to right due to forward and back tunneling paths, resulting in the same output current direction regardless of which electrode absorbs an X-ray photon. In an intermediate temperature range $k_{\mathrm{B}}T > eU_{\mathrm{b}}$, some thermally-excited electrons and holes tunnel in the directions opposite to the arrows, reducing the net current flow from right to left in Fig. 3.12. Therefore, in a subgap voltage region below $|\Delta_1 + \Delta_2|/e$, we must consider four dominant tunnel paths. The net subgap current I_{sub} of the four tunnel paths is expressed by integrating the multiplication of the BCS-DOS curves and the Fermi distribution function:

$$I_{\mathrm{sub}} = \frac{1}{eR_{\mathrm{n}}} \int_{-\infty}^{\infty} \frac{|E_k|}{\sqrt{E_k^2 - \Delta_1^2}} \frac{|E_k + eU_{\mathrm{b}}|}{\sqrt{(E_k + eU_{\mathrm{b}})^2 - \Delta_2^2}} \left[f(E_k) - f(E_k + eU_{\mathrm{b}})\right] \mathrm{d}E_k, \tag{3.21}$$

where the range of the integration excludes $|E_k| < |\Delta_1|$ and $|E_k + eU| < |\Delta_2|$. An approximation of the integration in Eq. (3.21) for $\Delta_1 = \Delta_2 = \Delta_{\mathrm{g}}$ and $k_{\mathrm{B}}T \ll \Delta_{\mathrm{g}}$ yields

$$I_{\mathrm{sub}} = \frac{2}{eR_{\mathrm{n}}} \exp\left(\frac{-\Delta_{\mathrm{g}}}{k_B T}\right) \sqrt{\frac{2\Delta_{\mathrm{g}}}{eU_{\mathrm{b}} + 2\Delta_{\mathrm{g}}}} \left(eU_{\mathrm{b}} + \Delta_{\mathrm{g}}\right) \sinh\left(\frac{eU_{\mathrm{b}}}{2k_{\mathrm{B}}T}\right) K_0\left(\frac{eU_{\mathrm{b}}}{2k_{\mathrm{B}}T}\right), \tag{3.22}$$

where K_0 is the zeroth-order modified Bessel function [146].

Subgap Structure and Saturation Current in a Temperature Range well below $0.1T_c$

Equations (3.21) and (3.22) exhibit the quasi-exponential temperature dependence of the subgap current. The BCS theory predicts a ratio I_{sub} (0.5 K)/I_{sub} (4.2 K) = 2×10^{-14} at a bias voltage of 0.5 mV in an ideal Nb/AlO_x/Nb junction. Experimentally, this has never been observed; there are two cases for large subgap saturation currents in an extremely low temperature range well below $0.1T_{\mathrm{c}}$.

The first case is observed for the junction with the 5-nm thick Al layers (denoted by "Al 5-nm" hereafter) in Fig. 3.10b. Incomplete Al layer coverage of the Nb surfaces results in a large subgap current of 8.7 μA at 0.5 mV and 0.5 K: I_{sub} (0.5 K)/ I_{sub} (4.2 K) = 5×10^{-2}, which is tremendously larger than the BCS prediction. In this large leak current, the *I*–*U* curves exhibit the linear increase of the subgap current with increasing bias voltage, which indicates that resistance shunts the top and bottom electrodes. The incomplete Nb-surface coverage tends to occur in Al layers thinner than 30 nm. A probable formation of normal-conducting NbO_x instead of AlO_x causes the shunt resistance. The dependence of I_{sub} (T)/I_{sub} (4.3 K) on inverse temperature for "Al 20-nm" in Fig. 3.13 also exhibits the shunt resistance case: the difference between 4.2 and 4.3 K is negligible. The experimental subgap currents denoted by the red open triangles are saturated in $1/T > 0.7$ ($T < 1.4$ K). The red solid curve is the BCS prediction: Eq. (3.22) is fitted to the experimetal data using the free parameter T_{c} and the two fixed parameters Δ_{g} (0 K) and I_{sat}, that is the saturation current at the lowest temperature. By adopting Δ_{g} (0 K) = 1.08 meV from the *I*–*U* curve in Fig. 3.15a and $I_{\mathrm{sat}}/I_{\mathrm{sub}}$ (4.3 K) = 5.9×10^{-3}, the fitted T_{c} value is 9.33 K. This is consistent with the bulk-Nb value of 9.23 K. The junctions shunted with resistance can detect X-ray photons, but energy resolution is considerably worse than that of Si detectors.

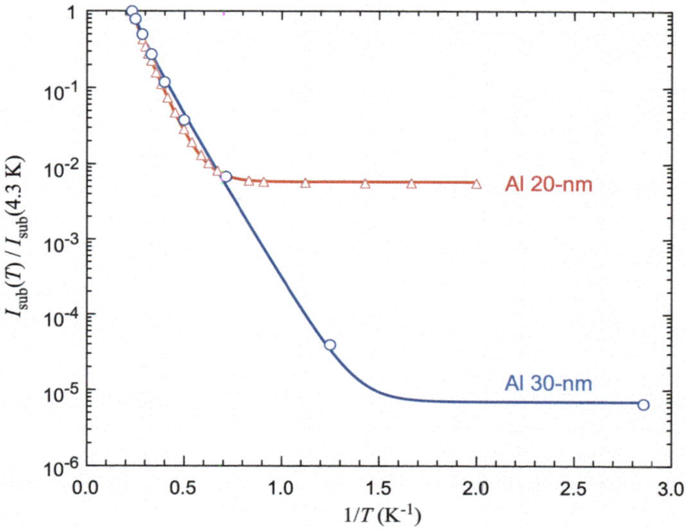

Fig. 3.13 Normalized subgap currents of Nb/Al square junctions as a function of inverse temperature. The junctions with a layer structure of Nb(200)/Al/AlO_x/Al/Nb(200)/Si have a size of 200 × 200 μm^2, a critical current of 200 A/cm^2: the thickness of the Al layers is either 20 nm or 30 nm. The junctions are biased at 1.0 mV for "Al 20-nm" and 0.6 mV for "Al 30-nm." The I_{sub} (4.3 K) for "Al 20-nm" is 3.4 mA. The solid lines in red and blue depict the BCS predictions fitted to the experimental data with saturation currents at low temperatures. The Δ_g (0 K) values are fixed at 1.08 meV for "Al 20-nm" and 0.80 meV for "Al 30-nm," which are determined from I–U curves. The junction quality necessary for X-ray detection requires Al layers with a thickness of 30 nm or thicker. The I_{sub} ($T < 0.5$ K)/I_{sub} (~ 4 K) values of high quality junctions fabricated at diverse institutions are commonly saturated at ~ 10^{-6}

The second case of the high subgap saturation current is shown by the blue open circles in Fig. 3.13. The junction with 30-nm thick Al layers "Al 30-nm" exhibits a I_{sub} (0.5 K)/I_{sub} (4.3 K) value of ~ 5×10^{-6}. The subgap saturation current is three orders of magnitude lower than that of "Al 20-nm," which implies that Al entirely covers the Nb surfaces. Although it is still higher than $2 \times 10^{-14}$of the BCS prediction, this relatively low saturation current is obtained in junctions with Al layers thicker than 30 nm. The lowest I_{sub} ($T < 0.5$ K)/I_{sub} (~ 4 K) values observed at diverse institutions are common at ~ 10^{-6}, which means either a limit of the tunnel barrier fabrication technology using surface oxidization of Al surfaces or a fundamental phenomenon without barrier defects such as pinholes and shorts. Junctions in this second case exhibit nonlinear dependence of subgap current on bias voltage [147]; especially, an abnormal current increase appears at $U_b = \Delta_g/e$, as shown in Fig. 3.14a. The nonlinear feature indicates that the subgap current is not due to a simple leak in the tunnel barrier, but rather to excess quasiparticles that are excited or injected into the superconductor electrodes. Empirically, we require junctions exhibiting the non-linear subgap current feature for detector applications. The latest 100–200-μm square junctions for analytical instruments have a subgap current below 10 nA.

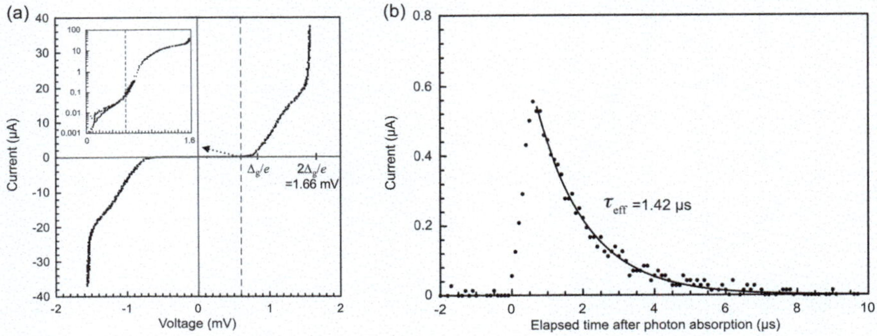

Fig. 3.14 Typical *I–U* curve at 0.47 K in (**a**) and output current pulse for 5895-eV X-ray photon in (**b**) at a bias voltage of 0.60 mV. The bias point is indicated by the vertical dashed lines in (**a**). The 200-μm square junction detector, which is indicated by "Al 30-nm" in Fig. 3.13, has a layer structure of Nb(200)/Al(30)/AlO$_x$/Al(30)/Nb(200) with thicknesses in nm in parentheses, an energy gap Δ_g(~ 0 K) of 0.83 meV, a normal resistance R_n of 23.1 mΩ, and a critical current density J_c of 200 A/cm^2. The dashed arrow in **a** denotes the readout circuit load line. The fitted solid line in (**b**) exhibits a quasiparticle effective lifetime τ_{eff} of 1.42 μs. Reproduced with permission from Ohkubo [144]. Copyright 2023 AIP Publishing LLC

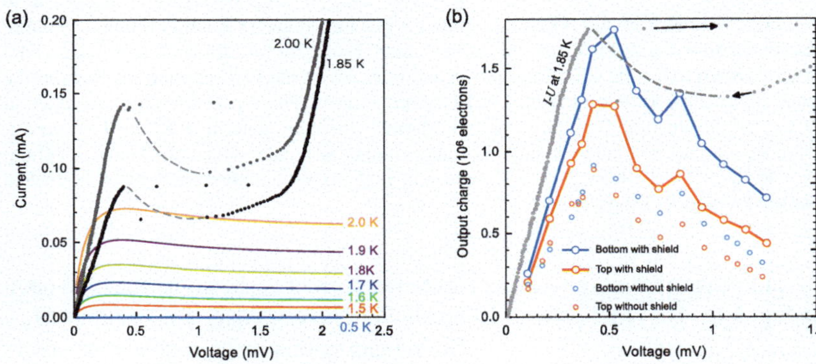

Fig. 3.15 Relation between static *I–U* curves and X-ray output amplitudes of 200-μm square junction with Δ_g (0 K) = 1.08 meV and R_n = 34.4 mΩ (13.8 μΩ cm^2): (**a**) experimental and calculated *I–U* curves at various temperatures and (**b**) bias-voltage dependence of output charge for 5895-eV X-rays at 0.33 K [169]. The junction has a layer structure of Nb(200)/Al(20)/AlO$_x$/Al(20)/Nb(200)/Si. The colored solid *I–U* curves in (**a**) are based on Eq. (3.22). The black dots denote the experimental data at 1.85 K and 2.00 K. The dashed lines for the experimental *I–U* curves in (**a**) and (**b**) show the expectation in the negative resistance region. The arrows in (**b**) indicate the current sweep directions. The output charge curves for the X-ray absorption events in either top (red open circle) or bottom (blue open circle) electrodes follow the experimental *I–U* curve at 1.85 K in (**b**) in a voltage bias region below 0.5 mV. The origin of the reproducible anomalous jumps at 0.84 mV is unknown. The maximum output charge for the bottom electrode is 1.7 × 10^6 electrons for cooling in a magnetic shield. Cooling in a geomagnetic field of ~ 35 μT without a magnetic shield results in the output charge reduction (blue and red open circles without a guide line) by a factor of 2 due to quasiparticle trapping by Abrikosov vortices (AVs)

What is the origin of the subgap saturation current with non-linear voltage dependence in a temperature range well below $T_c/10$? It resembles the excess quasiparticle density observed in qubits. Recently, one of the topics on superconductor qubits is that, at a few 10 mK that is the operating temperature, the quasiparticle density appears to exceed the thermally expected value by a factor of 10^{40} [148]. Such nonequilibrium excess quasiparticles cause decoherence of qubits, which is one of the serious problems on quantum computing. A plausible source for the excess quasiparticles is stray photons [148, 149]. If stray photon noise is the source of quasiparticle excitation, the ratio between photon energy E_0 and Δ_g of superconductors is close to E_0/E_g for an ultraviolet region in Si detectors. The theoretical treatment for a few e–h pair generation in Si is still an unsolved question. The situation in superconductor detectors is worse; currently, we have no data on the ε values of superconductors for meV photons both experimentally and theoretically. There are rooms for future studies on ε values in superconductor quantum detectors.

The exact origin of the nonequilibrium quasiparticles in a temperature range below $\sim T_c/10$ is still an open question, especially after the experiment on a microwave Al strip eliminating stray photons exhaustively [150]. The photon energy near $2\Delta_g$ is equivalent to a THz range for Nb and a microwave range for Al. They observed no dependence of quasiparticle density on RF input-power in the Al strip. Therefore, it seems that the possibility of stary photons is excluded, although a possibility of microwave noise cannot be completely excluded [150]. When we assume that some continuous excitation causes the excess quasiparticles at low temperatures, the energy distribution should deviate from thermal equilibrium distribution. However, there is a report showing that the excess quasiparticles have an energy distribution compatible with a thermal equilibrium [151]. The excess quasiparticle behavior is mysterious; further studies are required.

Cryostats for X-ray spectroscopy usually have at least two X-ray windows with aluminized parylene to protect STJDs from room temperature blackbody radiation. Because the X-ray windows cannot block visible light completely the saturation current at extremely low temperatures can be ascribable to stray photons of visible light when STJDs are used for material analysis. On the other hand, the data in Fig. 3.13 were measured with a cryostat closed completely without X-ray windows such that no blackbody radiation affects the subgap currents. Therefore, the saturation currents or the excess quasiparticles of the blue curve for “Al 30-nm” in an $1/T$ range above 1.5 should stem from an unknow mechanism other than stray photons. Nevertheless, in X-ray spectroscopy, the subgap saturation current due to the excess quasiparticles is negligibly small compared with current pulses ($\sim 1\ \mu$A) due to X-ray photon events.

When we postulate that the saturation current and the excess quasiparticles near 0 K stems from a physical origin instead of tunnel barrier defects and external noise, a plausible contribution to the excess quasiparticles is multiple quasiparticle tunneling (MQT) as discussed below. The MQT cycles multiply nonthermal hot quasiparticles that acquire the energy corresponding to a bias voltage on every tunnel cycle. Note that there is also a report claiming that the saturation currents can be explained by considering the imaginary part of Δ_g [152].

Figure 3.14a shows a typical I–U curve of a detector-quality junction with a I_{sub} (0.5 K)/I_{sub} (4.2 K) value of ~ 10^{-6} for "Al 30-nm." The subgap current below $2\Delta_g/e$ is highly nonlinear. Figure 3.14b shows a typical output current pulse for a 5895-eV X-ray photon, which is later discussed in Sect. 3.2.4.3. Figure 3.14a indicates that the Δ_g value in the Al layers near the Al/AlO_x interfaces rises to 0.83 meV from 0.17 meV, which is the intact Al value, due to the proximity effect with Nb. Accordingly, it decreases from the intact Nb value of 1.52 meV. The striking feature is the rapid subgap current increase at Δ_g/e, deviating from the BCS prediction. The observation of the deviation above Δ_g/e is a typical sign for high quality junctions with a low leak current.

One of the possible mechanisms for the current increase at a bias voltage of $2\Delta_g/2e$ as well as other subharmonic gap structures at $2\Delta_g/ne$, where n is integral, is multiple Andreev reflection (MAR) [153, 154]. The MAR process transfers multiple elementary charges ne at bias voltages $2\Delta_g/ne$, provided that probable pinholes exist in an AlO_x barrier layer, forming normal-superconducting (N-S) boundaries [155]. However, the subharmonic gap structures were observed in pinhole-free junctions in a variety of barrier transparency [156]. Furthermore, no pinholes were found in a 1-nm thick AlO_x layer, whose local thickness variation follows a Gaussian distribution [130]. Therefore, the origin of the subharmonic gap structures is an open question. The subharmonic gap structures with $n > 3$ clearly appear in junctions with a high transparency with $J_c >$ ~ kA/cm^2. On the other hand, the junction in Fig. 3.14 is classified as a low transparency with $J_c \cong 200$ A/cm^2 to avoid the shunt-resistance leakage. Pinholes are unlikely at this low J_c even for a large junction area of 50–200 μm^2; nevertheless, the anomalous current increase is clearly recognizable for $n = 2$ because it is the largest subgap structure. The subharmonic gap structures without pinholes can be explained by the MQT model proposed by Kozorezov, et al. [103, 157], although other possibilities such as Josephson radiation self-absorption still remain. Providing no quasiparticle loss other than self-recombination [158], the quasiparticles that tunnel via the path 1 and the path 2 in Fig. 3.12 acquire an energy of eU_b after tunneling. The forward-and-back multiple tunneling causes energy accumulation. This MQT process leads to current jumps at $2\Delta_g/\overline{N}_{tun}\, e$, where $\overline{N}_{tun}$ is the average repetition number of quasiparticle tunneling, in a similar way to n in the MAR model. When extra loss of quasiparticles occurs, the subharmonic gap structure becomes obscure. The MQT model requires no pinholes in the AlO_x tunnel barrier; therefore, the MQT model can explain the subgap structures. Quasiparticles are multiplied by energy accumulation due to the sequential MQT. Even when no stray photons exist, a few quasiparticles excited thermally or other perturbations at low temperatures can be multiplied by the MQT process, inducing the observable saturation excess current or non-equilibrium excess quasiparticles. The MQT model is applicable to junctions in operation.

Although some preliminary explanations are possible as mentioned above, the origin of the subharmonic gap structures and the excess quasiparticles are an open question [156]. The subgap structures associated with excess quasiparticles influence qubits as well as superconductor quantum detectors. We need further studies to completely understand the excess quasiparticles in an extremely low temperature range.

3.2.4.3 Response to Particle Incidence

STJD Output Pulse

We consider here the stage after the energetic electrons after X-ray absorption or molecule impact end up in generating quasiparticles near Δ_g as described in Fig. 3.3b and Sect. 3.2.3.2. These quasiparticles can be regarded as mobile carriers on statistical physics for tunnel junctions biased in a subgap region below Δ_g/e. The average number of quasiparticle pairs is approximately calculated by $E_0/\varepsilon \approx E_0/(1.7 \times 2\Delta_g)$ by using Eq. (3.20). The quasiparticles generated in either top or bottom electrode of a junction biased in a subgap region experience a random walk and eventually tunnel to the counter electrode through the tunnel paths in Fig. 3.12. The signal current induced by X-ray photons or other particles resembles a thermally excited subgap current, as described in Sub-sect. "Effective Temperature for 6-keV X-ray Excitation" in this section, although the particle-induced current returns to zero according to the quasiparticle recombination into Cooper pairs or loss processes in contrast to the steady current at an elevated temperature. In addition, it is expected that quasiparticle spectral distribution for particle incidence is different from that in a thermal equilibrium state. Nevertheless, the effective temperature for X-ray absorption is a meaningful concept of detector operation.

An increase in quasiparticles generated by a 6-keV X-ray photon results in the current pulse shown in Fig. 3.14b. The vertical dashed lines in Fig. 3.14a indicate the bias point at 0.60 mV, which is slightly lower than Δ_g/e. Although the bias point exceeds the optimum condition for the maximum pulse height, it is suitable to avoid bias-point jumps due to Fiske steps [159], which is a resonance between AC Josephson current and the SIS cavity structure. Furthermore, the bias point near Δ_g/e is effective in operating many pixels in an array STJD simultaneously and stably during material analysis. When a particle incidence event generates quasiparticles, which leads to a junction resistance reduction, the bias point shifts from almost zero current to a high current, following the 210-Ω load line indicated by the dotted arrow in Fig. 3.14a. The combination of a cryostat with a wiring-line resistance of 160 Ω and a readout circuit with an input resistance of 50 Ω determines the load line. The load line can be closer to the perpendicular direction by using a low wiring resistance and a preamplifier with constant voltage biasing [160]. The voltage shift affects quasiparticle tunneling processes, but experimental results on energy resolution and energy linearity described in this section show that it is negligible in an X-ray energy range below ~ 1 keV.

Figure 3.14b shows the current pulse for a 6-keV X-ray photon from a ^{55}Fe source. The current pulse is reproduced by differentiating an output pulse from a charge-sensitive preamplifier with a nominal conversion gain of 0.45 V/pC to reduce electronic noise. The X-ray event creates a jump of 714 nA and –0.15 mV from the bias point. The rise time is determined by the readout bandwidth, while the fall time represents the duration for which quasiparticles remain in the junction structure. The duration is equivalent to the effective quasiparticle lifetime τ_{eff} including quasiparticle recombination and phonon loss processes. The exponential decay curve depicted by the solid line in Fig. 3.14b indicates a τ_{eff} value of 1.42 μs. Currently, we observe no sign for charge imbalance, the possibility of which is discussed in Sect. 3.2.3.4. The charge imbalance can influence the output pulse shape in an interval of ns after the X-ray absorption; however, it disappears at the time scale of Fig. 3.14b. The integration of the current pulse represents the detected total charge or the particle kinetic energy E_0. The charge pulse for the 6-keV photon contains 6.0×10^6 electrons, which is surely higher than ~ 1600 electrons in Si detectors. Even though no strong charge collection mechanism exists in STJDs, the extremely high charge output ensures a high signal/noise (S/N) ratio and a high energy resolution statistically, following Eq. (3.2) or (3.4).

The effective quasiparticle lifetime τ_{eff} is primarily governed by recombination into Cooper pairs, phonon down-conversion, and phonon escape into a substrate. When two quasiparticles with energies slightly above Δ_g recombine into a Cooper pair, they emit a phonon with energy slightly exceeding $2\Delta_g$. This phonon can break another Cooper pair, so that τ_{eff} is considerably longer than a material-dependent intrinsic quasiparticle lifetime [121]. This is called the phonon trapping effect. The quasiparticle and phonon systems gradually approach a steady state, the relaxation time of which is governed by phonon energy down-conversion or phonon escape from a superconductor to surroundings. The Rothwarf-Taylor model describes this phenomenon [161]. In Fig. 3.14b, the τ_{eff} of 1.42 μs is significantly longer than the characteristic time values of 0.149 ns for Nb and 438 ns for Al. Phonon pair breaking mean free path in Nb is exceptionally shorter than those in other elemental superconductors: characteristic times of phonon pair-breaking are 4.17 ps for Nb, 22.7 ps for Ta, and 242 ps for Al [121]. Therefore, in Nb/Al/AlO_x/Al/Nb/Si junctions, the phonon escape into the Si substrate is negligible even when a Nb film thickness is 100 nm [162–164]. Other superconductors exhibit nonnegligible phonon escape into a substrate [102]. We have often observed that in Nb/Al junctions bottom electrodes have longer τ_{eff} values than top electrodes, resulting in a higher output-charge for the absorption events in the bottom electrodes. An explanation for this behavior is that quasiparticle trapping on the top Nb surface occurs into an oxidized metallic surface layer of NbO_x formed during fabrication processes [164]. The metallic NbO_x surface layer has a Δ_g value lower than that inside the Nb electrode.

Quasiparticle Loss Due to Abrikosov Vortices

Quasiparticle loss shortens τ_{eff} and reduces the output charge. Moreover, the quasiparticle loss induces additional fluctuation of carrier number N for each event, causing energy resolution degradation. One of the quasiparticle loss processes stems from Abrikosov vortices (AVs) trapped along the direction perpendicular to the junction surface. AV trapping occurs when cooling a junction from room temperature to ~

0.3 K in a residual magnetic field. Polycrystalline Nb is usually a Type-II superconductor, allowing AV trapping in Nb/Al junctions. Our observation also implies that an electric shock induces AV trapping, as shown in Fig. 3.28b [144]. The direction of a magnetic field applied to suppress the Josephson effect should be in parallel to the junction surface to avoid the electric-shock-induced AV trapping. AVs have a normal-conducting core at the center and a reduced Δ_g region surrounding the core. Therefore, AVs cause quasiparticle trapping and loss, shortening the quasiparticle lifetime [119] and thereby reducing the output charge [120].

The reduction of τ_{eff} due to perpendicular magnetic field was measured at 1.6 K by a low temperature scanning electron microscope (LTSEM) at Tübingen [119]. The LTSEM can also indirectly visualize AVs trapped in junctions by observing junction response to an electron beam [165]. Furthermore, a scanning SQUID microscope directly visualized magnetic flux distribution and identify the AVs trapped in junctions: each vortex carries one magnetic flux quantum Φ_0. AV patters of Nb/Al junctions cooled in a perpendicular magnetic flux density from 0 to 7.3 μT exhibited a junction edge region where no AVs exist; this is consistent with vortex nucleation based on the classical Bean and Livingstone surface barrier model [144, 166]. Due to this surface barrier, 200-μm square junctions have no AVs when a perpendicular magnetic flux density is less than ~ 1 μT. Correspondingly, complete exclusion of AVs requires a magnetic shield with an attenuation factor higher than ~ 50 in a geomagnetic field of ~ 35 μT.

The quasiparticle loss rate is expressed by

$$1/\tau_{\text{eff}} = 1/\tau_{\text{r}} + 1/\tau_{\text{AV}}, \tag{3.23}$$

where τ_{r} is the quasiparticle lifetime against recombination into Cooper pairs and phonon loss without AV trapping, and τ_{AV} is the quasiparticle loss time due to AVs. The AV quasiparticle loss rate $1/\tau_{\text{AV}}$ is given by $1.82 n_{\text{v}} \pi R_{\text{eff}}{}^2/\tau_0$, where n_{v} is the AV density, R_{eff} is the effective vortex radius, and τ_0 is the material dependent characteristic time [119, 121, 167]. We evaluate the τ_{AV} value for a 50-μm square junction with Nb(200)/Al(40)/AlO$_x$/Al(40)/Nb(200)/Si with thicknesses in nm in parentheses. The junction was cooled either in a magnetic shield at ~ 0 μT or in a geomagnetic field of 35 μT [120]. The τ_{AV} value evaluated from the measured values of τ_{r} in ~ 0 μT and τ_{eff} in 35 μT is 1.0 μs, which leads to $R_{\text{eff}} \approx 3\xi$, provided $n_{\text{V}} = 1.7 \times 10^{10}$ m^{-2} under 35 μT, $\tau_0 = 0.149$ ns [121], and $\xi = 10$ nm [169] in polycrystalline Nb films. This R_{eff} value is consistent with other reports [119, 167]. The charge output reduction after cooling in a geomagnetic field is 44%, which significantly degrades the detector performance. Therefore, for the best performance, we need to cool STJDs in a magnetic shield, which should meet the above-mentioned condition, to prevent AV trapping. For the details of AV trapping, see Refs. [120, 144, 166].

Effective Temperature for 6-keV X-ray Excitation

Previously, the bias point was fixed near Δ_g/e. However, the bias points significantly influence STJD operation, envisaged from the semiconductor-model description for STJDs in Fig. 3.12. The bias-point dependence is important to understand the detector physics. The bias dependence of STJD charge output for X-ray photons resembles the static *I–U* curve at an elevated temperature. Figure 3.15a shows a comparison between Eq. (3.22) and the experimental *I–U* curves at 1.85 and 2.00 K for a symmetric Nb/Al junction with 20-nm-thick Al layers and a normal resistance R_n of 34.4 mΩ. The proximity effect of the thin Al layers is moderate; the Δ_g value at 0 K is reduced from 1.52 meV of the intact Nb to 1.08 meV, suggesting that the deviation from the BCS DOS profile in the electrodes is also moderate. Exact calculations for junctions with significant proximity effects can be found in the literature, as in Refs. [132–134]. The curve calculated with Eq. (3.22) at 2.0 K has the maximum tunnel current at a bias voltage of 0.43 mV. As bias voltage increases further, tunnel current decreases, following the DOS reduction in Fig. 3.12. The curve calculated at 2.0 K is close to the experimental one at 1.85 K rather than 2.0 K. The experimental subgap current is about two times larger than the calculated one. However, it is tolerable, considering the large temperature dependence of subgap current in this temperature range. In addition, the experimental curves have sharper bumps at 0.45 mV than the calculated ones, which reflects a DOS profile of the weak proximity effect [132–134]. The negative resistance region between 0.4 and 1.1 mV cannot be traced experimentally; the dashed lines indicate anticipated curves. Another deviation from the calculation with Eq. (3.22) is the current increase above Δ_g/e (= ~ 1 mV). This excess current is independent of temperature. The MQT model can explain the subgap current increase at Δ_g/e [157], as discussed in Sub-sect. "Subgap structure and saturation current in a temperature range well below 0.1 T_c" in Sect. 3.2.4.2

Figure 3.15b superimposes the *I–U* curve at 1.85 K on the bias-voltage dependence of the output charge for 5895-eV X-rays at 0.33 K. The top and bottom electrodes respond in slightly different ways to the X-rays, producing a double peak in X-ray spectra [169]. Nonetheless, the bias-voltage dependence of the output charge follows the main feature of the *I–U* curve at 1.85 K, especially below 0.8 mV. Above ~ 1 mV, the monotonic decrease of the output charge stems from imperfect charge collection due to the junction resistance reduction in a bias region of $\Delta_g/e < U_b < 2\Delta_g/e$. The resistance reduction represents the tunnel current increase above Δ_g/e, the plausible mechanism of which is quasiparticle excitation due to the MQT process. It leads to discharge of the charge stored in the feedback capacitor of a charge-sensitive preamplifier through the junction in parallel with the feedback circuit. Therefore, it is reasonable that the X-ray data deviate downward from the static *I–U* curve above ~ 1 mV. The agreement below 0.8 mV in Fig. 3.15b indicates that the quasiparticle excitation by one 6-keV X-ray photon at 0.33 K is equivalent to that in the thermal equilibrium state at 1.85 K.

Strictly speaking, the quasiparticle energy distribution due to the energetic particle impact at 0.33 K is different from the Fermi distribution in the thermal equilibrium at 1.85 K. Furthermore, particle impact creates excitation locally, producing spatially inhomogeneous distribution of quasiparticles. In a thermal equilibrium state, on the other hand, the quasiparticle density is uniform throughout the top and bottom electrodes. Nonetheless, the voltage-bias dependence exhibits that the nonthermal excitation due to one 6-keV X-ray photon creates a similar state to the thermal excitation at ~ 2 K.

Energy Resolution

Figure 3.16 shows a pulse height spectrum acquired with the 100-μm square asymmetric junction having a layer structure of Nb(300)/Al(70)/AlO_x/Al(70)/Nb(100)/Si in nm. Monochromatized 400-eV photons with an uncertainty of well smaller than 0.1 eV from the KEK-PF synchrotron radiation source illuminated the entire area of the STJD chip, which is called full illumination. The absorption coefficient of the 300-nm-thick top Nb film is 99.7%, so that no double peak appears in pulse height spectra. The single peak with the tail toward the low energy side is also called the response function. The low-energy tail predominantly stems from spatial inhomogeneity of charge output: photon events near the four edges of the square junction generate lower output charges. A quasiparticle diffusion model explains the response function of Fig. 3.16, which is described in Sub-sect. "Quasiparticle diffusion model for spatial inhomogeneity" in Sect. 3.2.4.7.

The FWHM value of the 400-eV peak in Fig. 3.16 is 4.1 eV, which is 10 times better than 45 eV of the latest 10-mm^2 SDD [53]: the 45-eV value is estimated from the ΔE values for the C K and O K lines. Subtracting an electronic noise of 2.0 eV from 4.1 eV of the 100-μm square STJD, the full illumination intrinsic energy resolution is 3.6 eV. It is better than ~ 10 eV of typical WDS spectrometers for electron probe microanalyzer (EPMA). A better energy resolution in this soft X-ray range can be obtained with a Ta/Al-based junction developed at LLNL. The Ta/Al junction recorded 2.5 eV at 250 eV and 4 eV at 1 keV including an electronic noise of 2 eV [64]. Ta with a lower Δ_g outperforms Nb in energy resolution, following Eq. (3.4), although a lower operating temperature is required.

The energy resolution of STJDs based on Nb/Al or Ta/Al exceeds typical natural linewidths of characteristic X-rays from light elements embedded in a matrix: the natural linewidths are typically 10–20 eV in FWHM. The linewidths are considerably wider than those of isolated atoms: < 1 eV in $Z < 20$ and 1–10 eV in $20 < Z < 50$ [168]. For example, the carbon atoms in $CaCO_3$ have a K-line width of ~ 14 eV, measured with the array STJD which has an energy resolution of 6.0 eV. In addition, the Zn L_3-$M_{4,5}$ (Zn L) line from a trace of Zn atoms inside a dentin sample has a linewidth of ~ 15 eV. Broadening of characteristic lines exceeds the level widths of electronic shells due to chemical bonds with other atoms or electronic band structure of solids. Therefore, no further improvement in energy-resolution is necessary for elemental mapping analysis in a range below ~ 2 keV. An example of the elemental mapping with SEM equipped with an array STJD is described in Chap. 6, Sect. 6.3.1.3. A better energy resolution is required for measuring chemical states or chemical shift, which is called X-ray emission spectroscopy (XES) in Chap. 6, Sect. 6.3.3.4. For

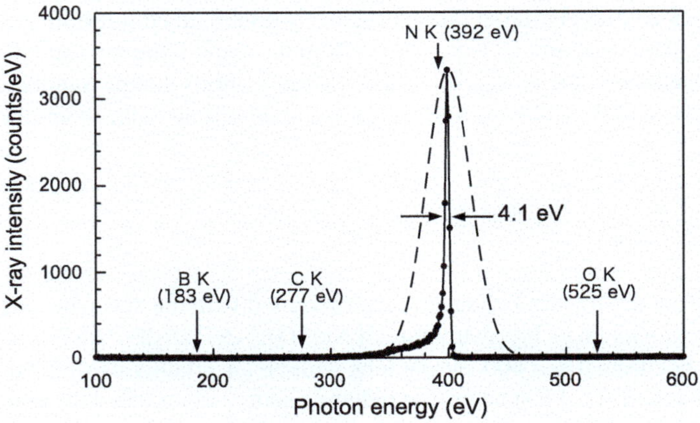

Fig. 3.16 Full-illumination pulse height spectrum of 100-μm square Nb/Al-junction for 400-eV photons with an uncertainty of less than 0.1 eV from synchrotron radiation at KEK-PF. The monochromatized X-ray photons illuminated the entire area of the STJD chip, which is full illumination. The junction has a layer structure of Nb(300)/Al(70)/AlO_x/Al(70)/Nb(100)/Si in nm: the junction is one of the 100-pixels in an array STJD. The bias point is near Δ_g/e. The energy bin for the x-axis is 1 eV. The y-axis indicates X-ray intensity expressed by the photon counts within each bin. This spectrum was acquired in real-time without off-line pulse processing. The 4.1-eV energy resolution at 400 eV includes an electronic noise of 2.0 eV, which means that the intrinsic resolution is 3.6 eV. Integrating all the 100 pixels, the array STJD has a ΔE value of 6.0 eV including electronic noise. The dotted line shows the peak with an FWHM value of 45 eV for the latest 10-mm^2 Si drift detector (SDD) [53]. The positions of the K lines for the light elements are indicated for reference. The peak shape is also called the response function. Reproduced with permission from Ohkubo [144]. Copyright 2023 AIP Publishing LLC

XES in a soft X-ray range below ~ 2 keV, the low energy tail of the STJD should be eliminated by adding an aperture structure in front of each pixel to improve energy resolution further.

The 100-pixel array Nb/Al-STJD including the pixel for Fig. 3.16 recorded an energy resolution of 6.0 eV including an electronic noise after integrating the pulse heigh spectra of all the pixels. With an X-ray polycapillary lens, the 100-pixel STJD is routinely utilized for imaging with SEM, covering a range below ~ 2 keV up to a count rate of 200 kcps (Fig. 3.33 in Sect. 3.2.4.8). A 25-mm^2 SDD with an energy resolution of 129 eV at 5.9 keV and a maximum count rate of 200 kcps simultaneously covers a range above ~ 2 keV. This STJD-SDD combination performs excellently, covering a wide element range from Li to U. In addition, a similar STJD system is in route use at the KEK-PF synchrotron radiation facility described in Chap. 6, Sect. 6.3.3.2.

The absorption efficiency of the top electrode decreases with increasing photon energy. Nonnegligible absorption events in the bottom electrode appear at 6 keV, as shown in Fig. 3.15b, resulting in a double peak [144, 162, 169]. The absorption events in the bottom electrode normally produce higher output pulses [162, 169]. The structural difference in grain sizes between the top and bottom electrodes in Fig. 3.9c

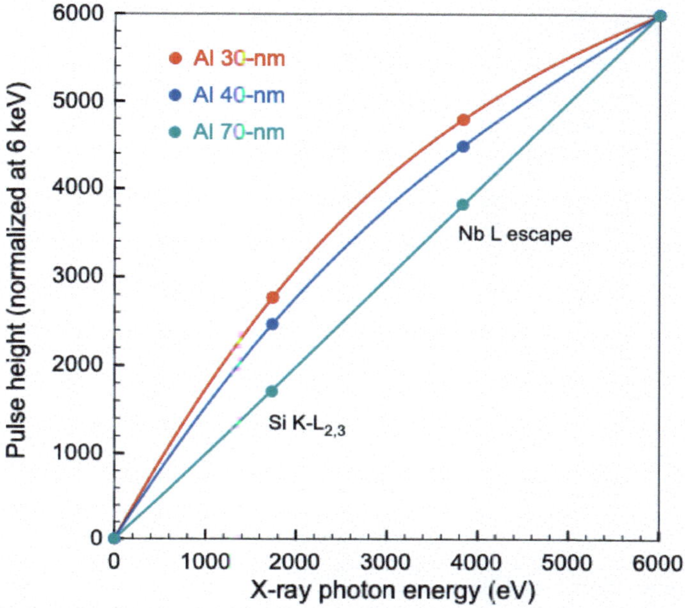

Fig. 3.17. Al-layer-thickness dependence of energy linearity (or nonlinearity) with respect to relation between absorbed energy (photon energy) E_0 and pulse height H, normalized at 6 keV. The superconductor tunnel junction detectors (STJDs) have a layer structure of Nb(200)/Al/AlO_x/Al/Nb(200)/Si in nm. The junction with 30-nm thick Al layers "Al 30-nm" has a size of 146-μm square, while other two junctions with "Al 40-nm" and "Al 70-nm" have a size of 200-μm square. The bias point is near Δ_g/e. The X-ray microbeam with a diameter of 5 μm irradiated the central area of each junction. The peak positions (pulse height) of the Si K-$L_{2,3}$ ($K\alpha_{1,2}$, 1.74 keV) line and the Nb-L escape peak (3.84 keV) are plotted in addition to that of the 6.00-keV full-energy peak. The solid lines are fitted to the data points with cubic polynomial functions of $H = 1.10 \times 10^{-8} E_0{}^3 - 2.24 \times 10^{-4} E_0{}^2 + 1.95 E_0$ for "Al 30-nm"; $H = 9.32 \times 10^{-9} E_0{}^3 - 1.71 \times 10^{-4} E_0{}^2 + 1.69 E_0$ for "Al 40-nm"; and with a quadratic polynomial function of $H = 2.2 \times 10^{-6} E_0{}^2 + 0.987 E_0$ for "Al 70-nm." The junction with the 70-nm thick Al layers has the perfect energy linearity

can affect the double peak phenomenon [89, 144, 169]. The double peak phenomenon is primarily due to a difference in τ_{eff} between the top and bottom electrodes. The top Nb electrodes are covered by a few nm thick NbO_x metallic layer, which acts as a quasiparticle trap and shorten τ_{eff} [164, 170]. Even in the symmetric layer structure, electric and phononic properties are asymmetric. The double peak is an obstacle to analyzing X-ray spectra: monochromatic X-rays produce a double peak. For 6-keV X-rays, ~ 90% of the incoming photons pass through entire STJ layer structure. In addition to the double peak, the high energy X-ray photons produce undesirable spectral features such as the Nb-L escape and the Si K-$L_{2,3}$ lines from the Si substrate, as shown in in Fig. 3.24 later in the explanation of spatial inhomogeneity. Therefore, STJDs covers a low X-ray energy region below ~ 2 keV. On the other hand, ions or

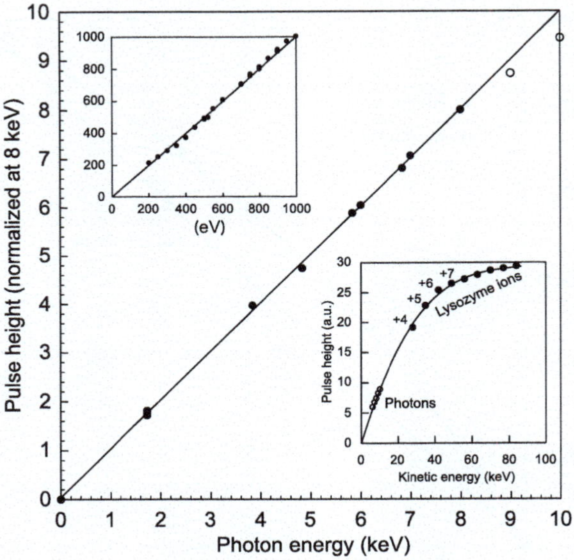

Fig. 3.18 Energy linearity of 200-μm square Nb(200)/Al(70)/AlO_x/Al(70)/Nb(200)/Si junction, measured with X-ray microbeams with energies above 1.74 keV and full illumination below 1 keV in the upper left inset. The thicknesses are in nm in parentheses. The synchrotron radiation microbeams with a diameter of 10 μm irradiated the center of the junction at 6, 7, 8, 9, and 10 keV. The data points include the Si K-$L_{2,3}$($K\alpha_{1,2}$, 1.74 keV) line and the Nb-L escape peak (3.84 keV). The lower right inset shows the data of the photons and the multiply charged ions of lysozyme (14 305 Da), accelerated at 7 kV. The slight deviation from the straight line below 1 keV is explained by the Nb M-edge absorption structure. The full illumination energy resolution at 400 eV is 28 eV with an electronic noise of 25 eV, which corresponds to an intrinsic energy resolution of 12 eV. Reproduced with permission from Ohkubo [144]. Copyright 2023 AIP Publishing LLC

molecule impact always produces a single peak regardless of the thickness of the top electrode because molecules stop on the top-Nb surface.

Energy Linearity

In addition to energy resolution, another important factor for spectroscopy is energy linearity; nonlinearity should be negligibly small to evaluate the energies of incident particles. Figure 3.17 shows the Al thickness dependence of energy nonlinearity normalized at 6.00 keV, measured with full illumination. The output charge values, which correspond to the pulse height values normalized at 6 keV, are plotted for three X-ray energies: 6.00-keV of the primary photons, 3.84 keV for the Nb-L escape peak, and 1.74 keV for the Si K-$L_{2.3}$, following a method in Ref. [171]. The latter two peaks, which are undesirable for X-ray spectroscopy, are useful for investigating energy linearity. In Fig. 3.17, the junctions with "Al 30-nm" and "Al

40-nm" exhibit strong nonlinearity. On the other hand, the junction with "Al 70-nm" exhibits a perfect linearity without any meaningful deviation from the straight line: the quadratic coefficient of 2.2×10^{-6} for "Al 70-nm" is negligibly smaller than 1.71×10^{-4} for "Al 40-nm." Therefore, it is apparent that Al layer thickness affects the energy linearity. In contrast to Fig. 3.17, it was reported that an excellent linearity was obtained for extremely thin Al layers at 7 and 0.7 nm [171]. At these extremely small thicknesses, it is considered that the Al layers exhibit almost no effect on quasiparticle dynamics and the Nb layers possess a superior energy linearity. However, junctions with thicker Al layers are preferable to spectroscopy for a better energy resolution; junctions with thin Al layers are not used for X-ray spectroscopy. Plausible mechanisms for the linearity change with Al thickness is discussed in Sect. 3.2.4.4.

The junction with "Al 70-nm" is suitable for spectroscopy in consideration of the superior energy linearity observed in full illumination. We precisely measured the energy linearity for "Al 70-nm" by using the microbeam with a diameter of 10 μm, which irradiated the junction center at 6, 7, 8, 9, and 10 keV. Figure 3.18 shows the plot of the Si K-$L_{2,3}$($K\alpha_{1,2}$) and Nb-L escape peaks as well as the data points for the primary X-ray photons. A slight deviation from the straight line is observed above 8 keV, indicated by two open circle data points. This STJD covers up to 30 keV with a tolerable energy nonlinearity as displayed in the lower right inset, which plots the pulse height values for the photons and multiply charged lysozyme ions accelerated at 7 kV. The data points for the charge states such as +4, +5, +6, +7 represent the kinetic energy E_0 values when the charge numbers are multiplied by 7 keV [172]. Lysozyme is an enzyme protein with a molecular weight of 14 305. Beyond ~ 30 keV, the strong nonlinearity prevents proper measurement of E_0. In spite of the excellent energy linearity up to 8 keV, a practical energy range with a meaningfully high X-ray absorption efficiency is below ~ 2 keV. The absorption efficiencies of the 200-nm thick Nb film are 55% at 1 keV and 95% at 500 eV [173].

The upper left inset shows the full illumination peak data below 1 keV without using X-ray microbeams. The full illumination energy resolution at 400 eV is 28 eV with an electronic noise of 25 eV, which corresponds to an intrinsic energy resolution of 12 eV. The energy resolution of this junction is worse than 3.6 eV of the junction in Fig. 3.16. The slight deviation from the perfect straight line stems from the X-ray absorption fine structure due to the Nb-M edge at 207 eV. The absorption efficiency decreases from 95% at 500 eV to 74% at 200 eV. This is opposite to a general tendency, where X-ray absorption efficiency decreases with increasing photon energy, due to the X-ray absorption fine structure. Except an energy range from ~ 300 to ~ 500 eV, a small number of photons are absorbed in the bottom electrode that produces slightly higher output pulses. This results in the slightly higher pulse height positions when we determine the peak centroid positions by fitting the Gauss function to the peaks. Regardless of this negligible deviation, we conclude that the STJD with "Al 70-nm" performs well in an energy range below ~ 2 keV with the adequate X-ray absorption.

In contrast to the excellent linearity in Fig. 3.18, it was reported that in an epitaxially grown Nb/Al/AlO_x/Al/Nb junction, remarkably large energy-nonlinearity was observed in a photon energy region between 100 and 500 eV; the ε value decreased rapidly by a factor of 43 [174]. This anomaly is ascribed to some local trap sites for quasiparticles; the assumed traps sites, where Δ_g is suppressed, are crystallographic defects. Beyond 500 eV, the ε value increased with increasing photon energy, which is assigned to self-recombination of quasiparticles. The increase in ε significantly degrades energy nonlinearity. Normally, we observe the output charge reduction or the ε increase due to the self-recombination as the photon energy increases; however, the rapid ε reduction in a low photon energy range is uncommon [120, 144]. Figure 3.18 shows that the Nb/Al junction with "Al 70-nm" has a constant ε value in a wide range from 200 to 8000 eV. The large reduction in ε resembles abnormally large output observed in a spatially resolved analysis of a junction with "Al 30-nm" in Sect. 3.2.4.7. This is explained by superconductivity breakdown due to insufficient DOS, based on the gap-reduction model in Sect. 3.2.4.4. Further studies are required to understand the disparity between the two junctions.

3.2.4.4 Quasiparticle Trapping and Gap Reduction Models on Superconducting Gap Engineering

In this section, we selectively use Δ for the order parameter and Δ_g for the bottom edges of quasiparticle DOS profiles. The order parameter Δ represents the Cooper pair potential, whereas the gap Δ_g is the minimum excitation energy for quasiparticles relative to E_F. For STJD operation, as mentioned earlier, we assume that the quasiparticles have a character of either pure electron or pure hole, considering the tunnel paths in the semiconductor-model description in Fig. 3.12.

The Δ_g values of superconductor electrodes significantly affect the STJD performance because ε is proportional to Δ_g. A Δ_g value determined from *I*–*U* curves of junctions is strongly dependent on Al thickness , which also influences DOS profiles as a function of quasiparticle excitation energy over a distance of several times longer than ξ along the direction perpendicular to a junction surface [134]. Polycrystalline Al films normally have a ξ value of 147 nm in Table 1.1. Usual Al layer thickness for gap engineering is in a range of 30–70 nm. In contrast to the strong dependence of Δ_g on Al layer thickness, the dependence on polycrystalline Nb layer thickness is relatively small. Typical Al layers are always thinner than ξ_{Al}, whereas typical Nb-layers (200–300 nm) are much thicker than ξ_{Nb}, which is 23 nm in polycrystalline films. Therefore, the Al layer thickness predominantly affects quasiparticle generation and collection processes, energy linearity, temporal response, and energy resolution. Indeed, quasiparticles tunnel from either Al layer to the counter Al layer in the Nb/Al/AlO_x/Al/Nb layer structure. The DOS profiles in the Al layers dominantly govern quasiparticle tunneling processes between the top and bottom electrodes.

It has been widely believed that quasiparticles become trapped in low-Δ_g Al-layers in proximized junction detectors, which typically have layer structures such as Nb/Al/AlO_x/Al/Nb or Ta/Al/AlO_x/Al/Ta: $\Delta_{g,Nb} > \Delta_{g,Al}$ or $\Delta_{g,Ta} > \Delta_{g,Al}$. In this belief, the Al layers act as layers trapping quasiparticles, while the high-*Z* Nb or Ta layers serve as a photon absorber. It is usually assumed that the high-*Z* layers maintain their intact Δ_g values because Nb or Ta layers are much thicker than ξ. Figure 3.19a illustrates this quasiparticle trapping model for a Nb(200)/Al(30)/AlO$_x$/Al(30)/Nb(200) junction with thicknesses in nm in parentheses. Note that the direction perpendicular to the junction surface corresponds to the x-axis in Fig. 3.19. The quasiparticle trapping model is certainly valid for detector designs such as a junction attached to a bulk superconductor [125] or a photon imaging spectrometer with two junctions attached to both ends of a superconductor strip absorber with a length of several hundreds of μm [126, 175, 176]. However, the quasiparticle trapping model is plausibly inappropriate for direct particle detection with thin-film junctions because the proximity effect influences quasiparticle DOS and Δ_g across a long distance presumably up to 6 μm. This supposition is based on an experiment on the region with a reduced Δ_g value in epitaxially-grown Ta strips in touch with Al [177]. The coherence length of bilayers such as Nb/Al or Ta/Al forming STJDs seems to be elongated to 500 nm–1.6 μm along the direction parallel to the film surface [178], which is called the long-range proximity effect. Furthermore, the temperature dependence of I_c in normal-metal/superconductor bilayers for TESDs indicates that the long-range proximity effect extends over several tens of μm [179, 180]. Long-range proximity superconductivity becomes pronounced along the direction parallel to the film surface when Δ_g of bilayers forming TESDs increases due to the presence of superconductor current leads with a higher T_c. Along the direction perpendicular to the film surface, Δ_g decreases due to normal metal posts to mount an normal-metal X-ray absorber on the TES bilayer. We use "horizontal" for the direction parallel to a film surface and "vertical" for the direction perpendicular to a film surface. Interestingly, both studies on STJDs and TESDs reported similar experimental results, which are consistent each other, claiming that the horizontal extent of the long-range proximity effect reaches several μm to several tens of μm. It is unclear whether the phenomenon is specific to the 2D cases of thin films; nevertheless, there seems to be no reason to distinguish the horizontal and vertical directions for the proximity effect. Therefore, the observed long-range proximity effect can be also valid along the vertical direction perpendicular to film surfaces. Although the theory of the proximity effect is normally applied to thin films with a thickness of much less than ξ [127], we need to reconsider the spatial change of Δ_g, DOS, and other superconducting properties beyond $10 \times \xi$. According to the above-mentioned experimental results, it is reasonable that Δ_g is constant throughout proximity bilayers along the direction perpendicular to film surfaces; still quasiparticle weak trapping into Al layers occurs due to the Andreev reflection associated with a sudden change of DOS profiles at Nb/Al or Ta/Al interfaces [178].

New model is termed as the gap-reduction model; the Nb absorber layer has a Δ_g value reduced from the intact $\Delta_{g,Nb}$ by an adjacent Al layer with a lower $\Delta_{g,Al}$. The Δ_g value is constant throughout Nb/Al proximity bilayers. The Δ_g reduction inside the Nb absorber layer causes an increase in the number of excited quasiparticles according to Eq. (3.20), and an improvement in energy resolution according to Eq. (3.4), when DOS near Δ_g inside the Nb layer is enough for accommodating quasiparticles excited by an incident particle. We compare the conventional and new models concerning quasiparticle generation and motion along the direction perpendicular to

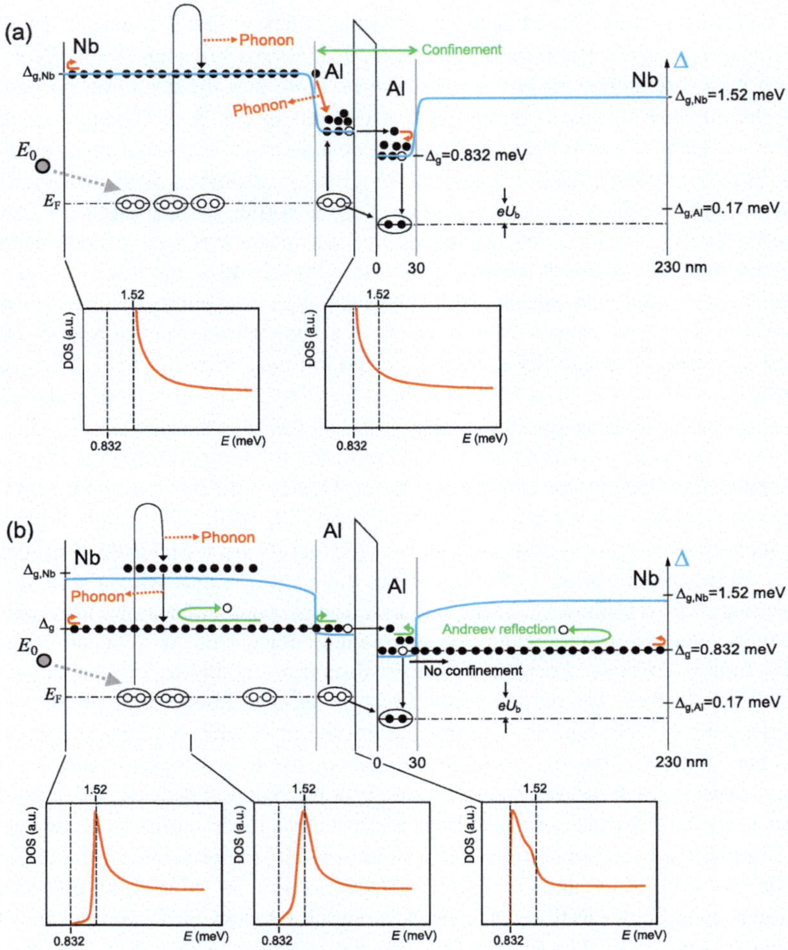

Fig. 3.19 Conventional quasiparticle trapping model and new gap-reduction model for proximized STJDs. The layer stricture of the junction is Nb(200)/Al(30)/AlO_x/Al(30)/Nb(200) in nm; the *I-U* curve and the output current pulse are shown in Fig. 3.14. The *x*-axis shows the distance from the AlO_x/Al interface along the direction perpendicular to the junction surface. The *y*-axis shows the quasiparticle energy in meV, displaying only the quasielectron part and omitting the quasihole part. The blue solid lines denote the order parameter or Cooper pair potential Δ. The Δ_g value of 0.832 meV is obtained from the voltage at which the slope of the *I-U* curve near $2\Delta_g/e$ crossed the abscissa. The trapping model in (**a**) assumes the BCS DOS profiles with the intact Δ_g value of 1.52 meV inside the Nb layer and 0.82 meV inside the Al layers. The gap-reduction model in (**b**) assumes that Δ_g is constant at 0.82 meV throughout the top and bottom Nb/Al bilayers and DOS profiles strongly depend on the distance from the AlO_x/Al interface. The DOS profiles are inferred from theories and experiments on the proximity effect [104, 133, 134, 181]. X-ray photons below ~ 2 keV are absorbed predominantly in the top/left Nb layer rather than the Al layer because of the higher atomic number. Macromolecules of a few 10 keV stop on the top/left Nb surface without penetration, which produces the Ω_D phonon bubble in the top Nb layer. Note that there are two ways to express eV: (1) Δ_g in J or eV and Δ_g/e in V, where e is 1.602×10^{-19} C, (2) Δ_g in J and Δ_g/e in eV, where e is just a unitless number of 1.602×10^{-19}. We use the former. Reproduced with permission from Ohkubo [144]. Copyright 2023 AIP Publishing LLC

the film surface, following the theoretical studies [104, 181] on Δ_g and DOS as well as the above-mentioned experimental evidences of the long-range proximity effect [177–180].

From the standpoint of the conventional quasiparticle trapping model for gap engineering in Fig. 3.19a, all quasiparticles that are excited inside the left Nb layer encounter the intact $\Delta_{g,Nb}$ value of 1.52 meV, except for those generated near the Nb/Al interface within a length of ~ ξ_{Nb}. Inside the Nb region near the interface, Δ_g steeply varies from 1.52 meV to 0.832 meV. The Δ_g value raised from 0.17 to 0.832 meV by the Nb proximity is constant within the Al layers because the Al thickness of 30 nm is considerably smaller than ξ_{Al} (= 147 nm). The value of 0.832 meV is determined from the intersection of the slope of the I–U curve near $2\Delta_g/e$ and the abscissa, as shown in Fig. 3.14. The excited quasiparticles subsequently encounter a random walk inside the left Nb layer. When quasiparticles enter the left Al layer, they lose their energies through phonon emission and become trapped in the Δ_g well at 0.832 meV. Once trapped in either Al layer, quasiparticles are confined within the Al/AlO$_x$/Al trilayer against the $\Delta_{g,Nb}$ walls at 1.52 meV. They tunnel back and forth with a high probability before recombining into Cooper pairs, which is an amplification of output charge due to multiple tunneling. The tunnel probability, which is inversely proportional to film thickness, becomes high because the Al layers are several times thinner than the Nb layers. Consequently, the quasiparticles excited inside the left Nb layer efficiently contribute to the tunnel current or the output charge. It is possible to explain output signals of STJDs by the quasiparticle trapping model with some fitting parameters. However, we demonstrate that the gap-reduction model better explains the experimental results rather than the quasiparticle trapping model.

The gap-reduction model in Fig. 3.19b begins from the same initial condition that all quasiparticles are excited inside the left/top Nb layer. The significant difference is that the Δ_g value in the Nb layer is not the intact $\Delta_{g,Nb}$ value but reduced to 0.832 meV due to the proximity effect of Al, holding the constant Δ_g throughout the left and right Nb/Al bilayers. Nonnegligible DOS exists between 0.832 meV and the intact $\Delta_{g,Nb}$ at 1.52 meV inside the Nb layers in accordance with the Usadel equations, in contrast to the BCS DOS profiles at the bottom of Fig. 3.19a [182]. The DOS profiles at different locations along the direction perpendicular to the junction surface are inferred from the theories and experiments on the proximity effect [104, 133, 134, 178–181]. The DOS values between 0.832 and 1.52 meV in the Nb layers gradually decrease with increasing distance from the Nb/Al interface, as shown in the bottom of Fig. 3.19b. Nevertheless, in principle, nonzero DOS extends to a scale substantially larger than the thickness of the Nb films. The proximity DOS increases with Al thickness.

Without perturbation or particle incidence, the Nb/Al bilayers hold the Δ_g value of 0.832 meV at any locations, while DOS profiles vary with the distance from the Nb/Al interface. On the other hand, it is plausible that a large perturbation destroys the proximity superconductivity or DOS below 1.52 meV. In other word, the presence of the proximity DOS depends on perturbation strength. In small perturbation, the experimental evidences of the long-range proximity effect indicate a spatial extent of the proximity-DOS in several μm for the injection of quasiparticles with an energy of

~ Δ_g in the 10-μm square junction with Nb(100)/Al(5–10)/AlO_x/Al(5–10)/Nb(100) in nm [178], and several tens of μm for the current injection at ~ 10 nA in the 29-μm square Mo/Au TES bilayer [180].

The high-energy quasiparticles relax to the edge of $\Delta_g = 0.832$ meV inside the Nb layer in the gap-reduction model. The relaxed quasiparticles tunnel to the counter electrode, spread throughout the counter Nb/Al bilayer because Δ_g is constant throughout bilayers. No quasiparticle trapping occurs, but the number of initial quasiparticles is 1.8 times higher than that of the trapping model. As an example, a Nb(100)/Al(90) bilayer holding a constant Δ_g of 0.5 meV has a DOS value of 1.43×10^6 states/meV/μm^3 at an energy of $0.3 \times \Delta_{g,Nb}$ on the Nb free surface [104]. This proximity DOS value is large enough for accommodating the quasiparticles excited by a 6-keV X-ray photon: $6000/(1.7\times0.832) \times 10^{-3} \approx 4 \times 10^6$ quasiparticles or 2×10^6 quasiparticle pairs. A quasiparticle cloud of 4×10^6 can be accommodated within an energy bin of 0.04 meV inside a Nb volume of $20 \times 20 \times 0.2$ μm^3, which is 100 times smaller than the Nb electrode volume. The quasiparticles can spread throughout the entire electrode volume, depending on diffusion constant; therefore, the quasiparticle density is considerably reduced and thereby it becomes far below the proximity DOS level. The upper limit of the allowable photon energy in the bilayer of Nb(200)/Al(30) is naturally lower than that of Nb(100)/Al(90). Nevertheless, even at the free surface of the 200-nm thick Nb layer in Fig. 3.19b, the DOS value is presumed to be sufficient to accept the quasiparticles excited by an X-ray photon up to ~ 300 eV. Therefore, the energy resolution of the junction with "Al 30-nm" is the same as that of the junction having more DOS with "Al 70-nm," as shown later in Sect. 3.2.4.6 and in Fig. 3.20. Consequently, we can treat the Nb(200)/Al(30) bilayers as uniform superconducting layers in an absorption energy range below ~ 300 eV. This enables a simple modeling for the output charge of proximized STJDs. Both tunnel directions from left to right and from right to left have the same quasiparticle tunnel probability P_{tun} in Fig. 3.19b. If the DOS at the Nb free surface is insufficient for excited quasiparticles, it induces nonlinear effects such as Δ_g reduction and even breakdown of the proximity superconductivity. In this case, Δ_g is not constant anymore during the event duration.

In the gap-reduction model in Fig. 3.19b, quasiparticle trapping into the Al layers occurs due to the difference in DOS between the Nb and Al layers, although it is extremely weak compared with the strong trapping in Fig. 3.19a. In addition, Andreev reflection can occur at the Nb/Al interfaces because the DOS profile change is abrupt [134, 178]. Concisely, a quasielectron or quasihole is reflected into an Al layer as the counter quasiparticle such as a quasihole or quasielectron, adding or removing a Cooper pair in a Nb layer. Furthermore, the Andreev reflection occurs even inside the Nb layer because there are different DOS profiles along the direction perpendicular to the junction surface, as depicted in Fig. 3.19b [104]: this direction corresponds to the horizontal axis in Fig. 3.19. The Andreev reflection contributes to moving quasiparticles toward the AlO_x barrier and thereby can enhance quasiparticle tunneling. The exact quasiparticle distribution in proximity bilayers depends on diverse factors: particle energy, relaxation process to the edge of Δ_g, Andreev reflection, quasiparticle

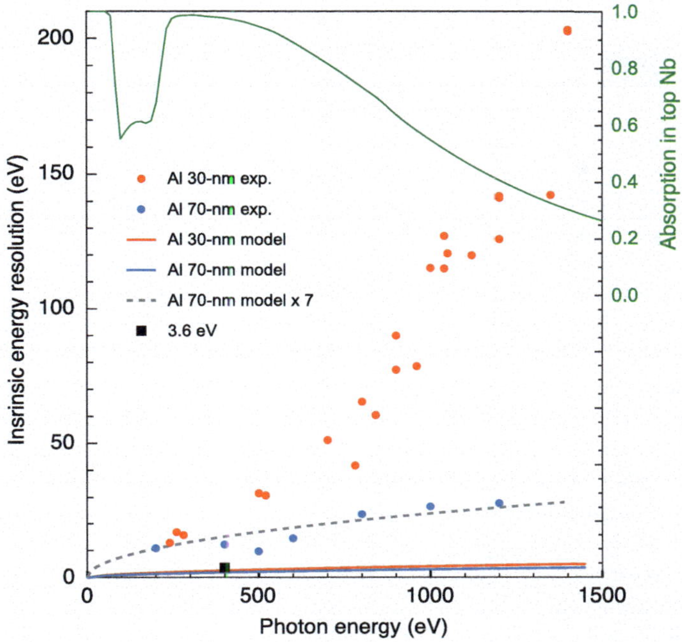

Fig. 3.20 Comparison in ΔE between the gap-reduction model with multiple quasiparticle tunneling and the experimental data below 1.4 keV. The red and blue solid lines represent the model calculations; the red and blue dots are the experimental data. For multiple tunneling, the assumed parameters are $G = 1/P_{\text{tun}} = 1.5$ and $\Delta_g = 0.832$ meV for the junction with "Al 30-nm"; and $G = 1.4$ and $\Delta_g = 0.44$ meV for "Al 70-nm"; and $F = 0.107$ with $\varepsilon = 1.72 \times 2\Delta_g$ [63, 144]. A theoretical ΔE calculation gives 1.9 eV at 400 eV (N K-$L_{2,3}$ line) and 2.2 eV at 525 eV (O K-$L_{2,3}$ line) for "Al 70-nm," following $\Delta E(\text{eV}) = 2.355\sqrt{E_0\varepsilon(F+G)} = 0.094\sqrt{E_0}$. The dashed line represents the curve obtained by multiplying the theoretical curve for "Al 70-nm" by a factor of 7. For the experimental data, monochromatized X-rays from synchrotron radiation irradiated the entire junction area, which is full illumination [187]. The electrical noises of 6.5 eV for "Al 30-nm" and 7.1 eV for "Al 70-nm" were subtracted from the measured ΔE data. The experimental intrinsic ΔE of 3.6 eV (4.1 eV with the electrical noise) for the asymmetric STJD in Fig. 3.16 is also plotted by the solid black square at 400 eV. The upper green curve represents the photon absorption efficiency of the 200-nm thick top Nb layer, exhibiting the Nb M-edge structure around 207 eV and the reduction of absorption efficiency above ~ 500 eV Reproduced with permission from Ohkubo [144]. Copyright 2023 AIP Publishing LLC. Experimental data reproduced with permission from Ohkubo [187]. Copyright 2004 Elsevier B.V

reflection at Nb free surfaces, dynamic processes concerning quasiparticle-density dependent DOS, and DOS profile variation along the horizontal axis.

The quasiparticle energy down-conversion processes are compared between the BCS-DOS in Fig. 3.19a and the proximity DOS profiles Fig. 3.19b [181]. The quasiparticle generation efficiency in the proximity DOS profile is lower than that in the BCS-DOS profile, when they have the same Δ_g. This low efficiency stems from the indirect energy relaxation to Δ_g (= 0.832 meV) via the intact states at $\Delta_{g,\text{Nb}}$

(= 1.52 meV); a short trap of quasiparticles occurs at the DOS peak at 1.52 meV [181]. Nevertheless, the Δ_g of proximized STJDs is significantly lower than that of non-proximized STJDs that have no thick Al layers. In addition, thick Al layers prolong τ_{eff}, resulting in a higher output. Consequently, the reduction of Δ_g and the prolongation of τ_{eff} surpass the inefficient quasiparticle generation in proximity bilayers.

Which model better explains the experimental data: the quasiparticle trapping model or the gap-reduction model? The first experimental data set for this comparison was obtained with Nb/Al junctions with an Al thickness in a range of 5–70 nm and a low critical current J_c of 200 A/cm^2 [144]. The average number of quasiparticle tunnel times $\overline{N}_{\mathrm{tun}}$ for 5.9-keV X-rays varies from 0.26 to 7.5 on the trapping model and from 0.24 to 2.2 on the gap-reduction model, depending on Al thickness [144]. The highest average tunnel repetition of 7.5 per quasiparticle is unrealistic at 200 A/cm^2, which indicates a low tunnel probability. Therefore, it is concluded that the $\overline{N}_{\mathrm{tun}}$ variation of 0.24–2.2 on the gap-reduction model is more reasonable for the low J_c junctions rather than 0.26–7.5 on the quasiparticle trapping model [144].

Another experimental data set is the response to 5.9-keV X-rays of a junction with very thick Al layers: Nb(150)/Al(200)/AlO$_x$/Al(200)/Nb(240) tested at LLNL [183]. The measured $2\Delta_g$ value of the junction is 0.34 meV at 0.95 K. Therefore, the Δ_g value inside the Al layers is 0.17 meV, which is equal to $\Delta_{g,\mathrm{Al}}$ of the intact Al. It is reasonable that the proximity effect of the Nb layers thinner than the Al layers is negligibly small. In the quasiparticle trapping model, the junction with Δ_g = 0.17 meV has a $\overline{N}_{\mathrm{tun}}$ value of 20, assuming that the Nb X-ray absorber has the intact $\Delta_{g,\mathrm{Nb}}$ value of 1.52 meV. This high $\overline{N}_{\mathrm{tun}}$ value is inconsistent with the junction layer structure; the 200-nm thick Al layers have no strong quasiparticle confinement effect because the tunnel probability is inversely proportional to film thickness. In the gap-reduction model, on the other hand, the Δ_g values are constant at 0.17 meV throughout the Nb/Al bilayers. The number of initial quasiparticle-pairs is evaluated at 1.0×10^7, which corresponds to $\overline{N}_{\mathrm{tun}} = 2.3$. This $\overline{N}_{\mathrm{tun}}$ value in the gap-reduction model is more plausible than 20 in the quasiparticle trapping model.

Consequently, the gap-reduction model better explains the experimental data. The most striking feature of the gap-reduction model is not quasiparticle trapping but the Δ_g reduction in Nb layers, leading to an increase of the number of excited quasiparticles and thereby an improvement in energy resolution. We are necessitated to perform an experimental method to precisely measure the number of excited quasiparticles and to determine ε of superconductors in a similar way to semiconductors. This is also important for excess quasiparticles in qubits in an extremely low temperature range. For such experiment, STJDs are unsuitable because they have many unknown factors such as interface transparency of two superconductors, tunnel probability, imperfect quasiparticle collection, and charge amplification due to multiple tunneling. MKIDs with bilayer structure will play a crucial role to verify the gap-reduction model better than STJDs. MKIDs simply utilize surface resistance change due to quasiparticle excitation in superconductor strips. In addition, we can also use superconductor RF filters with bilayer structure. Fundamental physics on quasiparticle generation and relaxation in proximity bilayers and the long-range proximity effect are crucial for both STJ and TES detectors. These are experimentally and theoretically unsolved questions.

3.2.4.5 Optimum Al-Layer Thickness

Let us consider the optimum Al thickness for STJDs within the framework of the gap-reduction model. In Table II in Ref. [144], τ_{eff} of Nb/Al junctions monotonically increases with increasing Al thickness. This is reasonable because Al has an extremely longer characteristic time τ_0 than Nb: $\tau_{0,Al} = 438$ ns and $\tau_{0,Nb} = 0.149$ ns [121]. On the other hand, $\overline{N}_{tun}$ increases from 0.24 for "Al 5-nm" to the maximum value of 3.1 for "Al 40-nm," then slightly decreases to 2.2 for "Al 70-nm." At 40 and 70 nm, the output charge is considerably enhanced by the two mechanisms: lowering Δ_g in the Nb absorber layers and holding the high $\overline{N}_{tun}$ of 3.1 and 2.2. At 70 nm, $\overline{N}_{tun}$ slightly decreases because the thicker Al layers lead to a lower tunnel probability and a mild DOS profile change inside the Nb layer leads to a mild quasiparticle movement toward the AlO_x tunnel barrier. Nevertheless, the output charge for 5.9 keV monotonically increases with increasing Al thickness due to the gap reduction. When we consider energy linearity in Fig. 3.17 and Fig. 3.18, operating temperature, and spatial uniformity discussed in Sect. 3.2.4.7, we adopt 70 nm for the optimum Al thickness. For Ta/Al junctions, the optimum Al thickness should be similar, although Ta/Al junctions require an operating temperature range below 0.1 K that is lower than <0.5 K for Nb/Al junctions.

3.2.4.6 Multiple Quasiparticle Tunneling Model

With the Gray effect in Fig. 3.12, quasiparticles can tunnel forth and back. The output charge amplification due to multiple tunneling is advantageous in enhancing the S/N ratio and energy resolution. Note that the multiple tunneling model here is different from the MQT model [157] that explains excess quasiparticles in $T \ll 0.1T_c$ and the sub-harmonic gap structure in Sect. 3.2.4.2. In the gap-reduction model, the quasiparticle trapping with proximity bilayers is negligibly weak; nevertheless, $\overline{N}_{tun}$ is larger than 1.0 in Al layers thicker than 20 nm. The multiple tunneling occurs when the tunnel time τ_{tun} is shorter than τ_{eff}. Detected charge Q_c after an absorption event in the left/top electrode in Fig. 3.19b is given by

$$Q_c = Q_0 \overline{N}_{tun} = Q_0 \frac{P_t(1 + P_b)}{1 - P_t P_b}, \tag{3.24}$$

where P_t is the probability of top-to-bottom quasiparticle tunneling, and P_b represents the tunnel direction opposite to P_t [184, 185]. When an absorption event occurs in the bottom electrode, the subscripts "t" and "b" are exchanged. The tunnel probability P_t is given by

$$P_{\mathrm{t}} = \frac{\tau_{\mathrm{tun,\,t}}^{-1}}{\tau_{\mathrm{tun,\,t}}^{-1} + \tau_{\mathrm{eff,t}}^{-1}}, \tag{3.25}$$

where $\tau_{\mathrm{tun,t}}$ is the top-to-bottom tunnel time, and $\tau_{\mathrm{eff,t}}$ is the effective quasiparticle lifetime for the top electrode. For the bottom electrode, the subscript "t" is replaced by "b." Symmetric layer structure with uniform superconductor electrodes holds $\tau_{\mathrm{tun,t}} = \tau_{\mathrm{tun,b}} = \tau_{\mathrm{tun}}$ and $P_{\mathrm{t}} = P_{\mathrm{b}} = P_{\mathrm{tun}}$, which transforms Eq. (3.24) to $Q_{\mathrm{c}} = Q_0\,\overline{N}_{\mathrm{tun}} = Q_0 P_{\mathrm{tun}}/(1\text{-}P_{\mathrm{tun}})$ with $P_{\mathrm{tun}} = \overline{N}_{\mathrm{tun}}/\left(1 + \overline{N}_{\mathrm{tun}}\right)$. Based on the gap-reduction model, the P_{tun} values for the Nb/Al junctions with Al layers with a thickness of 5–70 nm are listed in Table II in Ref. [144]: 0.19 at 5 nm, 0.52 at 30 nm, 0.66 at 40 nm, and 0.69 at 70 nm for the differing Al thicknesses in the Nb/Al/AlO$_x$/Al/Nb layer structure. At 30 nm or thicker, more than half of the quasiparticles inside either top or bottom electrode tunnel to the counter electrode.

Multiple tunneling causes the uncertainty of output charge and thereby degrades energy resolution: it is a statistical process with the average number of tunnel times $\overline{N}_{\mathrm{tun}}$ and its variance expressed by [185]

$$\sigma_{\mathrm{tun}}^2 = \frac{P_{\mathrm{t}} - P_{\mathrm{t}}^2 + 3P_{\mathrm{t}}P_{\mathrm{b}} + P_{\mathrm{t}}^2 P_{\mathrm{b}}}{(1 - P_{\mathrm{t}}P_{\mathrm{b}})^2}. \tag{3.26}$$

In units of the number of quasiparticles, the variance becomes $\frac{E_0}{\varepsilon}\sigma_{\mathrm{tun}}^2$. Furthermore, for the contribution of tunnel fluctuation to energy resolution in FWHM, we need to convert the standard deviation $\sqrt{E_0/\varepsilon}\sigma_{\mathrm{tun}}$ into an expression in eV by using the average partition energy $E_0/\left(Q_0/e \cdot \overline{N}_{\mathrm{tun}}\right)$ per quasiparticle tunneling event and the relation $Q_0/e = E_0/\varepsilon$. Finally, we obtain the FWHM tunneling fluctuation in eV:

$$\Delta E_{\mathrm{tun}} = 2.355\sqrt{E_0/\varepsilon}\sigma_{\mathrm{tun}} \frac{eE_0}{Q_0 \overline{N}_{\mathrm{tun}}} = 2.355\sqrt{E_0 \varepsilon G},$$
$$\text{where } G = \left(\frac{\sigma_{\mathrm{tun}}}{\overline{N}_{\mathrm{tun}}}\right)^2 = \frac{P_{\mathrm{t}} - P_{\mathrm{t}}^2 + 3P_{\mathrm{t}}P_{\mathrm{b}} + P_{\mathrm{t}}^2 P_{\mathrm{b}}}{P_t^2 (1 + P_b)^2}. \tag{3.27}$$

The factor G corresponds to the relative standard deviation of multiple tunneling uncertainty. For symmetric junctions with uniform Δ_{g} in the gap-reduction model for proximized juncitons, G simply equals $1/P_{\mathrm{tun}}$, where $P_{\mathrm{tun}} = P_{\mathrm{t}} = P_{\mathrm{b}}$. This condition is the same as the model for the junction with the 200-nm thick Al layers [183]. The propagation law combining the uncertainty of Eqs. (3.4) and (3.27) yields

$$\Delta E = 2.355\sqrt{E_0 \varepsilon (F + G)}. \tag{3.28}$$

Another noise source of quasiparticle tunneling originates from reverse tunneling of quasiparticles highly excited above $\Delta_{\mathrm{g}} + eU_{\mathrm{b}}$ due to bias voltage [186]. However, it is not straightforward to evaluate the exact amount of the number of highly excited quasiparticles [157], and it is negligible compared with the multiple tunneling noise.

Therefore, Eq. (3.28) provides practical guidance in evaluating the energy resolution limit of proximized STJDs, omitting the reverse tunneling noise.

The junction with "Al 70-nm" has parameters $P_{\mathrm{tun}} = 0.69$ and $G = 1.4$, both of which can be defined for the spatially uniform Δ_g in the Nb/Al bilayers under the gap-reduction model. When we use $F = 0.214$ obtained by the Monte Carlo simulation [63] for single quasiparticles in Nb and consider that the number of quasielectron-quasihole pairs represents output charge, $F = 0.214$ is transformed to $F = 0.107$ for quasiparticle pairs [144]. Two quasiparticles are simultaneously generated, so that the number of pairs governs the statistical fluctuation. A precise probability distribution function for carrier generation is analytically unknown; Monte Carlo simulations provide a reasonable evaluation of the variance ratio F of the putative precise probability distribution function to the Poisson distribution function. Equation (3.28) indicates that the most fluctuation originates from the multiple tunneling noise rather than the Fano factor. In contrast, semiconductor detectors with carrier collection by a strong electric field hold $\overline{N}_{\mathrm{tun}} = 1$, $P_{\mathrm{t}} = 1$, $P_{\mathrm{b}} = 0$, and thereby $G = 0$. The high G value is one of the disadvantages of STJDs. Nonetheless, extremely small Δ_g values of superconductors in meV compensate the disadvantage. Experimentally, STJDs outperform semiconductor detectors in energy resolution by a factor of ~ 10, as shown in Fig. 3.16 previously. Additionally, high charge output due to multiple tunneling is advantageous in noisy experimental conditions, for example, at beamlines of synchrotron radiation facilities, where electronic noise often significantly exceeds the intrinsic energy resolution of semiconductor detectors.

Figure 3.20 shows the comparison between the experimental data of intrinsic energy resolution for X-ray photons below 1400 eV and the theoretical ΔE values using Eq. (3.28) based on the gap-reduction model. The red and blue solid lines, which almost lie on the horizontal axis, represent the theoretical curves for the two junctions, each with an Al layer thickness of either 30 nm or 70 nm. The theoretical curves adopt, $\Delta_g = 0.832$ meV and $G = 1.5$ for "Al 30-nm" and $\Delta_g = 0.44$ meV and $G = 1.4$ for "Al 70-nm" with the modified Monte Carlo simulation factors: $\varepsilon = 1.72 \times 2\Delta_g$ and $F = 0.107$ for quasiparticle pairs in Nb [63]. The experimental data points in Fig. 3.20 represent the ΔE values in FWHM of the two 100-μm square junctions with "Al 30-nm" or "Al 70-nm" [187]. The monochromatic X-ray photons from synchrotron radiation uniformly irradiated the entire junction area. The full-energy peaks for the primary photons and higher order photons are used to plot the data points as a function of photon energy. The electronic noise values of 6.5 eV for "Al 30-nm" and 7.0 eV for "Al 70-nm" are subtracted from the measured ΔE values.

The dashed line is the theoretical ΔE curve for "Al 70-nm," multiplied by 7. The experimental ΔE data are 7 times larger than the theoretical values. The best experimental intrinsic ΔE value for "Al 70-nm" is 9.7 eV at 500 eV, which is worse than the theoretical value of 2.1 eV. One of the reasons for the factor 7 is that the dip structure of the blue experimental data points between 400 and 600 eV stems from X-ray transmission through the top 200-nm thick Nb in an energy range near the Nb M-edge. The absorption coefficient in the top Nb layere is depicted by the green solid line. In the energy range between 400 and 600 eV, almost all the X-ray photons are absorbed by the top Nb layer. On the other hand, for the data points below 400 eV and above 800 eV, 30% or more of the incoming photons transmit the top Nb layer and thereby are absorbed in the bottom Nb layer [173]. As observed in the double peak phenomenon at 6 keV, the absorption events in the bottom Nb layer produce slightly higher output pulses. This causes the degradation in ΔE on the occasion of Gaussian fitting to full-energy peaks. This is similar to the case of the energy linearity in Fig. 3.18. Another reason for the factor 7 is spatial inhomogeneity that slightly remains even for the junction with "Al 70 nm": spatial distribution of charge output is shown in Sect. 3.2.4.7. Nonetheless, the experimental ΔE data for "Al 70 nm" follows the $\sqrt{E_0}$ dependence as expected from the general behavior of quantum detectors. The black square at 400 eV in Fig. 3.20 plots an intrinsic ΔE value of 3.6 eV (4.1 eV with electronic noise) for the asymmetric junction with Nb(300)/Al(70)/AlO_x/Al(70)/Nb(100)/Si in nm in Fig. 3.16. This ΔE value is close to the theoretical value of 1.9 eV.

It is intriguing to examine the data of the junction with "Al 30-nm;" the ΔE values increase abruptly above 500 eV, deviating from the $\sqrt{E_0}$ dependence. It exceeds a typical ΔE value of ~ 50 eV in SDDs at ~ 800 eV. This abrupt ΔE degradation is a sign of spatial inhomogeneity, which is the absorption-position dependence of output charge [129, 188, 189]. The ΔE values for "Al 30-nm" are roughly proportional to E, suggesting a large spatial inhomogeneity [129]. The spatial inhomogeneity is the topic of Sect. 3.2.4.7 in the next section. The linear ΔE increase is typical for junctions with an Al layer thickness of 30 nm or smaller. Noticeably, in an energy range below ~300 eV, this junction has the same ΔE values as those of the junction with "Al 70-nm." This implies that the spatial inhomogeneity depends on photon energy as well as Al thickness. High energy photons can cause the enhancement of self-recombination of a large number of quasiparticles near the junction edges, assisted by the quasiparticle confinement effect due to reflection at the edges; therefore, spatial inhomogeneity is enhanced [188]. Furthermore, the gap-reduction model with constant Δ_g and position-dependent DOS profiles in Fig. 3.19b explains this behavior as follows. At the free surface of the left/top Nb layer, a DOS value for the Nb(200)/Al(70) bilayer is higher than that for the Nb(200)/Al(30). Correspondingly, the top Nb layer of the Nb(200)/Al(70) bilayer has a higher allowance for quasiparticle density without collapse of the proximity superconductivity, holding constant Δ_g throughout the bilayers up to 1.2 keV. The bilayer of Nb(200)/Al(30) can also accept a low density of quasiparticles excited in a photon energy range below ~ 300 eV. However, beyond ~ 300 eV, the quasiparticle density begins to exceed a threshold level for linear response near the junction edges due to the quasiparticle reflection. In a higher energy range, the DOS between 1.52 and 0.832 meV even vanishes due to a high density of quasiparticles near the junction edges; the Δ_g increases from 0.832 to 1.52 meV inside the Nb layers. Therefore, output charge decreases near the junction edges above ~ 300 eV, leading to the ΔE degradation. Direct measurement of the spatial inhomogeneity in a high photon energy range is shown in Figs. 3.23 and 3.29 in the following Sect. 3.2.4.7.

3.2.4.7 Spatial Inhomogeneity of STJD

Microscopes for Spatially Resolved Measurement

Besides the multiple quasiparticle tunneling uncertainty, the most serious source of energy resolution degradation is spatial inhomogeneity. Charge output for monochromatic X-rays varies depending on the location where a photon is absorbed within the junction's sensitive area. The spatial distribution of charge output is not necessarily flat, which is spatial inhomogeneity. Analytical science requires detectors with a large sensitive area to complete measurement for a short time; a large junction size is preferable to material analysis. Correspondingly, particles such as X-rays, ions, and molecules uniformly illuminate an entire detector chip, which is called full illumination. The ΔE degradation induced by spatial inhomogeneity often surpasses other noise sources such as fundamental statistical fluctuation and electronic noise. For example, in the junction with "Al 30-nm" in Fig. 3.20, the signifiant spatial inhomogeneity appears in an photon energy range above ~ 300 eV, as inferred from the ΔE degradation following $\Delta E \propto E_0$ in the red dots in Fig. 3.20, instead of $\Delta E \propto E_0{}^{0.5}$ in an ideal quantum detector. A microscopic method is indispensable for directly observing the spatial inhomogeneity of 100–200-μm square junctions to approach the theoretical energy resolution predicted for superconductor quantum detectors.

One powerful imaging instrument used to directly visualize the spatial distribution of detector output is a low-temperature scanning electron microscope (LTSEM) at the University of Tübingen [90, 190]. The LTSEM has a high spatial resolution – significantly better than 1 μm. In addition, electron-beam short pulses enable temporal response measurement. The decay time of detector output pulses for the electron-beam pulses represents τ_{eff}. However, one disadvantage of the LTSEM is a high base-temperature of ~ 1.5 K reached by pumping liquid ^{4}He. In addition, a large open hole between the electron gun and the sample stage is indispensable; infrared radiation precludes proper junction operation below ~0.5 K. At the LTSEM base temperature, the thermal excitation of quasiparticles is nonnegligible. The T/T_c value is ~0.5 for the proximized Nb/Al junction with "Al 70-nm," which exhibits a significant reduction on Δ_g due to thermally excited quasiparticles, as shown in Fig. 3.6. Therefore, the quasiparticle recombination is not the self-recombination but the thermal recombination. The self-recombination means that quasiparticle pairs excited by energetic particles recombine into Cooper pairs [158], whereas the thermal recombination occurs between excited quasiparticles and thermally excited quasiparticles. The self-recombination rate is proportional to the density of excited quasiparticles; the energy nonlinearity becomes significant at a high density. On the other hand, the thermal recombination rate is determined by the density of thermally excited quasiparticles. When we maintain the beam current at a low level such that the density of excited quasiparticles is negligibly smaller than that of thermally excited quasiparticles, the recombination rate becomes independent of the electron beam current. We usually operate the LTSEM under this condition. According to LTESM experiments, top electrodes normally exhibit a shorter τ_{eff} than bottom electrodes, which is attributed to quasiparticle trapping into a metallic NbO_x layer on the Nb surface [164]. Top Nb-electrode surfaces are easily oxidized during fabrication processes. The longer τ_{eff} in the bottom electrode yields a higher LTSEM signal than the top electrode. This behavior is consistent with the bias-voltage dependence of X-ray full-energy peak positions in Fig. 3.15, which shows that the bottom electrode generates higher output charge values than the top electrode.

LTSEM signal-amplitude imaging does not completely imitate X-ray absorption events at ~0.3 K, where thermally excited quasiparticles are absent. However, LTSEM visualizes the spatial profiles of detector response at a sub-μm scale, which enables optimization of STJD structural design. Figure 3.21a shows an example of the LTSEM images of a Nb/Al circular junction with the large contact hole for the top electrode and the wide wiring leads for both top and bottom electrodes. The focused electron-beam pulses with an energy of 10 keV and a pulse width of 100 ns scanned the junction area and surroundings at 2.1 K. The amplitude of electron-beam-induced voltage pulses are scaled in order of blue, green, yellow, orange, red, and white. The circular shape and the unusually wide leads are devoted to investigating their effects on junction response. The cross section in Fig. 3.21b shows the layer structure of the junction, contact hole, and wiring leads. Naturally, the signal is the highest in the junction area in red, while it is the lowest in the Si substrate area in blue. In this junction design, the most junction area is exposed except the narrow rim region covered by SiO_2 layer for insulating the top and bottom electrodes. The junction rim covered by the SiO_2 insulation layer generates intermediate signals: the 1-μm wide SiO_2 region on the junction is barely recognizable in yellow and the 2-μm wide SiO_2 region outside the junction area is apparent in light green in Fig. 3.21a. The intermediate signal amplitudes in yellow and light green stem from indirect quasiparticle excitation through phonons generated inside the insulating SiO_2 surface in contrast to the direct quasiparticle excitation in the junction area. The 30-μm square large contact hole region in dark-to-light yellow in Fig. 3.21a also generates intermediate amplitudes. Full illumination of the whole area with monochromatic X-rays can be imitated by integrating signal heights at all the locations. The lower signals in the contact hole and the wide leads produce artifacts besides full-absorption peaks: tailing toward the low energy side in pulse height spectra. Consequently, Fig. 3.21 suggests a better STJD design; the contact hole should be as small as possible, the SiO_2 insulator region at the junction rim and the wiring leads should be as narrow as possible. Modern STJDs have a 1-μm square contact hole and a-few-μm wide wiring leads to eliminate undesirable low output signals: Fig. 3.9a shows the optimized design.

The white line curve shows LTSEM signal amplitudes along the light green arrow in Fig. 3.21a. LTSEM amplitude images of STJDs normally exhibite a dome shape, which stems from the amplitude reduction near the circlar junction rim, similar to the edges of square junctions [170]. The low signals near the junction rim imply that the rim functions as a quasiparticle loss site, which has a quasiparticle reflection rate less than 1.0 due to quasiparticle loss or trapping. The dome shape is explained by a quasiparticle diffusion model [188]. In addition to the junction rim, the signal amplitudes near the wiring leads in the junction area are lower than those at the circular junction center in Fig. 3.21a. This reduction is caused by quasiparticle escape into the

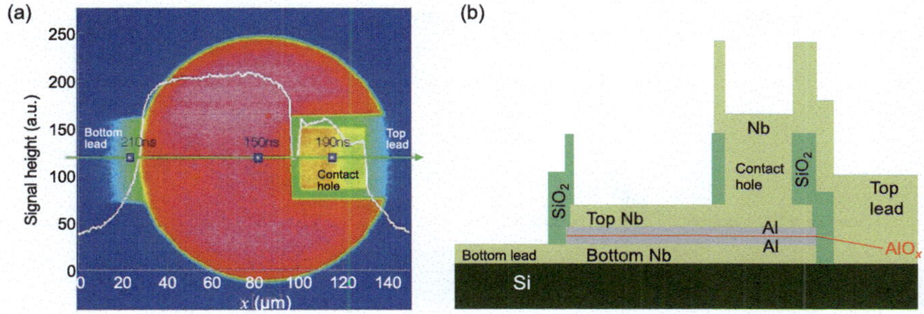

Fig. 3.21 LTSEM imaging of circular STJD with a layer structure of Nb(200)/Al(20)/AlO_x/Al(20)/Nb(200)/Si in nm: (**a**) rainbow color image of output signal amplitude scaled in order of blue, green, yellow, orange, red, and white, and (**b**) cross-sectional layer structure along the light green arrow in (**a**). The junction with a diameter of 113 μm has the 40-μm wide wiring leads, and the very-large 30-μm square contact hole for the top electrode. The 10-keV electron pulses with a duration of ~100 ns scanned the junction area and surroundings at 2.1 K. The white solid line in (**a**) exhibits the output signal amplitude along the light green arrow. The pulse fall time, which corresponds to $\tau_{\rm eff}$, depends on locations: 210 ns on the bottom wiring lead, 150 ns at the junction center, and 190 ns at the contact hole center on the top wiring lead. Reproduced with permission from Ohkubo [144]. Copyright 2023 AIP Publishing LLC

wiring leads. Alternatively, the absorption events in the wiring lead for the bottom electrode generate low output signals because quasiparticles reach the junction area by diffusion even at a position 30-μm apart from the junction rim. The exponential decay curve inside the bottom lead area in Fig. 3.21a corresponds to a quasiparticle diffusion length of ~ 10 μm. From a $\tau_{\rm eff}$ value of 210 ns in the bottom wiring lead, the quasiparticle diffusion constant is evaluated at 2.4 cm^2/s, which is consistent with 1.0 cm^2/s in Ref. [170], considering the difference in experimental conditions.

Although LTSEM has a sub-μm high spatial resolution, the base temperature is considerably higher than a normal operating temperature of ~ 0.3 K. Another microscope using an X-ray microbeam with a diameter of 5–10 μm with synchrotron radiation directly yields spatial distribution of X-ray signals at ~ 0.3 K. This microscope proposed by Pressler is called the low temperature scanning synchrotron microscope (LTSSM) [191, 192]. The LTSSM is equipped with a ^{3}He sorption cryostat and X-ray windows that block infrared radiation, which ensures a detector chip temperature of ~ 0.3 K. Before and after the LTSSM studies, location-selective X-ray irradiation experiments were performed [193–196]. However, the LTSSM realizes full imaging analysis, similar to the LTSEM. Spatial profiles depend on bias voltage (or current), junction size, photon energy, AV trapping, and magnetic field strength to suppress the DC Josephson effect. Although theories cannot fully explain LTSSM images, these findings are important for optimizing STJ operation condition. LTSSM images of X-ray pulse height and $\tau_{\rm eff}$ support the gap-reduction model as discussed in the subsequent sub-sections.

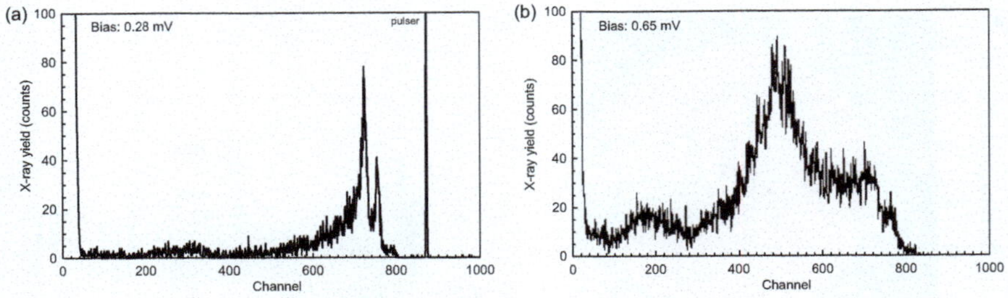

Fig. 3.22 Full illumination pulse height spectra of 200-μm square junction for ^{55}Fe X-ray source at bias points of 0.28 mV in (**a**) and 0.65 mV in (**b**). The junction has a layer structure of Nb(200)/Al(30)/AlO_x/Al(30)/Nb(200)/SiO_2(600)/Si in nm and a critical current density of 200 A/cm^2: the SiO_2 layer does not affect detector performance. A magnetic field of 12 mT parallel to the junction surface suppresses the DC Josephson effect. The 5.9-keV X-ray photons illuminated the whole junction area. The maximum output charge is normalized at the 800 channel. At 0.28 mV, the response function is consistent with a dome shape profile of output charge, based on the quasiparticle diffusion model [188]. On the other hand, at 0.65 mV, the response function is unexplainable by the diffusion model [191]. The fundamental properties of this junction are shown in Figs. 3.14, 3.17, and 3.20. The Δ_g value obtained from the *I*–*U* curve is 0.832 meV

Abnormal Response Function for Full Illumination

Before LTSSM images, it is instructive to examine abnormal pulse height spectra of a 200-μm square junction with "Al 30-nm". The X-rays from a ^{55}Fe source illuminated the entire detector chip. Although the ^{55}Fe source emits two lines near 5.9 keV (89.5%) and one line at 6.5 keV (10.5%), we can approximate the former two lines as a weighted-average single line at 5.895 keV. A pulse height spectrum to monochromatic X-rays is called the response function. The fundamental properties of the 200-μm square junction are shown in Figs. 3.14, 3.17, and 3.20. Figure 3.22 shows two pulse height spectra of ^{55}Fe full illumination at bias voltages of 0.28 mV in (a), which is far lower than a normal bias point near $\Delta_g/e = 0.832$ mV, and of 0.65 mV in (b), which is close to Δ_g/e . At 0.28 mV in Fig. 3.22a, the double peak appears with tailing toward the lower energy side, which is consistent with the quasiparticle diffusion model with the edge loss of quasiparticles [188]. The diffusion model predicts that the response function with low-energy tailing stems from low output signals near the junction edges. The double peak is due to the difference in response to X-rays between the top and bottom electrodes, as explained in Sect. 3.2.4.3. When the bias voltage is raised to 0.65 mV close to Δ_g/e, the response function exhibits a drastic change as shown in Fig. 3.22b. This bias dependence has been a mystery for a long time before using LTSSM.

Figure 3.23 shows LTSSM images of the same 200-μm junction as that in Fig. 3.22. The junction has a layer structure of Nb(200)/Al(30)/AlO_x/Al(30)/Nb(200). The 10-μm microbeam of 6-keV X-rays scanned the junction area and surroundings at a step of 10 μm in (a) and 1 μm in (b) while recording a pulse height spectrum at each coordinate [191]. Figure 3.24 shows one of the pulse height

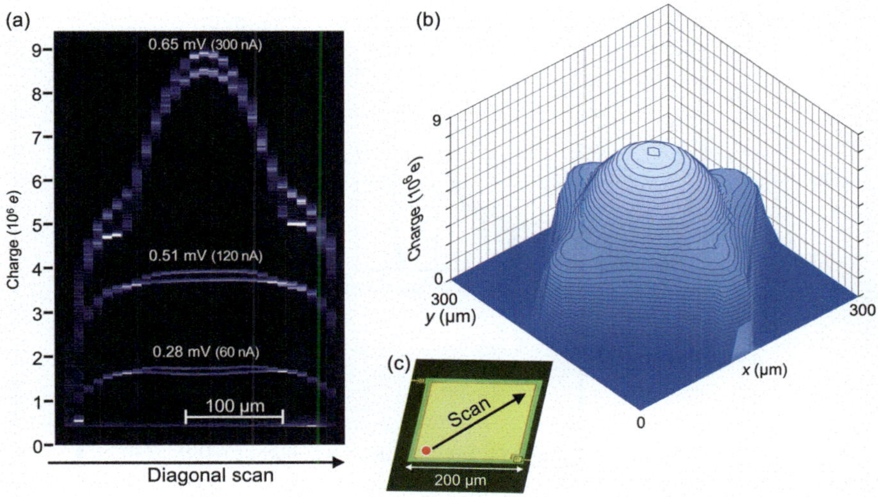

Fig. 3.23 LTSSM images of junction response to 6-keV X-ray photons in a parallel magnetic field of 12 mT: (**a**) bias voltage dependence of diagonal scans from one corner to opposite corner for the absorption events in the top and bottom electrode, (**b**) 3D plot of output charge values for the absorption events inside the top electrode at a bias voltage of 0.65 mV, and (**c**) microscope image of 200-μm square junction along with 6-keV X-ray microbeam illustrated by the red solid circle with a diameter of 10 μm, and the arrow indicating the diagonal scan direction. The 200-μm square junction has a layer structure of Nb(200)/Al(30)/AlO_x/Al(30)/Nb(200) in nm. The Δ_g value obtained from the *I*–*U* curve is 0.832 meV as shown in Fig. 3.14a. The pulse height spectra for full illumination are shown in Fig. 3.22. Scan steps are 10 μm in (**a**) and 1 μm in (**b**). In (**a**), the event numbers per charge bin or channel bin as a function of coordinate along the diagonal direction are represented by a gradient with black for the lowest count, blue for the middle, and white for the highest. For the ordinate, the channel numbers are converted to output charges. In (**b**), the output charges represent the centroids of the full-energy peaks for the events inside the top electrode, scaled from blue to white with the contour lines. Reproduced with permission from Pressler et al. [191].

spectra near the junction center, although it is another 146-μm square junction with the same layer structure as the 200-μm square junction. There is almost no spatial inhomogeneity within the microbeam area, so that the peaks are sharp compared with those in Fig. 3.22. The double peak comes from absorption events in the top and bottom electrodes. It appears as the double curve in Fig. 3.23a, which plots the number of detected photons in a scale from black to white at each coordinates along the diagonal direction. Due to this large spatial inhomogeneity in $E_0 \gg \sim 300$ eV, the STJD with "Al 30-nm" exhibits the linear ΔE degradation in Fig. 3.20. For the ordinate of Fig. 3.23a, the channel numbers in Fig. 3.24 are converted to the total charges of STJD output pulses. In Fig. 3.23b, the output charge values representing the centroids of the full-energy peaks for the events in the top electrode are scaled from blue to white with the contour lines.

The diagonal scans at different bias points in Fig. 3.23a exhibit a drastic change of the spatial profiles. Below 0.51 mV, the spatial profiles resemble the dome shape

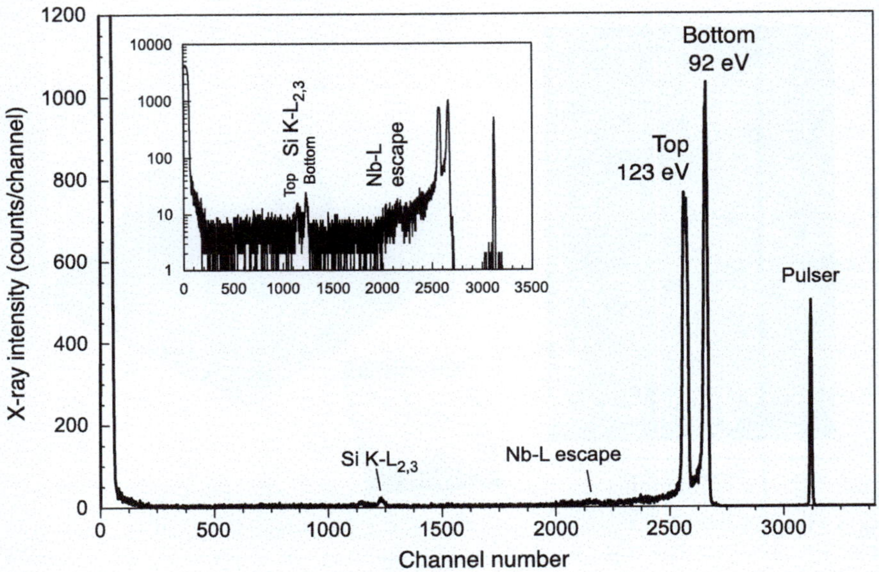

Fig. 3.24 Pulse height spectrum near the center of 146-μm square junction with a layer structure of Nb(200)/Al(30)/AlO_x/Al(30)/Nb(200). A 6-keV X-ray microbeam with a diameter of 5 μm illuminated a small area near the junction center. The double peak for the 6 keV X-rays appears in a channel range between 2500 and 2700, which originates from absorption events in the top and bottom electrodes. The channel numbers represent output charge values. The ΔE values in FWHM are 123 eV for the top Nb events and 92 eV for the bottom Nb events, considering energy nonlinearity. The higher peak for the bottom electrode has a smaller peak area than that of the lower peak for the top electrode, which is consistent with the X-ray absorption ratio. The ΔE values are calibrated by the energy nonlinearity using the positions of the Si K-$L_{2,3}$, Nb-L escape, and full-energy peaks in the same way as Fig. 3.17. The pulser peak represents an electronic noise of 20 eV. Reproduced with permission from Ohkubo [144]. Copyright 2023 AIP Publishing LLC

consistent with the quasiparticle diffusion model, although the profile at 0.28 mV is slightly concave at the center. The top and bottom electrodes exhibit similar but slightly different spatial profiles. The output charges of the top and bottom electrodes at the center are reversed with increasing bias voltage. At 0.65 mV, the abnormally large dome feature appears on the terrace feature: the output charge at the center is increased by a factor of 6.5. Figure 3.23b shows the full 3D image at 0.65 mV. The integration of the output charges at all the locations is identical to the ^{55}Fe-full-illumination pulse height spectrum in Fig. 3.22b.

Regardless of the abnormal spatial inhomogeneity at 6 keV in Fig. 3.22 and Fig. 3.23, the junction with "Al 30-nm" has the same ΔE values as those of "Al 70-nm" below ~ 300 eV in Fig. 3.20. This implies that the output charges of the junction with "Al 30-nm" are homogeneous only in a range below ~ 300 eV. The excitation below ~ 300 eV is a small perturbation such that the junction operation is within a linear regime. The abnormally large spatial inhomogeneity is observed when detector operation satisfies two conditions: a photon energy is above ~ 300 eV and a bias voltage is close to Δ_g/e. It is noteworthy that this voltage corresponds to the bias voltage, at which the subgap current begins to rapidly increases due to the plausible MQT quasiparticle multiplication near Δ_g/e [156, 157]. The bias voltage of 0.65 mV requires 5 times higher current than that at 0.28 mV, which means that the number of the MQT excess quasiparticles is also 5 times larger. Therefore, the center dome is related to the MQT excess current near Δ_g/e. An explanation appears in the fourth sub-section ahead "An explanation of abnormal spatial inhomogeneity."

Quasiparticle Diffusion Model for Spatial Inhomogeneity

The response function with low energy tailing in Fig. 3.22a is consistent with that of the quasiparticle diffusion model [188]. The diffusion model postulates that quasiparticles excited in either top or bottom electrode encounter diffusion inside that electrode, reflection at the junction edges, tunneling to the counter electrode, back tunneling, loss at the junction edges, and bulk loss inside the electrodes. Figure 3.25 shows a model system by Cristiano et al., who analytically solved the coupled equations of diffusion and tunneling with quasiparticle loss within a 2D system [188]. For a symmetric layer structure with uniform electrodes, which is applicable even to proximized junctions under the gap-reduction model, we can define a single tunneling rate for the tunnel directions of both top-to-bottom and bottom-to-top ($\gamma = \gamma_{\text{t-b}} = \gamma_{\text{b-t}} = 1/\tau_{\text{tun,t}} = 1/\tau_{\text{tun,b}}$). For symmetric square junctions, three important parameters are defined by

$$\beta = \frac{L(1-R)}{2l},$$
$$\Lambda = \left(\frac{D}{\gamma + \gamma_{\text{loss}}}\right)^{\frac{1}{2}}, \text{ and}$$
$$\Gamma = \frac{\gamma}{\gamma + \gamma_{\text{loss}}}, \tag{3.29}$$

where L is the junction side length, R is the reflection probability at junction edges, l is the mean free path, D is the quasiparticle diffusion constant, γ is the tunnel rate to the counter electrode, γ_{loss} is the quasiparticle loss rate. The γ_{loss} represents the bulk loss rate inside the electrodes, phonon escape into surroundings, and other loss processes. Quasiparticle thermal recombination loss is disregarded because the density of thermally excited quasiparticles is negligibly low in an operation temperature range. Instead, we assume a constant γ_{loss} ($= 1/\tau_{\text{eff}}$) everywhere in both top and bottom electrodes, disregarding the phonon escape asymmetry between the top and bottom electrodes [189]. Quasiparticle recombination at an operating temperature is self-recombination, emitting $2\Delta_g$ phonons that break Cooper pairs. Due to this phonon trapping effect, the quasiparticle bulk loss rate is equal to the phonon decay rate into energies below $2\Delta_g$, enhanced by the phonon escape rate. The solution of the coupled equations is not simple, but yields

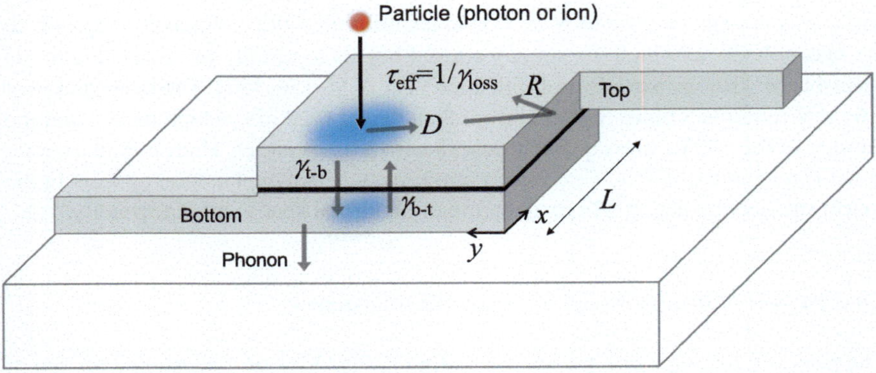

Fig. 3.25 Schematic junction structure for quasiparticle diffusion model [188]. The black layer is the insulating AlO_x tunneling barrier. An SiO_2 insulating layer, which is omitted, isolates the top and bottom electrodes. We define the quasiparticle diffusion D, the tunnel rate from top to bottom $\gamma_{t\text{-}b}$, the tunnel rate from bottom to top $\gamma_{b\text{-}t}$, the reflection rate at the junction edges R, the quasiparticle bulk loss rate γ_{loss} ($= 1/\tau_{eff}$). Symmetric junctions hold $\gamma = \gamma_{t\text{-}b} = \gamma_{b\text{-}t}$. The blue regions depict the quasiparticle clouds after absorbing particle energy, tunneling, and diffusion. Based on the gap-reduction model within small perturbation, proximity bilayers are regarded as uniform superconductor electrodes with a constant Δ_g. Hence, calculations can be performed within a 2D condition, although the junction has the 3D structure

$$Q_c(x, y) = Q_0\Gamma \sum_{n=1}^{\infty}\sum_{m=1}^{\infty} \frac{A_n(x)B_m(y)}{\left[1 + \frac{\Lambda^2(u_n^2+v_m^2)}{L^2} - \Gamma\right]}, \tag{3.30}$$

where A_n, B_m, u_n, v_m are the functions of β (See Ref. [188]). This 2D model reproduces a dome-shaped spatial inhomogeneity. Kozorezov et al. performed a sophisticated fully temporal and spatial analysis by using generalized Rothwarf-Taylor balance equations [189].

The 146-μm square junction exhibited a simple dome profile compatible with the diffusion model at any bias points [192]. The RRR value of the polycrystalline Nb films for this junction corresponds to a l value of 10 nm, which is considerably shorter than 23 nm in Table 1.1 due to small grains. The τ_{eff} value is fixed at 1.4 μs from the decay time of the output pulse in Fig. 3.14b. The fitting parameters are R, D, and γ in Eq. (3.29). Figure 3.26 shows a comparison between calculation curves and experimental data point of the bottom electrode events along the diagonal direction at bias points of 0.15 mV (15 nA) and 0.30 mV (30 nA) in a bias range significantly lower than Δ_g/e (=0.841 meV) avoiding the effect of the large MQ T excess current. The photon energy of the X-ray microbeam is 5 keV, which does not make any meaningful difference from 6 keV in Fig. 3.23. The best fits to the experimental data points are obtained at $D = 1.8$ cm^2/s, $R = 0.999952$, $\gamma = 1/0.594$ μs^{-1} for 0.15 mV and $D = 4.5$ cm^2/s, $R = 0.999916$, $\gamma = 1/0.343$ μs^{-1} for 0.30 mV. The R values are slightly less than 1.0, which indicates that the small quasiparticle-loss at the junction edges causes the convex dome shape. The model curves agree well with the

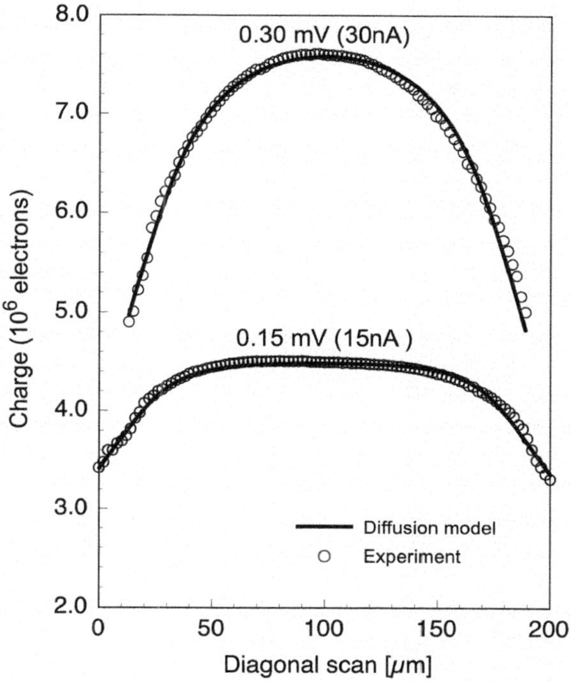

Fig. 3.26 Comparison between LTSSM diagonal scan curves of output charge in units of 10^6 e with 5 keV X-ray microbeam and theoretical curves based on quasiparticle diffusion model [192]. The 146-μm square junction with 30-nm thick Al layers and a Δ_g value of 0.841 meV is operated at the bias voltages of 0.15 and 0.30 mV in a parallel magnetic field between 10 and 16.7 mT. The LTSSM data points (open circles) represent the positions of the full-energy absorption peaks of the bottom electrode events. The fixed parameters for fitting are $\gamma_{loss} = 1/1.4$ μs and $l = 10$ nm. The best fits yield $\beta = 0.353$, $\Lambda = 8.65$, $\Gamma = 0.703$ ($D = 1.8$ cm²/s, $R = 0.999952$, $\gamma = 1/0.594$ μs^{-1}) for 0.15 mV and $\beta = 0.616$, $\Lambda = 11.26$, $\Gamma = 0.803$ ($D = 4.5$ cm²/s, $R = 0.999916$, $\gamma = 1/0.343$ μs^{-1}) for 0.30 mV

experimental data points at both bias voltages. In a bias range below the maximum tunnel current in Fig. 3.15, a net tunnel probability of quasiparticles increases with increasing bias voltage according to the BCS-DOS profile. Therefore, it is reasonable that the γ values are different at 0.15 and 0.30 mV. The diffusion constant of 1.8 cm²/s at 0.15 mV is consistent with 2.4 cm²/s obtained from the LTSSM image in Fig. 3.21 and 1.0 cm²/s for polycrystalline Nb in the literature [170]. The D value of 4.5 cm²/s at 0.30 mV is also within an agreement: the quasiparticle diffusion constant is a matter of large uncertainty. However, the diffusion constant should be essentially unaffected by the bias points. Therefore, we should consider some factors other than the quasiparticle diffusion to explain the bias-voltage dependence of spatial inhomogeneity at 5–6 keV.

Deviation from the quasiparticle diffusion model becomes large in the 200-μm square junction at 6 keV in Fig. 3.23. The abnormal inverted tulip shape in Fig. 3.23b is partly consistent with an advanced nonlinear model with the generalized Rothwarf-Taylor balance equations [170, 189]. The advanced nonlinear model involves all possible processes such as diffusion, recombination, multiple tunneling, phonon coupling between two electrodes, and loss processes. The model can reproduce the pulse height spectra in Fig. 3.22b. However, the inverted tulip shape requires highly asymmetric different parameters for the top and bottom electrodes [189], which is unrealistic in the symmetric Nb/Al junction. Therefore, the bias voltage dependence and the inverted tulip shape originates from an unknown mechanism differing from the quasiparticle diffusion; we need to consider the the MQT process [157] and the dynamics of Josephson vortices (JVs).

It is surprising again that, in a photon energy range below 300 eV in Fig. 3.20, the junction with "Al 30-nm" has the same energy resolution as the junction with "Al 70-nm," where another junction recorded the best energy resolution of 4.1 eV. This signifies that the junction with "Al 30-nm" operates in a linear regime in a photon energy range below 300 eV, similar to the junction with "Al 70-nm." To reveal the origin of the abnormal spatial inhomogeneity observed at a bias voltage near Δ_g/e and photon energies near 6 keV, we examine the dependence of spatial inhomogeneity on magnetic field strength and Al thickness in the following discussion.

Dependence of Spatial Inhomogeneity on Magnetic Field

The dome shape of the 146-μm square junction with "Al 30-nm" at 0.30 mV in Fig. 3.26 resembles the central dome enhancement in the 200-μm square junction in Fig. 3.23. It seems that the dome shape has a specific length of ~ 150 μm. When the abnormal dome enhancement is observed, τ_{eff} becomes shorter. This is inconsistent with the common signal generation due to quasiparticle tunneling, where pulse height is roughly proportional to τ_{eff}. The dependence of spatial inhomogeneity on magnetic field provides insight into what happens in proximized junctions.

Figure 3.27 shows the magnetic field dependence of diagonal scan curves of pulse height between 13.5 and 31.6 mT at two bias points: 0.3 and 0.6 mV. The junction with "Al 30-nm" is the same one in Fig. 3.23. The magnetic field range is higher than a normal magnetic field of ~ 10 mT. The upper panels in Fig. 3.27a, b show the diagonal scan curves of double full-energy peak. The lower panels show the diagonal scan curves of charge-sensitive-preamplifier rise-time values, equivalent to τ_{eff}. The bias voltage of 0.3 mV (60 nA) in (a) is significantly lower than Δ_g/e (= 0.832 mV), whereas 0.6 mV (150 nA) in (b) is close to Δ_g/e within the skirt of the excess current due to the MQT current enhancement. The line scan curves at 0.3 mV are flatter than those at 0.6 mV. The diagonal scans show a tendency of continuous output charge reduction from left corner to right one. This reduction is consistent with the rise time change, which indicates that the quasiparticle current partly contributes to the signal generation. However, the magnetic field dependence exhibits unexplainable behaviors. Firstly, the pulse height order of the top and bottom electrodes near the junction center changes with increasing magnetic field in Fig. 3.27a. Secondly, at a magnetic field of 31.6 mT, the convex dome shape changes to the inverted dome one. At the high bias point of 0.6 mV, the inverted tulip shape in Fig. 3.27b almost disappears by increasing the magnetic field to 31.6mT. More surprisingly, the rise time profiles apparently exhibit the opposite dependence to the output charge curves. These behaviors, especially magnetic field dependence, certainly indicate that an unknown mechanism related to Josephson vortices (JVs) governs the dome and inverted tulip shapes in addition to the conventional quasiparticle tunneling current. The number of JVs increases with increasing magnetic field. Moreover, the bias point dependence suggests the effect of the MQT excess current near Δ_g/e.

It is important to consider that the bias current at 0.6 mV (150 nA) is 2.5 times larger than that at 0.3 mV (60 nA). As shown in Fig. 3.14a, the bias current rapidly increases near Δ_g/e, which can attribute to an excess current due to the MQT process. The MQT injects quasiparticles into the states significantly higher than Δ_g due to the bias voltage, which induces the quasiparticle multiplication. Although it is not shown in Fig. 3.27, the MQT current further increases to 300 nA at 0.65 mV from 150 nA at 0.6 mV. The 300-nA current is 42% of the current pulse peak, extrapolated to the time zero, for the 6 keV X-ray. The significant reduction in rise time or τ_{eff} at the junction center from ~3.1 μs in the far left panel in Fig. 3.27a to ~2.2 μs in the far left panel in Fig. 3.27b is consistent with enhanced quasiparticle recombination due to a high quasiparticle density indued by the MQT.

An accidental sudden change of spatial inhomogeneity in Fig. 3.28 provides another insight into the signal generation mechanism. Additional LTSSM measurement run after Fig. 3.23 is shown in Fig. 3.28: (a) displays the top view of the same inverted tulip shape as in Fig. 3.23b. The sudden change from the inverted tulip shape to the dome shape occurs as shown in Fig. 3.28b. It is attributed to AV trapping near the top corner due to an electric shock. When a junction traps AVs perpendicular to the junction surface, the output charge significantly decreases: SQUID AV imaging and charge output reduction are shown in Figs. 13 and 14 in Ref. [144]. Interestingly, the inverted tulip shape profile simultaneously disappears just as an unstable resonance is disturbed, transformed into a simple dome shape except the top corner, where AVs exist. This phenomenon implies that the AVs disturb the JV motion, so that the abnormal inverted tulip shape disappears.

The signal generation relates to two mechanisms: the conventional tunneling current of excited quasiparticles and the resonance of JV motion. High magnetic field suppresses the abnormally large inhomogeneity. This does not occur in a bias range significantly lower than Δ_g/e or in a photon energy range below ~ 300 eV. The inverted tulip shape appears under the following conditions: a Al layer thickness of ~ 30 nm, a large perturbation due to ~ 6-keV X-rays, a low magnetic field of ~ 10 mT, and a high bias position near Δ_g/e.

Dependence of Spatial Inhomogeneity on Al Thickness

To further investigate the abnormal spatial inhomogeneity in a high photon energy range of the nonlinear regime (> ~ 300 eV), we measured the dependence on Al thickness. Figure 3.29 shows the spatial homogeneity improvement at a bias point

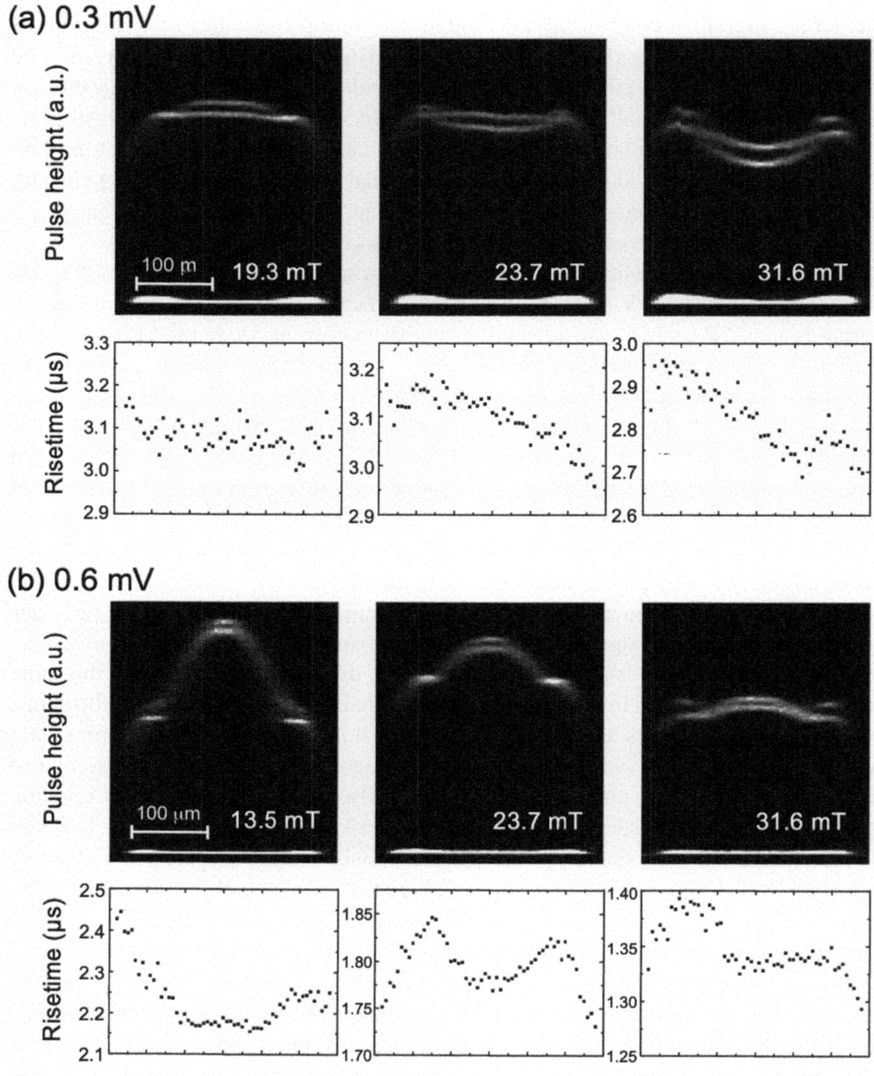

Fig. 3.27 Magnetic field dependence of diagonal scans at bias voltages of 0.3 mV (60 nA) and 0.6 mV (150 nA) for 200-μm square Nb(200)/Al(30)/AlO_x/Al(30)/Nb(200) junction with Δ_g = 0.832 meV. It is the same junction as in Fig. 3.23. The 6-keV X-ray microbeam scanned the junction surface at a step of 1 μm. The magnetic field is applied in parallel to the junction surface along the diagonal direction. No dependence of magnetic field direction is observed. The rise time of a charge-sensitive preamplifier is equivalent to quasiparticle effective lifetime τ_{eff}. The number of photons per energy bin, in units of counts/channel, as a function of coordinate along the diagonal direction is represented by a gradient, with black indicating the lowest count and white indicating the highest

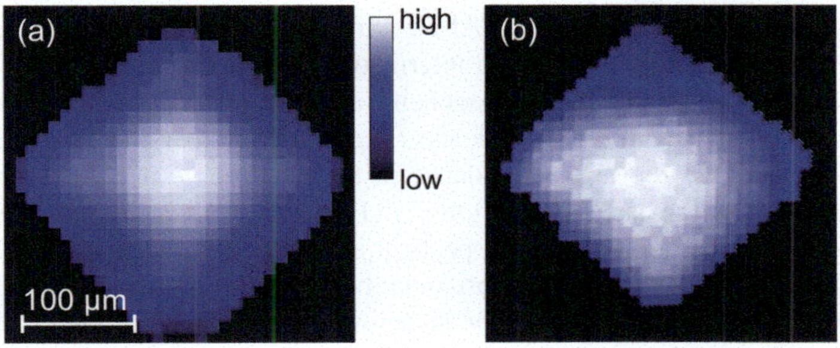

Fig. 3.28 LTSSM scan images after measuring Fig. 3.23 in the same operation condition. A 6-keV X-ray microbeam with a diameter of 10 μm scanned the 200-μm square junction at a step of 10 μm. The output charges are represented by black for the lowest charge to white for the highest via blue. The 2D output charge image in (**a**) reproduces the same inverted tulip shape as in Fig. 3.23, whereas an unintentional sudden change occurred in (**b**) after measuring the image in (**a**). The output charge reduction at the top corner stems from AV trapping due to an electric shock. The inverted tulip shape disappears, and is transformed into the normal dome shape consistent with the quasiparticle diffusion model except the top corner

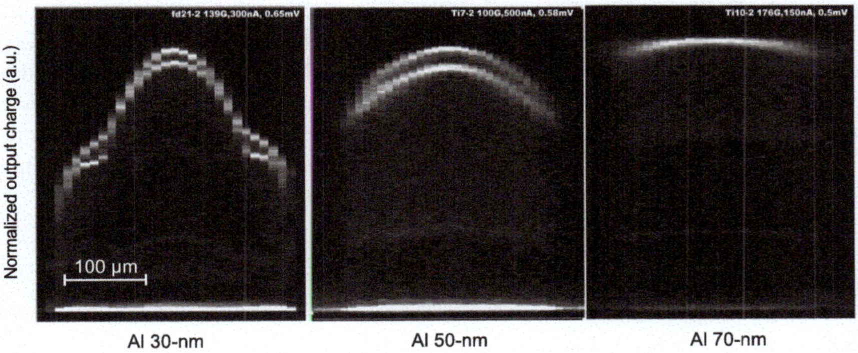

Fig. 3.29 Al layer thickness dependence of 6-keV LTSSM diagonal scans for 200-μm square junctions with an Al layer thickness between 30 and 70 nm at a bias point near Δ_g/e [197]. The microbeam size and the scan step are 10 μm. The parallel magnetic fields range from 10 to 17.6 mT: the effect on the spatial profiles is negligible. The numbers of photon absorption events are scaled from black to white. The *y*-axis is normalized at maximum output charges. The spatial uniformity is significantly improved by increasing Al thickness. The events near zero output charge are due to substrate events. At 70 nm, lower limit discrimination cut the substrate events. Reproduced with permission from Ukibe et al. [197]. Copyright 2004 Elsevier B.V.

near Δ_g/e by increasing Al thickness [197]. When the Al thickness increases from 30 to 50 nm, the inverted tulip shape disappears; it is transformed into a simple dome shape with slight remains of the terrace feature near the both ends of the diagonal scan. At "Al 70-nm," the spatial uniformity is further improved, although it is not perfectly flat. This experimental observation explains the difference in photon energy dependence of ΔE between "Al 30-nm" and "Al 70-nm" in Fig. 3.20. The junction with "Al 50-nm" exhibits an intermediate feature between 30 and 70 nm. The abnormally large spatial inhomogeneity and output charge enhancement are observed in junctions with an Al thickness range at 50 nm and thinner.

Figure 3.30 shows the bias dependence of the junction with "Al 70-nm" for 6-keV X-rays. Unlike the case of "Al 30-nm" in Fig. 3.23, neither inverted tulip shape nor abnormal central dome growth appears at any bias points even beyond Δ_g/e (= 0.44 mV). At 0.4 mV or above, the horizontal bands of the full-energy peak become broad compared with those at 0.3 mV or below. This stems from the double peak phenomenon, in which top and bottom electrodes generate slightly different output charges. Bias positions affect the split between the top and the bottom peaks and the pulse-height order [169]. In Fig. 3.30, the split becomes prominent at 0.4 mV and above. Below ~ 1 keV, the double peak disappears due to a high X-ray absorption efficiency of the top Nb electrode; the ΔE value of 4.1 eV indicates that the spatial profile is flat.

The spatial profiles without the inverted tulip shape have a connection with the gap-reduction model; the proximity DOS values between 0.44 and 1.52 meV inside the Nb layers in "Al 70-nm" are considerably higher than those in "Al 30-nm." In addition, the DOS profile change against the distance from the AlO_x barrier should be milder than that in "Al 30-nm." This results in more uniform proximity bilayers and more uniform quasiparticle distribution inside the Nb/Al bilayers; therefore, the spatial inhomogeneity becomes less prominent. On the other hand, in the junction with "Al 30-nm," quasiparticles tend to be confined near the AlO_x barrier due to more DOS near Δ_g inside the Al layers. Such quasiparticles interact with quasiparticles injected by the MAR process, which induces the inverted tulip shape.

An Explanation of Abnormal Spatial Inhomogeneity

Although complete understanding of the STJD signal generation related to the inverted tulip shape is a challenging task, a model with the MQT process and the JV motion can explain the following experimental results.

1. The junctions with Al $\leqq$ 50 nm exhibit the abnormally large spatial inhomogeneity for photons above ~ 300 eV at bias points near Δ_g/e. One of the mechanisms of the bias current increase near Δ_g/e is the MQT process [157].
2. In a photon energy range below 300 eV, even the junction with "Al 30-nm" has the same superior energy resolution as that of the junction with "Al 70-nm" in full illumination, ensuring that spatial inhomogeneity is negligibly small.
3. No abnormal spatial inhomogeneity appears in the junctions with "Al 70-nm" at any bias points.
4. When abnormal spatial inhomogeneity appears for the junctions with Al $\leqq$ 50 nm, output charge is inversely proportional to τ_{eff}, which indicates that an unknow

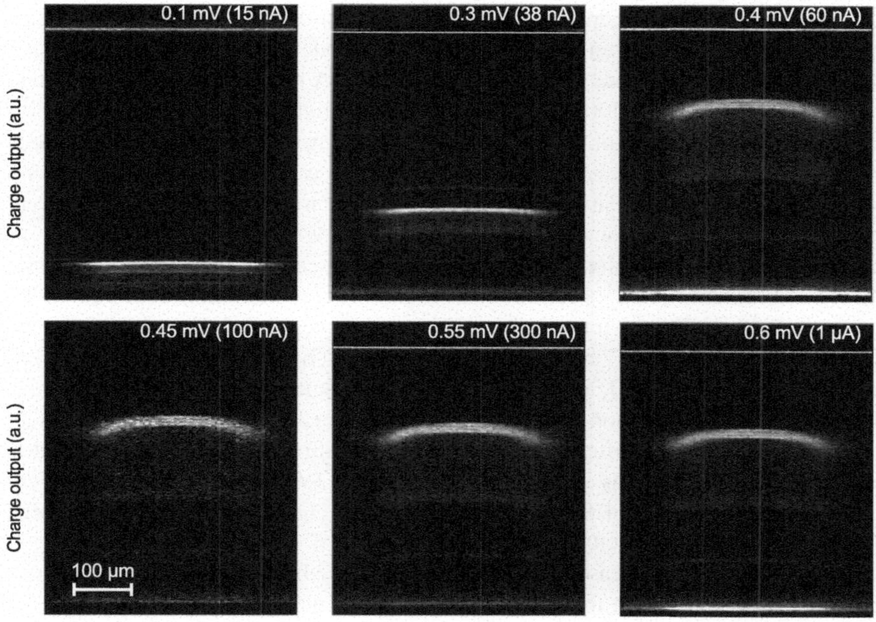

Fig. 3.30 Bias dependence of diagonal scans for junction with "Al 70-nm." The 6-keV X-ray microbeam scanned the 200-μm square Nb(200)/Al(70)/AlO$_x$/Al(70)/Nb(200) junction with $\Delta_g/e = 0.44$ mV in a magnetic field of 9.68 mT that is equivalent to the lowest field case that the apparent inverted tulip shape appears in Fig. 3.27. The microbeam size and the scan step are 1 μm. The horizontal thin lines at the upper parts are due to a pulser, representing electronic noise. The junction exhibits neither a central dome feature nor any growth of such a feature with increasing bias voltage, even beyond Δ_g/e. The spatial profiles are compatible with the quasiparticle diffusion model at any bias points

mechanism other than simple quasiparticle tunneling governs the signal generation. The abnormal spatial inhomogeneity accompanies the fivefold enhancement of output charge.

5. The abnormal spatial inhomogeneity or inverted tulip shape is unstable. AV trapping transforms the inverted tulip shape into the normal dome shape compatible with the quasiparticle diffusion model [188].
6. In Ref. [174], the junctions with 10 nm $\leqq$ Al $\leqq$ 20 nm exhibit a constant responsivity (Q_c/E_0) or ε^{-1} value below ~ 50 eV, the maximum responsivity at ~ 500 eV, and then gradual decrease with photon energy. The energy linearity is extremely poor beyond ~ 50 eV. The responsivity enhancement resembles the dome shape growth in Figs. 3.23 and 3.26.
7. High magnetic field suppresses the spatial inhomogeneity, approaching a flat spatial profile.

Figure 3.31 depicts a conceptual model to explain the above-mentioned experimental observations, combining the gap-reduction model and the MQT excess quasiparticle injection at bias points near and above Δ_g/e. A model of proximized square junctions has an area of 200 × 200 μm^2 and a layer structure of Nb(200)/Al(30)/AlO_x/Al(30)/Nb(200) in nm. Under MQT, the tunneling current stepwisely increases at every bias point expressed by $2\Delta_g/ne$ in the same way as MAR. Junctions with a low critical current density J_c of 200 A/cm^2 exhibit an apparent abrupt current increase only for $n = 2$ in Fig. 3.14a; the steps for $n \geq 3$ are not discernible. A magnetic field of 12 mT along the direction parallel to a junction side and the junction surface induces 180 JVs inside the 200–μm square junction, providing that the magnetic field penetrates into the layer with a thickness of $(2\,\lambda + d) = 155$ nm, where $\lambda = \lambda_{Nb} + t_{f,Al} = 77$ nm with $\lambda_{Nb} = 47$ nm and $t_{f,Al} = 30$ nm for the Al layer, and $d = 1$ nm for the AlO_x layer. The number of JVs in a magnetic field of 12 mT along the diagonal direction is 254. In both cases, the JV density is approximately 1 JV/μm. Therefore, a difference between the two directions of magnetic field is negligibly small. At a bias voltage of 0.28 mV, the JVs or the alternating current pattern moves at 0.05% of the vacuum light speed. Figure 3.31 depicts only three JVs for the sake of simplicity.

Figure 3.31a shows a case of $E_0 <$ ~300 eV at any bias points. For "Al 30-nm," the DOS values between the reduced Δ_g (= 0.832 meV) and the intact $\Delta_{g,Nb}$ (= 1.52 meV) inside the Nb layers are considerably lower than those for "Al 70-nm." Even in this low DOS, as shown in Fig. 3.31a, the junction with "Al 30-nm" has a rather uniform spatial variation of output charges below ~ 300 eV. The signal generation is based on conventional quasiparticle tunnel current. This results in the superior energy resolution at any bias positions. At a low bias of 0.28 mV, no excess quasiparticles due to MQT exist, because the significant MQT excess current is observed exclusively in a bias range near and above Δ_g/e (= 0.832 mV) as shown in Fig. 3.14a. We assume that in a photon energy range below ~ 300 eV, a photon absorbed in the Nb layer generates a quasiparticle density lower than the critical level, at which the proximized DOSs between Δ_g and $\Delta_{g,Nb}$ cannot afford quasiparticles. Therefore, the simple quasiparticle tunnel current exclusively generates output charge with a negligibly small spatial inhomogeneity for photons below ~ 300 eV.

Figure 3.31b shows a case of $E_0 =$ ~6 keV at bias points significantly lower than Δ_g/e ($\leq$ 0.28 meV).When the photon energy exceeds ~ 500 eV and approach 6 keV while keeping a low bias at 0.28 meV, the junction operation begins to shift to the concave spatial profile. It is assumed that photon energy or quasiparticle density is sufficiently high to break down the proximity superconductivity with $\Delta_g = 0.832$ meV near the junction edges inside the Nb layer. The density of the excited quasiparticles near the edges must be higher than that near the junction center, considering a reflection efficiency above 0.99 and a diffusion length ($\sqrt{2D\tau_{eff}}$) of ~ 23 μm. The dashed black line in (b) depicts the expected quasiparticle density profile. It is assumed that the quasiparticle density near the edges exceeds the critical level and thereby the proximity superconductivity is disrupted. The Δ_g value changes from 0.832 meV to 1.52 meV near the junction edges inside the Nb layer, as depicted by the concave dashed blue line in the upper panel in Fig. 3.31b. Near the center, the Nb/Al bilayers

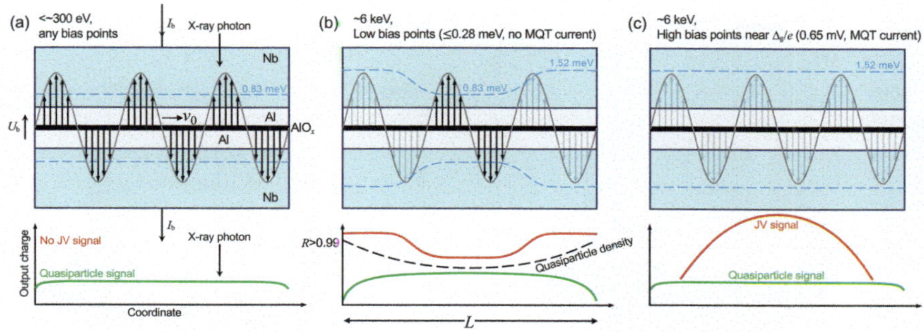

Fig. 3.31 Conceptual model to explain abnormal spatial inhomogeneity of 200-μm square proximized Nb/Al junctions with Al ≦ 50 nm: (**a**) a photon energy range below ~ 300 eV at any bias points, (**b**) ~ 6 keV at low bias points below 0.28 meV, (**c**) ~ 6 keV at high bias points near and above Δ_g/e (= 0.832 meV for "Al 30-nm"). This model junction has a layer structure of Nb(200)/ Al(30)/AlO_x/Al(30)/Nb(200) in nm. A magnetic field B is applied perpendicular to the surface of the paper, which shows the junction cross-section with length L. The L is 200 μm for a magnetic field in parallel to the junction side and 283 μm for the diagonal direction. The junction cross-section is enlarged along the vertical direction. The pairs of upward-and-downward black solid arrow-groups represent three supercurrent pairs or three Josephson vortices (JVs) for the sake of simplicity. At B = 12 mT and U_b = 0.28 mV, a JV density is approximately 1 JV/μm and a flow velocity of 0.05% of the vacuum light speed in both parallel and diagonal directions. The blue dashed lines represent expected Δ_g variations after X-ray photon absorption. The faint arrow-group pairs indicate that excited quasiparticles disturb the Josephson supercurrent pairs. The lower panels show the spatial variation of output charges due to the conventional quasiparticle tunnel current in green and the JV current in red, assisted with the MQT quasiparticle multiplication process at bias points near and above Δ_g/e. In the lower panel in (**b**), additionally, the dashed black line depicts an expected density variation of quasiparticles excited by an X-ray photon, based on the quasiparticle diffusion model with a quasiparticle reflection rate R of >0.99 at the junction edges

hold Δ_g = 0.832 meV. The quasiparticles near the edges between 0.832 and 1.52 meV return to normal electrons, which are unable to contribute to generating output charge. Therefore, the conventional quasiparticle current generated by events near the edges decreases, forming the convex dome shape indicated by the green solid line in the lower panel in Fig. 3.31 b. This situation is held in a photon energy range between ~ 300 and ~ 1.5 keV. In addition to the conventional quasiparticle current, additional JV current contributes to charge output. We assume that, near the edges, the Nb and Al layers return to the intact films without the proximity effect; the Nb film has an intact $\Delta_{g,Nb}$ value of 1.52 meV and the Al layer has an intact $\Delta_{g,Al}$ value of 0.17 meV. This transformation disturbs a small number of JVs near the junction edges, as depicted by the faint arrows in the upper panel in Fig. 3.31b. The disturbance reduces the number of moving JVs, leading to the reduction of bias voltage U_b across the top and bottom electrodes. Preamplifier feedback circuits cannot temporally respond to this fast change, so that the voltage is reduced. The voltage reduction causes a negative current flow in the external readout circuit, following the load line in Fig. 3.14a. Therefore, the JV current for photon absorption near the junction edges is higher than that near the center, forming the concave spatial profile depicted by the red

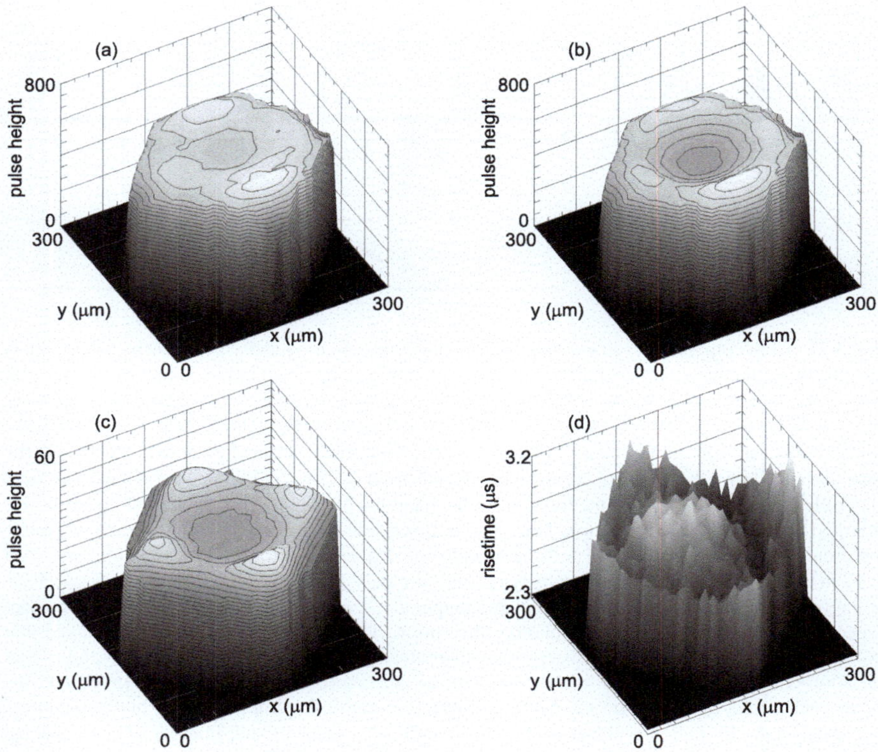

Fig. 3.32 LTSSM scan images for 200-μm square junction with "Al 30-nm", shown in Fig. 3.23, at a bias point of 0.16 mV: (**a**) output charge profile of the top electrode events, (**b**) output charge profile of the bottom electrode events, (**c**) output charge profile of the substrate events, and (**d**) rise time profile of a Charge-sensitive preamplifier. The bias point is lower than 0.28 mV, which is the lowest bias voltage in Fig. 3.23a. The X-ray microbeam energy with a diameter of 10 μm is 5 keV, which is essentially the same as 6 keV. The concave spatial profiles are consistent with the model in Fig. 3.31b. The rise time is equivalent to τ_{eff}, if conventional quasiparticle tunnel current exclusively governs the signal generation mechanism. Alternatively, the opposite curvatures in a–d implies that the contribution of the JV current mechanism is significant. Reproduced with permission from Pressler et al. [192]. Copyright 2001 IEEE

solid line in Fig. 3.31b. The total current summing up both quasiparticle current and JV current results in the concave spatial profile in Fig. 3.27a. Full 3D LTSSM scan images at a bias voltage of 0.16 mV, which is lower than 0.28 meV, are shown in Fig. 3.32 [192]. The concave profiles are consistent with Fig. 3.31 (b). The convex rise time profile in Fig. 3.32d is opposite to the concave output-charge profiles. If only the quasiparticle tunnel current governs charge output, the rise time profiles should follow the concave output-charge profiles. Therefore, the opposite profiles indicates that the JV current significantly contributes to the signal generation in addition to the conventional quasiparticle current.

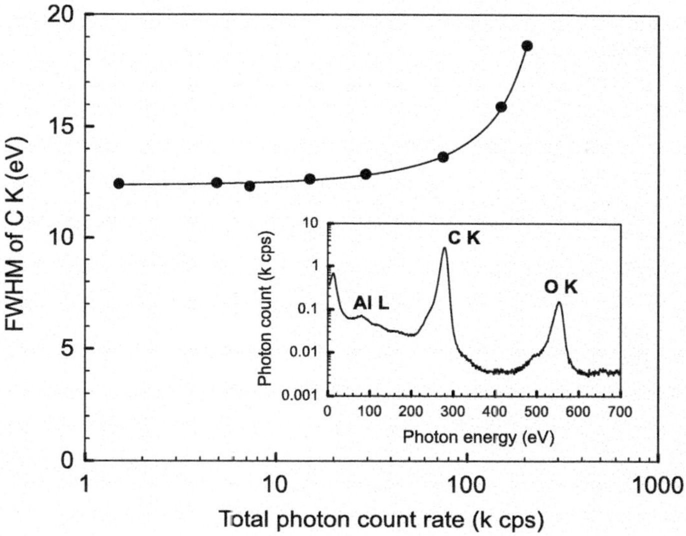

Fig. 3.33 Count rate performance of STJD array detector with 100 STJ-pixels of 100-μm square junctions with a layer structure of Nb(300)/Al(70)/AlO$_x$/Al(70)/Nb(100) in Fig. 3.16. The array detector has a ΔE value of 6.0 eV including an electronic noise of 2.0 eV after integrating all the pixels. The FWHM values for the C K-L$_{2,3}$ line (C Kα) are plotted as a function of photon count rate. A carbon nanotube X-ray source emits Bremsstrahlung and the characteristic X-rays of C K, O K, Al L-lines. The inset shows the X-ray spectrum with an energy bin of 1 eV at 100 kcps. The X-rays illuminated the whole area of the 100-pixle STJ array. By subtracting the STJD energy resolution of 6 eV from the 12.5-eV FWHM value below 10 kcps for the C K line, the carbon atoms in the tape have a natural linewidth of 11.0 eV. The count-rate upper limit for practical material analysis is ~200 kcps. Reproduced with permission from Ohkubo [144]. Copyright 2023 AIP Publishing LLC

Figure 3.31c shows a case of E_0 = ~6 keV at a high bias point near Δ_g/e (0.65 mV). When the bias voltage approaches Δ_g/e (= 0.832 mV), a 6-keV photon disrupts the proximity superconductivity everywhere in the junction area, assisted by the MQT quasiparticles. Even at the junction center, the total quasiparticle density exceeds the critical level for holding the proximity superconductivity. A photon absorption event disturbs the JVs everywhere. At the junction center, the collapse of the superconductivity occurs in an area of ~ 50 μm in diameter, considering the quasiparticle diffusion, so that many JVs are disturbed. This generates the maximum output current at the center. An event near the junction edges influences a smaller number of JVs because the quasiparticle diffusion area is limited by the junction edges. Therefore, the JV current assisted with the MQT quasiparticles generates a large dome shape. The sum of the conventional quasiparticle current in green and the JV current in red results in the dome on the terrace as shown in Fig. 3.23, which is the inverted tulip shape.

The magnetic field dependence in Fig. 3.27 reveals the close relationship between the abnormal signal enhancement and JVs. A high magnetic field of 31.6 mT represents a density of 2.4 JV/μm. One cycle of sinusoidal alternation is ~ 400 nm, which approach 147 nm of the coherence length of polycrystalline Al. JVs or pairs of alternating current groups begins to collapse. Therefore, the JV current decreases, leading to a flatter spatial curve exclusively due to quasiparticle tunnel current, as shown in Fig. 3.27b. In addition, the instability of the inverted tulip shape in Fig. 3.28 stems from the AVs pinned at the top corner. The AVs block the flow of JVs such that no JV current is generated everywhere in the junction area. Therefore, the inverted tulip shape transforms to the convex one in Fig. 3.28b: the convex shape is compatible with the quasiparticle diffusion model.

In the junction with “Al 70-nm,” the effect of the MQT quasiparticles on spatial inhomogeneity is less important because DOS values between 0.44 and 1.52 meV inside the Nb layers are sufficiently higher than those in "Al 30-nm. We assume that the Nb layers with a reduced Δ_g value of 0.44 meV has a large DOS margin even up to ~ 8 keV. The proximity Nb/Al bilayers hold a constant Δ_g value of 0.44 meV even when quasiparticles are excited by a 8-keV photon under the quasiparticle injection due to MQT at bias points near Δ_g/e. In other word, the junction with “Al 70 nm” operates within a linear regime up to ~ 8 keV at any bias points, as shown in Fig. 3.18, resulting in spatially uniform response or better energy resolution. The above-mentioned explanation is qualitative; more quantitative consideration is necessary for future investigations on the unsolved question.

3.2.4.8 Count Rate Performance

Regardless of the large spatial inhomogeneity in a high energy range in general, STJDs operate effectively in a soft X-ray range especially below ~ 1 keV and moderately up to ~ 2 keV without double peak. The energy resolution is 10 times better than that of the latest SDDs, enabling separation of closely spaced peaks of characteristic X-ray lines and even line shape analysis. In addition to energy resolution, most analytical applications require a high count-rate. The count rate represents the number of single photons or particles measured per second (counts per second: cps). Especially, for imaging analysis of advanced materials, a high count-rate is crucial for a tolerable measurement time to scan sample’s surfaces. Figure 3.33 shows a count rate performance of the 100-pixel STJD having a layer structure of Nb(300)/Al(70)/AlO_x/Al(70)/Nb(100). The thicker top Nb layer has a higher X-ray absorption efficiency than the bottom electrode, eliminating the double peak phenomenon. All the STJ pixels were uniformly irradiated by X-rays generated with a carbon nanotube X-ray source [198], which emitted Bremsstrahlung and the characteristic X-rays of C K, O K, and Al L-lines by electron impact onto a conductive carbon tape on an Al plate. Figure 3.33 shows the FWHM values for the C K-line as a function of photon count rate for all the characteristic X-rays and Bremsstrahlung. The inset shows an X-ray spectrum at 100 kcps. A pulse height spectrum of one pixel in the STJ array detector for monochromatic 400-eV photons is shown in Fig. 3.16 exhibiting a ΔE

value of 4.1 eV. Summing up all the pixels without off-line pulse processing, the array detector has an FWHM value of 6.0 eV including an electronic noise of 2.0 eV. The observed FWHM value of 12.5 eV does not originate from the detector energy resolution but the natural linewidth (11.0 eV) of carbon atoms embedded in the molecule structure of the tape.

Signal readout circuitry consists of the arrays of charge-sensitive preamplifiers, flash ADCs, and digital signal processing (DSP) circuits that have pulse shaping processors and multichannel analyzer memories for the 100 STJ-pixels [144]. The bias point is just after the rapid subgap-current increase near Δ_g/e as usual, which provides simultaneous and stable operation of all the pixels. Each pixel has its own gain, whose relative standard deviation is 4% (standard deviation divided by average peak position in pulse height spectra). A control software package automatically executes energy calibration for each pixel by using known characteristic X-ray peaks.

The energy resolution begins to rapidly degrade at ~ 200 kcps, which is the maximum count rate for analytical science. The count rate of 200 kpcs is slightly lower than several 100 kcps of SDDs; nevertheless, it is the highest record among superconductor detector types. Currently, insufficient bias current supply to each junction at a high-count rate region limits the count rate: a bias position change causes the ΔE degradation. A single junction can be operatable up to ~ 50 kcps without significant degradation of energy resolution [199, 200]. Therefore, ~ 5 Mcps is achievable in principle. STJ arrays with 200 pixels or more will achieve a higher count rate and a larger sensitive area [122]. A detector chip with 1024 pixels can be fabricated [147], but a typical complete system for analytical instruments has 100-to-200 readout channels due to a limit on parallel readout.

3.2.4.9 Optimum STJ Design

Figure 3.34 shows a close-packed square arrangement for the 100 STJ-pixels of 100-μm square junctions [122]. The arrangement has a gap space of 20 μm between the adjacent pixels and a filling factor of 70%. This high filling factor is realized by the 3D layer structure using chemical mechanical polishing (CMP) after fabricating the embedded Nb wiring leads underneath a SiO_2 insulation layer. The tunnel junction layer is deposited on the planarized flat surface of the SiO_2 layer. The lower panel shows the cross-sectional layer structure. The sensitive area of the array detector is still 25 times smaller than 25-mm^2 SDDs; therefore, a polycapillary focusing lens that is placed in front of the array detector compensates the small sensitive area [201]. STJ array detectors are employed for X-ray absorption fine structure (XAFS) spectroscopy at synchrotron radiation facilities and SEM. An application to particle induced X-ray emission (PIXE) analysis is under progress [202, 203]. An energy coverage of STJD spectrometers for X-rays is generally between 50 eV and 2 keV, holding an energy linearity and a practically high absorption efficiency or detection efficiency. A combination with a semiconductor detector covers up to ~ 20 keV on SEM. A 100 STJ-pixel array detector with 200-μm square junctions is employed

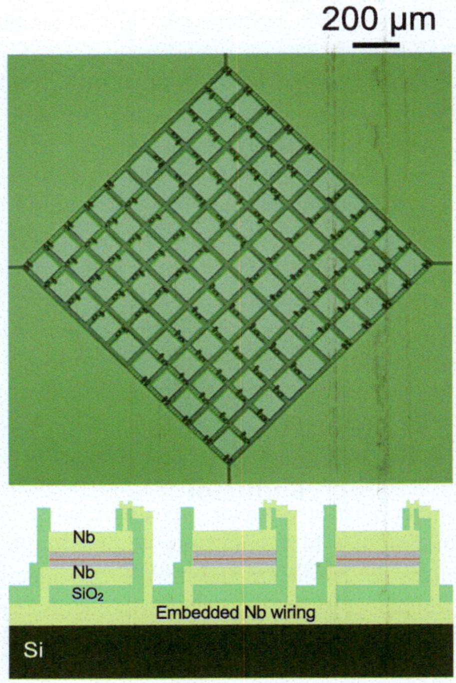

Fig. 3.34 Nb/Al STJ array detector with 100 pixels of 100-μm square junctions [122]: upper panel for optical top-view image and lower panel for schematic illustration of cross-sectional layer structure. Chemical mechanical polishing (CMP) enables embedding the wiring leads, which run underneath the junction layer and separated by a SiO_2 layer. Due to CMP, the close-packed pixel arrangement realizes a filling factor of 70%. The gap between the adjacent pixels is 20 μm. An example of X-ray spectra acquired by the best pixel is shown in Fig. 3.16. Reproduced with permission from Fujii et al. [122]. Copyright 2015 IOP Publishing

for large molecule detection in mass spectrometry (MS) in an energy range of 3–30 keV [204–207]. Due to spatial inhomogeneity, the energy resolution of 200-μm square junctions is larger than 10 eV, but it is acceptable for MS, in which molecule impact causes a large fluctuation in energy absorbed by a junction.

3.2.5 *Microwave Kinetic Inductance (MKI) Type*

3.2.5.1 Structure of MKI Type and Operation Principle

The second readout scheme for superconductor quantum detection utilizes a surface resistance change due to quasiparticle excitation in a RF range [81]. The MKI type consists of superconductor *LC* resonance circuits coupled to a RF excitation and readout line as illustrated in Fig. 3.35a. Either capacitive or inductive coupling between the resonance circuits and the external line is accessible. Figure 3.35b depicts a circuit diagram of capacitive coupling with C_c for one pixel. In a steady state, the surface resistance R_s is close to zero; the resonance frequency f_0 is expressed by

$$f_0 = \frac{1}{2\pi\sqrt{LC}}, \tag{3.31}$$

where L is the sum of surface inductance L_s and other stray contribution L_i, and C varies as C_1, C_2, C_3, etc., such that each pixel has a unique resonance frequency like f_1, f_2, f_3, etc. The ratio L_s/L_i is typically 0.04 [81]. The L_s inductor segment in Fig. 3.35 functions as an absorber.

Photon or continuous particle flux break Cooper pairs, generating quasiparticles in the inductor segment and inducing an increase of L_s and R_s. This change results in a resonance frequency shift and a Q-value reduction as depicted by the blue and red curves in Fig. 3.35c. The shift from f_0 is expressed by

$$\frac{\Delta f}{f_0} = \frac{\Delta L}{2L} \cong -\frac{1}{2}\frac{L_s}{L_i}\frac{\Delta L_s}{L_s}. \tag{3.32}$$

This resonance frequency shift results in an increase of RF transmission power when excitation RF is fixed at f_0 as shown in Fig. 3.35c. The C_1, C_2, C_3, etc. for pixels are adjusted such that the resonance frequencies are separated by a regular interval and the whole frequency range fits to a bandwidth of the RF circuit.

Kinetic inductance due to Cooper pairs in superconductors has a nonnegligible contribution to the total inductance in an RF range. A change of complex surface impedance $Z_s = R_s + i\omega L_s$, where ω is the angular frequency, governs output signals. In a temperature range below ~ $T_c/10$, a condition of $R_s \ll L_s$ is held because the number of thermally excited quasiparticles is negligibly small. Therefore, we need to exclusively consider L_s. Energy storage with L_s has two components: magnetic field within a penetration depth and kinetic inertia of Cooper pairs. Both magnetic inductance and kinetic inductance are proportional to London penetration depth λ, which is inversely proportional to the square root of Cooper pair density. When a photon breaks Cooper pairs, λ increases due to a decrease of Cooper pair density, leading to an increase of L_s. Simultaneously, R_s increases due to quasiparticles that dissipate

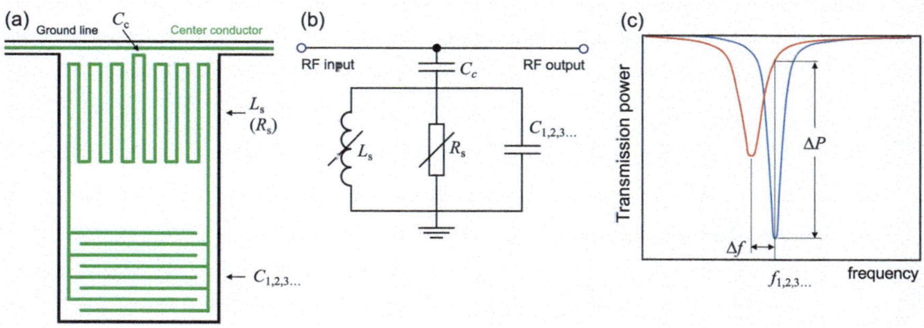

Fig. 3.35 Microwave kinetic inductance detector (MKID): (**a**) conceptual arrangement of circuit elements in one pixel, (**b**) equivalent circuit of one pixel, and (**c**) RF transmission power change due to particle incidence. Thousands of pixels dangle from the one transmission line. Each pixel consists of an inductive part with L_s and a capacitive part with C_i: $C_{1,2,3}$, etc. The LC circuit is coupled to the center conductor of the transmission line via a coupling capacitor C_c. Inductive coupling is also possible. Reproduced with permission from Ohkubo [144]. Copyright 2023 AIP Publishing LLC

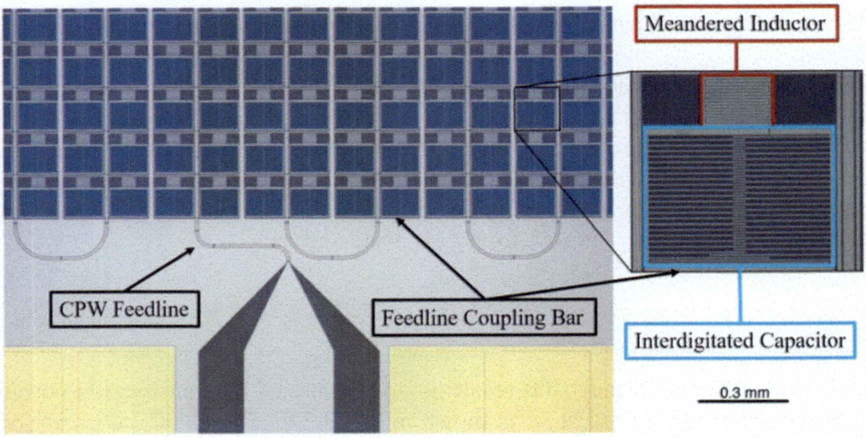

Fig. 3.36 20,000-pixel MKID similar to camera installed on Subaru telescope [82]. The enlarged image on the right-hand side shows the inductor and the capacitor of one pixel. The superconductor used in the circuit is PtSi with $T_c = 0.8$ K on a sapphire substrate. The operating temperature is 90 mK. Reproduced with permission from Walter et al. [82]. Copyright 2020 IOP Publishing

energy. The increase of L_s and R_s induces the resonance curve change as shown in Fig. 3.35c. The absorbed energy is determined by measuring the resonance amplitude change or the degree of phase shift.

The transmission power change ΔP for each element at a unique frequency provides a method of multiplexing thousands of pixels. Instead of the C adjustment, it is also possible to vary L_s. One excitation and readout line is sufficient for 10,000 pixels or more. Frequency-domain multiplexing provides a imaging capability in reading out a large number of pixels with sophisticated hardware and software [208]. Figure 3.36 shows an example of kilo-pixel MKID arrays with a superconductor PtSi with a T_c value of 0.8 K [82]. This MKID operating at 90 mK is similar to the MKID exoplanet camera covering a wavelength band of 800–1400 nm at a spectral resolution ($\lambda/\Delta\lambda$) of 5–7, installed on the Subaru telescope. In this wavelength range (1.55–0.886 eV), $E_0/\Delta E$ ranges 5–7.

3.2.5.2 Energy Resolution and Covering Energy Range

The energy resolution limit of MKIDs is the same as other quantum detectors using an energy gap. Therefore, the explanations and formula in the preceding sections on STJDs are also applicable to MKIDs; for example, $E = 2.355\sqrt{FE_0\varepsilon}$ in Eq. (3.4). Although the exact probability function for quasiparticle pair generation in superconductors is analytically unknown, the binominal function suitably explains the statistical process, as discussed in Sect. 3.1.4. A good guidance is numerical calculations based on the Monte Carlo method; we use the estimated F and ε values [63] by adjusting them to quasiparticle pairs. MKIDs utilize a simple superconductor thin film; therefore, we can skip the complicated multiple tunneling factor G for STJDs in Eq. (3.28). The theoretical ΔE values for 1-eV photon absorption, which is within a typical covering range of MKIDs, are 0.056 eV ($E_0/\Delta E = 18$) for Nb, 0.040 eV ($E_0/\Delta E = 25$) for Ta, and 0.018 eV ($E_0/\Delta E = 54$) for Al. When we use $\Delta_g = 0.07$ meV [209] for PtSi thin films in Fig. 3.36 and assume $F = 0.1$, it yields a ΔE value of 0.011 eV ($E_0/\Delta E = 87$) for 1 eV photons. These theoretical limits of $E_0/\Delta E$ are considerably larger than experimental values of 5–7. This disagreement is natural, considering many uncertain factors in a complex impedance change, a relationship between quasiparticle density and resonance curve shift, and measurement uncertainty of ΔP or Δf. Nonetheless, we can skip the complicated quasiparticle dynamics such as the multiple tunneling effect, the MQT multiplication of quasiparticles, and JV motion in STJDs. Therefore, with MKID, we can determine the relationship between ε and Δ_g in elemental superconductors [210]. In addition, the gap-reduction model in Fig. 3.19b can be verified by elucidating the number of quasiparticles in MKIDs with a proximity bilayer.

Besides photon detection near 1 eV, X-ray photon detection with MKIDs was also pursued, based on the conventional quantum scheme and a newly contrived thermal scheme [211]. One example of the quantum scheme is a 1D X-ray imager with two Al-MKIDs at both ends of a 200 × 35 μm^2 Ta strip [212], in a similar way to 1D STJ imagers using quasiparticle diffusion [126]. For events at the center of the Ta strip, the energy resolution for 5.9-keV X-rays is 62 eV, which is better than ~ 120 eV of Si detectors, as shown in Fig. 3.37. However, the X-ray energy range appears to be inappropriately high for MKIDs, as expected from the experimental results on STJDs, which are adequate to an energy range below ~1 keV.

The MKI type is suitable for a range of near-infrared to ultraviolet for astrophysics [213, 214], and used for cosmic microwave background (CMB) radiation (0.1–4 meV) [215, 216]. For optical/IR astronomy, the figures of merit of MKID are compared with TESD and STJD [217]. The MKI type is the most promising due to a tremendous number of pixels and a large sensitive area. One of the latest MKID applications is the 20,440-pixel camera in Fig. 3.36 for the eight-meter Subaru telescope at Mauna Kea [82]. The MKID camera has a photon detection capability of arrival time, location, and photon energy, which cannot be realized by CCD cameras for a near-infrared range of 800–1400 nm. The performance, physics, and applications appear in a review and the recent LTD workshop proceedings [218–220].

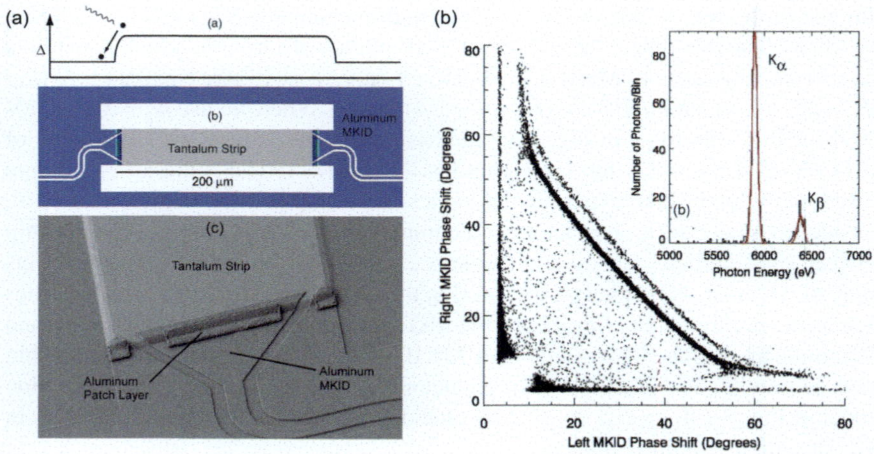

Fig. 3.37 X-ray photon imagers with two Al-MKIDs at both ends of 200 × 35 μm^2 Ta strip [212]: (**a**) imager structure and (**b**) scatter plot of absorption events detected by the right and left MKIDs. The inset in (**b**) shows a pulse heigh spectrum for the events at the strip center; the energy resolution is 62 eV in FWHM for 5.9 keV X-rays. Reproduced with permission from Mazin et al. [212]. Copyright 2006 AIP Publishing LLC

References

1. A.G. Kochur, V.A. Popov, Rad. Phys. Chem. **75**, 1525 (2006)
2. K.M. Anagnost, E. Lee, Z. Wang, J. Liu, E.R. Fossum, Sensors **21**, 7566 (2021)
3. U. Fano, Phys. Rev. **72**, 26 (1947)
4. G. F. Knoll, *Radiation Detection and Measurement*, 4th ed. (Wiley, 2010), p. 117
5. W.E. Drummond, J.L. Moll, J. Appl. Phys. **42**, 5556 (1971)
6. K. Ramanathan, N. Kurinsky, Phys. Rev. D **102**, 063026 (2020)
7. W. Roosbroeck, Phys. Rev. **139**, A1702 (1965)
8. R. Devanathan, L.R. Corrales, F. Gao, W.J. Weber, Nucl. Instrum. Methods in Phys. Res. A **565**, 637 (2006)
9. D.V. Jordan et al., Nucl. Instrum. Methods in Phys. Res. A **585**, 146 (2008)
10. D. Durnford, Q. Arnaud, G. Gerbier, Phys. Rev. D **98**, 103013 (2018)
11. H.R. Zulliger, D.W. Aitken, IEEE Trans. Nucl. Sci. **17**, 187 (1970)
12. I.V. Kotov, H. Neal, P.O' Gonnor, Nucl. Inst. Methods in Phys. Res. A **901**, 126 (2018)
13. D. Rodrigues et al., Nucl. Instrum. Methods in Phys. Res. A **1010**, 165511 (2021)
14. P. Varshni, Physica **34**, 149 (1967)
15. K.P. O'Donnell, X. Chen, Appl. Phys. Lett. **58**, 2924 (1991)
16. M. Brigida et al., Nucl. Instrum. Methods in Phys. Res. A **533**, 322 (2004)
17. M.J. Harrison, D.S. McGregor, F.P. Doty, Phys. Rev. B **77**, 195207 (2008)
18. C. Klein, J. Appl. Phys. **39**, 2029 (1968)
19. R.C. Alig, S. Bloom, Phys. Rev. Lett. **35**, 1522 (1975)
20. R.C. Alig, S. Bloom, C.W. Struck, Phys. Rev. B **22**, 5565 (1980)
21. R.C. Alig, Phys. Rev. B **27**, 968 (1983)
22. R.H. Pehl, F.S. Goulding, D.A. Landis, M. Lenzlinger, Nucl. Inst. Methods **59**, 45 (1968)
23. T. Yamaya, R. Asano, H. Endo, K. Umeda, Nucl. Inst. Methods **159**, 181 (1979)
24. A. Owens, A. Peacock, Nucl. Inst. Methods in Phys. Res. A **531** 18 (2004)
25. A.M. Barnett, J.E. Lees, D.J. Bassford, Appl. Phys. Lett. **102**, 181119 (2013)

26. M.D.C. Whitaker, G. Lioliou, A.B. Krysa, A.M. Barnett, Semicond. Sci. Technol. **35**, 095026 (2020)
27. R. Devanathan, L.R. Corrales, F. Gao, W.J. Weber, Nucl. Instrum. Methos Phys. Res. A **565**, 637 (2006)
28. G. Bertuccio, D. Maiocchi, J. Appl. Phys. **92**, 1248 (2002)
29. G.W. Fraser, A.F. Abbey, A. Holland, K. McCarthy, A. Owens, A. Wells, Nucl. Instrum. Methods Phys. Res. A **350**, 368 (1994)
30. W. Shockley, Solid-State Electron. **2**, 35 (1961)
31. A.G. Kozorezov, A.F. Volkov, J.K. Wigmore, A. Peacock, A. Poelaert, R. den Hartog, Phys. Rev. B **61**, 11807 (2000)
32. D.S. Kim, H.L. Smith, J.L. Niedziela, C.W. Li, D.L. Abernathy, B. Fultz, Phys. Rev. B **91**, 014307 (2015)
33. G. Klemens, Phys. Rev. **148**, 845 (1966)
34. B.C. Daly, K. Kang, Y. Wang, D.G. Cahill, Phys. Rev. B **80**, 174112 (2009)
35. C.R. Crowell, S.M. Sze, Appl. Phys. Lett. **9**, 242 (1966)
36. V. Laporta, L.D. Pietanza, G. Colonna, Nucl. Instrum. Methods in Phys. Res. A **636**, 67 (2011)
37. D. von der Linde, R. Lambrich, Phys. Rev. Lett. **42**, 1090 (1979)
38. F. Güthoff, B. Hemion, C. Herzig, W. Petry, H.R. Schober, J. Trampenau, J. Phys. Condens. Matter **6**, 6211 (1994)
39. P. Pandey, S.K. Pandey, J. Phys. Condens. Matter **36**, 165602 (2024)
40. S. Poncé, E.R. Margine, C. Verdi, F. Giustino, Comput. Phys. Commun. **209**, 116 (2016)
41. A.R. Jani, N.E. Brener, J. Callaway, Phys. Rev. B **38**, 9425 (1988)
42. W.L. McMillan, Phys. Rev. **167**, 331 (1968)
43. S. Baroni, S. de Gironcoli, A.D. Corso, P. Giannozzi, Rev. Mod. Phys. **73**, 515 (2001)
44. J. Sjakste, K. Tanimura, G. Barbarrino, L. Perfetti, N. Vast, J. Phys. Condens. Matter **30**, 353001 (2018)
45. L. Waldecker, R. Bertoni, R. Ernstorfer, J. Vorberger, Phys. Rev. X **6**, 021003 (2016)
46. E. Pop, R.W. Dutton, K.E. Goodson, J. Appl. Phys. **96**, 4998 (2004)
47. M.V. Fischetti, P.D. Yoder, M.M. Khatami, G. Gaddemane, M.L. Van de Put, Appl. Phys. Lett. **114**, 222104 (2019)
48. M. Bernardi, D. Vigil-Fowler, C.S. Ong, J.B. Neaton, S.G. Louie, Proc. Proc. Natl. Acad. Sci. U. S. A. **112**, 5291 (2015)
49. R. Sen, N. Vast, J. Sjakste, Appl. Phys. Lett. **120**, 082101 (2022)
50. A. Owens, G.W. Fraser, A.F. Abbey, A. Holland, K. McCarthy, A. Keay, A. Wells, Nucl. Instrum. Methods in Phys. Res. A **382** 503 (1996)
51. J.A. Detwiler, R.G.H. Robertson, Phys. Rev. C **107**, L042501 (2023)
52. F. Scholze, H. Rabus, G. Ulm, J. Appl. Phys. **84**, 2926 (1998)
53. D.M. Schlosser et al., Nucl. Instrum. Methods in Phys. Res. A **624**, 270 (2010)
54. G.H. Wood, Master thesis, *An Investigation of the Possibility of Charged Particle Detection Using Tunneling Between Superconductors* (University of British Columbia, 1965). https://doi.org/10.14288/1.0085916
55. C.H. Wood, B.L. White, Appl. Phys. Lett. **15**, 237 (1969)
56. J. Matisoo, Appl. Phys. Lett. **9**, 167 (1966)
57. D. Van Vechten, K.S. Wood, Phys. Rev. B **43**, 12852 (1991)
58. K.S. Wood, D. Van Vechten, Nucl. Instrum. Methods in Phys. Res. A, **314**, 86 (1992)
59. M. Kurakado, H. Mazaki, Phys. Rev. B **22**, 168 (1980)
60. M. Kurakado, Nucl. Instrum. Meth. **196**, 275 (1982)
61. N. Rando, A. Peacock, A. van Dordrecht, C. Foden, R. Engelhardt, B. Taylor, P. Gare, J. Lumley, C. Pereira, Nucl. Instrum. Methods Phys. Res. A **313**, 173 (1992)
62. N. Rando, A. Peacock, C. Foden, A. v. Dordrecht, J. Lumley, C. Pereira, J. Appl. Phys. **73**, 5098 (1993)
63. C. Bray, L.J. Hiller, K.G. Leach, S. Friedrich, J. Low Temp. Phys. **209**, 857 (2022)
64. S. Friedrich, F. Ponce, M.H. Carpenter, R. Cantor, Synchrotron Radiation News **27**, 28 (2014)

65. A. Peacock, P. Verhoeve, N. Rando, A. van Dordrecht, B.G. Taylor, C. Erd, M.A.C. Perryman, R. Venn, J. Howlett, D.J. Goldie, J. Lumley, M. Wallis, Nature **381**, 135 (1996)
66. T. Peacock, P. Verhoeve, N. Rando, C. Erd, M. Bavdaz, B.G. Taylor, D. Perez, Astron. Astrophys. Suppl. Ser. **127**, 497 (1998)
67. S. Shiki, H. Sato, Y. Takizawa, M. Kurakado, H. Shimizu, Publ. Astron. Soc. Jpn. **56**, L19 (2004)
68. B. Cabrera, R.M. Clarke, P. Colling, A.J. Miller, S. Nam, R.W. Romani, Appl. Phys. Lett. **73**, 735 (1998)
69. D. Twerenbold, Nucl. Instrum. Method Phys. Res. A **370**, 253 (1996)
70. D. Twerenbold, J.-L. Vuilleumier, D. Gerber, A. Tadsen, B. van den Brandt, P.M. Gillevet, Appl. Phys. Lett. **68**, 3503 (1996)
71. H. Hilton, J.M. Martinis, D.A. Wollman, K.D. Irwin, L.L. Dulcie, D. Gerber, P.M. Gillevet, D. Twerenbold, Nature **391**, 672 (1998)
72. R.J. Wenzel, U. Matter, L. Schultheis, R. Zenobi, Anal. Chem. **77**, 4329 (2005)
73. S. Shiki, M. Ukibe, Y. Sato, S. Tomita, S. Hayakawa, M. Ohkubo, J. Mass Spectrom. **43**, 1686 (2008)
74. M. Ohkubo, S. Shiki, M. Ukibe, S. Tomita, S. Hayakawa, Int. J. Mass Spectrom. **299**, 94 (2011)
75. H. Hayakawa, Phys. Today **39**, 46 (1986)
76. R.B. Laibowitz, J.J. Cuomo, J. Appl. Phys. **41**, 2148 (1970)
77. R.B. Laibowitz, A.F. Mayadas, Appl. Phys. Lett. **20**, 254 (1972)
78. M. Gurvitch, M.A. Washington, H.A. Huggins, Appl. Phys. Lett. **42**, 472 (1983)
79. S. Morohashi, F. Shinoki, A. Shoji, M. Aoyagi, H. Hayakawa, Appl. Phys. Lett. **46**, 1179 (1985)
80. R.F. Broom, S.I. Raider, A. Oosenbrug, R.E. Drake, W. Walter, IEEE Trans. Electron. Devices **ED-27**, 1998 (1980)
81. P.K. Day, H.G. LeDuc, B.A. Mazin, A. Vayonakis, J. Zmuidzinas, Nature **425**, 817 (2003)
82. A.B. Walter et al., Publ. Astron. Soc. Pac. **132**, 125005 (2020)
83. L. Strüder, A. Niculae, P. Holl, H. Soltau, Microscopy Today **28**, 46 (2020)
84. J. Bardeen, L.N. Cooper, J.R. Schrieffer, Phys. Rev. **108**, 1175 (1957)
85. G.W. Webb, F. Marsiglio, J.E. Hirsch, Physica C **514**, 17 (2015)
86. J.E. Hirsch, M.B. Maple, F. Marsiglio, Physica C **514**, 1 (2015)
87. T. Imamura, S. Hasuo, Appl. Phys. Lett. **58**, 645 (1991)
88. C. Peroz, C. Villard, Phys. Rev. B **72**, 014515 (2005)
89. Y.E. Chen, M. Ukibe, K. Odaka, A. Kurokawa, M. Ohkubo, Physica C **469**, 1680 (2009)
90. R. Gross, D. Koelle, Rep. Prog. Phys. **57**, 651 (1994)
91. J.R. Schrieffer, D.J. Scalapino, J.W. Wilkins, Phys. Rev. Lett. **10**, 336 (1963)
92. G.B. Scott, M. Springford, Proc. Roy. Soc. Lond. A **320**, 115 (1970)
93. D.P. Karim, J.B. Ketterson, G.W. Crabtee, J. Low Temp. Phys. **30**, 389 (1978)
94. T. Van Duzer, C.W. Turner, *Principles of Superconductive Devices and Circuits* (Elsevier, 1981), p. 61
95. M. Tinkham, *Introduction to Superconductivity*, 2nd ed. (McGraw-Hill, 1996), p. 406
96. V.V. Schmidt, *The Physics of Superconductors* (Springer, 2010), Fig. 7.2 on p. 171
97. N.N. Bogoliubov, Nuovo Cimento **7**, 794 (1958)
98. D.F. Agterberg, P.M.R. Brydon, C. Timm, Phys. Rev. Lett. **118**, 127001 (2017)
99. A.L. Kuzemsky, Int. J. Mod. Phys. B **24**, 835 (2010)
100. Y. Ronen, Y. Cohen, J.-H. Kang, A. Haim, M.-T. Rieder, M. Heiblum, D. Mahalu, H. Shtrikman, Proc. Natl. Acad. Sci. U. S. A. **113**, 1743 (2016)
101. J. Niu et al., Phys. Rev. Lett. **132**, 076001 (2024)
102. A. Kozorezov, J.K. Wigmore, D. Martin, P. Verhoeve, A. Peacock, Phys. Rev. B **75**, 094513 (2007)
103. A. Poelaert, R. den Hartog, A. Peacock, A. Kozorezov, J.K. Wigmore, Nucl. Instrum. Methods in Phys. Res. A **442**, 275 (2000)

104. G. Brammertz, A. Poelaert, A.A. Golubov, P. Verhoeve, A. Peacock, H. Rogalla, J. Appl. Phys. **90**, 355 (2001)
105. J. Clarke, Phys. Rev. Lett. **28**, 1363 (1972)
106. M. Tinkham, J. Clarke, Phys. Rev. Lett. **28**, 1366 (1972)
107. W.K. Skocpol, M.R. Beasley, M. Tinkham, J. Low Temp. Phys. **16**, 145 (1974)
108. J. Demsar, J. Low Temp. Phys. **201**, 676 (2020)
109. A.M. Gulian, D.V. Vechten, Modern. Phys. Lett. B **10**, 329 (1996)
110. A.M. Gulian, D.V. Vechten, Nucl. Instrum. Methods in Phys. Res. A **370**, 107 (1996)
111. M. Kalenkov, A.D. Zaikin, Phys. Rev. B **92**, 014507 (2015)
112. M. Johnson, Phys. Rev. Lett. **67**, 374 (1991)
113. V.V. Schmidt, *The Physics of Superconductors* (Springer, 2010), p. 172
114. F.C. Wellstood, C. Urbina, J. Clarke, Phys. Rev. B **49**, 5942 (1994)
115. R. Yagi, Phys. Rev. B **73**, 134507 (2006)
116. Y. Takane, J. Phys. Soc. Jpn. **75**, 023706 (2006)
117. K. Tsuboi, R. Yagi, J. Phys. Soc. Jpn. **79**, 013704 (2010)
118. J.C. Hübler, J.C. Lemyre, D. Beckmann, H.V. Löhneysen, Phys. Rev. B **81**, 184524 (2010)
119. J.B. le Grand, J. Martin, R.P. Huebener, A.W. Hamster, G.C.S. Brons, J. Flokstra, J. Appl. Phys. **81**, 7413 (1997)
120. M. Ohkubo et al., Nucl. Instrum. Methods in Phys. Res. A **444**, 237 (2000)
121. S.B. Kaplan, C.C. Chi, D.N. Langenberg, J.J. Chang, S. Jafarey, D.J. Scalapino, Phys. Rev. B **14**, 4854 (1976)
122. G. Fujii, M. Ukibe, M. Ohkubo, Supercond. Sci. Technol. **28**, 104005 (2015)
123. M. Maezawa, M. Aoyagi, H. Nakagawa, I. Kurosawa, S. Takada, Appl. Phys. Lett. **66**, 2134 (1995)
124. E. Gatti, P. Rehak, Nucl. Instrum. Methods in Phys. Res. **225**, 608 (1984)
125. N. Booth, Appl. Phys. Lett. **50**, 293 (1987)
126. H. Kraus, F. Feilitzsch, J. Jochum, R.L. Mössbauer, Th. Peterreins, F. Pröbst, Phys. Lett. B **231**, 195 (1989)
127. W.L. McMillan, Phys. Rev. B **175**, 537 (1968)
128. J. Du, A.D.M. Charles, K.D. Petersson, E.W. Preston, Supercond. Sci. Technol. **20**, S350 (2007)
129. P. Verhoeve, N. Rando, J. Verveer, A. Peacock, A. van Dordrecht, P. Videler, M. Bavdaz, D.J. Goldie, T. Lederer, F. Scholze, G. Ulm, R. Venn, Phys. Rev. B **53**, 809 (1996)
130. L.J. Zeng, S. Nik, T. Greibe, P. Krantz, C.M. Wilson, P. Delsing, E. Olsson, J. Phys. D Appl. Phys. **48**, 395308 (2015)
131. L. Zeng, D.T. Tran, C.W. Tai, G. Svensson, E. Olsson, Sci. Rep. **6**, 29679 (2016)
132. E.P. Houwman, J.G. Gijsbertsen, J. Flokstra, H. Rogalla, J.B. le Grand, P.A.J. de Korte, A.A. Golubov, IEEE Trans. Appl. Super. **3**, 2170 (1993)
133. A.A. Golubov, E.P. Houwman, J.G. Gijsbertsen, V.M. Krasnov, J. Flokstra, H. Rogalla, M.Y. Kupriyanov, Phys. Rev. B **51**, 1073 (1995)
134. A. Zehnder, Ph. Lerch, S.P. Zhao, Th. Nussbaumer, E.C. Kirk, H.R. Ott, Phys. Rev. B **59**, 8875 (1999)
135. I. Giaever, Phys. Rev. Lett. **5**, 464 (1960)
136. D. Twerenbold, Eur. Phys. Lett. **1**, 209 (1986)
137. J.G. Gijsbertsen, E.P. Houwman, B.B.G. Klopman, J. Flokstra, H. Rogalla, D. Quenter, S. Lemke, Physica C **249**, 12 (1995)
138. C. Nappi, R. Cristiano, L. Frunzio, S. Pagano, M.P. Lisitkii, J. Appl. Phys. **80**, 3401 (1996)
139. K. Kikuchi, H. Myoren, T. Iizuka, S. Takada, Jpn. J. Appl. Phys. **39**, 5828 (2000)
140. S. Friedrich, M.F. Cunningham, M. Frank, S.E. Labov, A.T. Barfknecht, S.P. Cramer, Nucl. Instrum. Methods Phys. Res. A **444**, 151 (2000)
141. K. Yamaguchi, H. Nakagawa, M. Aoyagi, M. Naruse, H. Myoren, T. Taino, Physica C **530**, 90 (2016)
142. L. Frunzio, L. Li, D.E. Prober, I.V. Vernik, M.P. Lisitskii, C. Nappi, R. Cristiano, Appl. Phys. Lett. **79**, 2103 (2001)

143. S. Imai, S. Morita, A. Ishikawa, Y. Takeuti, N. Mikoshiba, IEEE Trans. Mag. **21**, 906 (1985)
144. M. Ohkubo, J. Appl. Phys. **134**, 081101 (2023)
145. K.E. Gray, Appl. Phys. Lett. **32**, 392 (1978)
146. T. Van Duzer, C.W. Turner, *Principles of Superconductive and Circuits* (Elsevier, New York, 1981), p.87
147. G. Fujii, M. Ukibe, S. Shiki, M. Okubo, J. Low Temp. Phys. **184**, 194 (2016)
148. C.H. Liu et al., Phys. Rev. Lett. **132**, 017001 (2024)
149. M. Houzet, K. Serniak, G. Catelani, M.H. Devoret, L.I. Glazman, Phys. Rev. Lett. **123**, 107704 (2019)
150. P.J. de Visser, J.J.A. Baselmans, P. Diener, S.J.C. Yates, A. Endo, T.M. Klapwijk, Phys. Rev. Lett. **106**, 167004 (2011)
151. T. Connolly et al., Phys. Rev. Lett. **132**, 217001 (2024)
152. T. Noguchi, T. Suzuki, A. Endo, T. Tamura, Physica C **469**, 1585 (2009)
153. A.F. Andreev, Sov. Phys. JETP **19**, 1228 (1964)
154. Y. Uzawa, Z. Wang, Phys. Rev. Lett. **95**, 017002 (2005)
155. A.W. Kleinsasser, R.E. Miller, W.H. Mallison, G.B. Anrnold, Phys. Rev. Lett. **72**, 1738 (1994)
156. T. Greibe, M.P.V. Stenberg, C.M. Wilson, T. Bauch, V.S. Shumeiko, P. Delsing, Phys. Rev. Lett. **106**, 097001 (2011)
157. A.G. Kozorezov, J.K. Wigmore, A. Peacock, R. den Hartog, D. Martin, G. Brammertz, P. Verhoeve, N. Rando, Phys. Rev. B **69**, 184506 (2004)
158. A. Zehnder, Phys. Rev. B **52**, 12858 (1995)
159. D.D. Coon, M.D. Fiske, Phys. Rev. **138**, A744 (1965)
160. S. Friedrich et al., IEEE Trans. Appl. Super. **7**, 3383 (1997)
161. A. Rothwarf, B.N. Taylor, Phys. Rev. Lett. **19**, 27 (1967)
162. W. Eisenmenger, K. Laßmann, H.J. Trumpp, R. Krauß, Appl. Phys. **11**, 307 (1976)
163. S.B. Kaplan, J. Low Temp. Phys. **37**, 343 (1979)
164. M. Ohkubo, J. Martin, K. Drachsler, R. Gross, R.P. Huebener, I. Sakamoto, N. Hayashi, Phys. Rev. B **54**, 9484 (1996)
165. J. Martin, R.P. Huebener, J.B. le Grand, C.A. Mears, S.E. Labov, A.T. Barfknecht, Appl. Phys. Lett. **73**, 378 (1998)
166. M. Ohkubo, K. Suzuki, K. Tanabe, H. Pressler, M. Ukibe, A.I.P. Conf, Proc. **605**, 55 (2002)
167. A.A. Golubov, E.E. Houwman, Physica C **205**, 147 (1993)
168. S.I. Salem, P.L. Lee, Atom. Data Nucl. Data Tables **18**, 233 (1976)
169. M. Ohkubo, D. Fukuda, I. Sakamoto, H. Hayashi, J. Martin, F. Panteleit, R.P. Huebener, J. Appl. Phys. **85**, 595 (1999)
170. R. den Hartog, A.G. Kozorezov, J.K. Wigmore, D. Martin, P. Verhoeve, A. Peacock, A. Poelaert, G. Brammertz, Phys. Rev. B **66**, 094511 (2002)
171. N. Rando, A. Peacock, A. van Dordrecht, P. Hübner, P. Videler, J. Salmi, I. Suni, J. Appl. Phys. **76**, 2490 (1994)
172. M. Ohkubo, A.I.P. Conf, Proc. **1185**, 381 (2009)
173. Berkeley Lab. *X-ray Interactions with matter*. https://henke.lbl.gov/optical_constants
174. A. Poelaert, A.G. Kozorezov, A. Peacock, J.K. Wigmore, Phys. Rev. Lett. **82**, 1257 (1999)
175. S. Friedrich, K. Segall, M.C. Gaidis, C.M. Wilson, D.E. Prober, A.E. Szymkowiak, S.H. Moseley, Appl. Phys. Lett. **71**, 3901 (1997)
176. L. Li, L. Frunzio, C. Wilson, D.E. Prober, A.E. Szymkowiak, S.H. Moseley, J. Appl. Phys. **90**, 3645 (2001)
177. R. den Hartog, A. Golubov, D. Martin, P. Verhoeve, A. Poelaert, A. Peacock, M. Krumrey, Nucl. Instrum. Methods Phys. Res. A **444**, 28 (2000)
178. R. den Hartog, A. Golubov, P. Verhoeve, A. Poelaert, D. Matin, A. Peacock, A. van Dordrecht, D.J. Goldie, Phys. Rev. B **63**, 214507 (2001)
179. J.E. Sadleir, S.J. Smith, S.R. Bandler, J.A. Chervenak, J.R. Clem, Phys. Rev. Lett. **104**, 047003 (2010)
180. J.E. Sadleir, S.J. Smith, I.K. Robinson, F.M. Finkbeiner, J.A. Chervenak, S.R. Bandler, M.E. Eckart, C. Kilbourne, Phys. Rev. B **84**, 184502 (2011)

181. R.A. Hijmering, P. Verhoeve, D.D.E. Martin, A.G. Kozorezov, J.K. Wigmore, R. Venn, P.J. Groot, I. Jerjen, J. Appl. Phys. **105**, 123906 (2009)
182. K.D. Usadal, Phys. Rev. Lett. **25**, 507 (1970)
183. C.A. Mears, S.E. Labov, A.T. Barfknecht, Appl. Phys. Lett. **63**, 2961 (1993)
184. P. Brink, D.J. Goldie, C. Patel, N.E. Booth, G.L. Salmon, J. Low Temp. Phys. **93**, 555 (1993)
185. D.J. Goldie, P.L. Brink, C. Patel, N.E. Booth, G.L. Salmon, Appl. Phys. Lett. **64**, 3169 (1994)
186. K. Segall, C. Wilson, L. Frunzio, L. Li, S. Friedrich, M.C. Gaidis, D.E. Prober, A.E. Szymkowiak, S.H. Moseley, Appl. Phys. Lett. **76**, 3998 (2000)
187. M. Ohkubo, M. Ukibe, T. Zama, T. Ikeuchi, M. Katagiri, S. Ichimura, Nucl. Instrum. Methods Phys. Res. A **520**, 231 (2004)
188. R. Cristiano, E. Esposito, L. Frunzio, C. Nappi, G. Ammerndola, L. Parlato, G. Pepe, H. Kraus, P. Walko, J. Appl. Phys. **86**, 4580 (1999)
189. A.G. Kozorezov, J.K. Wigmore, R. den Hartog, D. Martin, P. Verhoeve, A. Peacock, Phys. Rev. B **66**, 094510 (2002)
190. R.P. Huebener, Rep. Prog. Phys. **47**, 175 (1984)
191. H. Pressler, M. Ohkubo, M. Koike, M. Ukibe, T. Zama, Appl. Phys. Lett. **77**, 4055 (2000)
192. H. Pressler, M. Ohkubo, M. Koike, T. Zama, D. Fukuda, N. Kobayashi, IEEE Trans. Appl. Super. **11**, 696 (2001)
193. H. Kraus, F. Pröbst, F. V. Feilitzsch, T.H. Peterreins, in Superconductive particle detectors, ed. A. Barone (World Scientific, Singapore, 1988). p. 66
194. F. Pröbst, H. Kraus, T.H. Peterreins, F.V. Feilitzsch, Nucl. Instrum. Methods in Phys. Res. A **280**, 251 (1989)
195. B. Beckhoff, R. Fliegauf, G. Ulm, Spectrochimica Acta B **58**, 615 (2003)
196. S. Bechstein, B. Beckhoff, R. Fliegauf, J. Weser, G. Ulm, Spectrochim. Acta B **59**, 215 (2004)
197. M. Ukibe, T. Ikeuchi, T. Zama, M. Ohkubo, Nucl. Instrum. Methods in Phys. Res. A **520**, 260 (2004)
198. H. Kato, B.E. O'Rourke, R. Suzuki, J. Wang, T. Ooi, H. Nakajima, Nucl. Instrum. Methods in Phys. Res. A, **807**, 41 (2016)
199. M. Frank, L.J. Hiller, J.B. de Grand, C.A. Mears, S.E. Labov, M.A. Linderman, H. Netel, D. Chow, A.T. Barfknecht, Rev. Sci. Instrum. **69**, 25 (1998)
200. S. Friedrich, J. Synchrotron Rad. **13**, 159 (2006)
201. G. Fujii, M. Ukibe, S. Shiki, M. Ohkubo, X-ray Spectrom. **46**, 325 (2017)
202. A. Yamazaki et al., Nucl. Instrum. Methods in Phys. Res. B **404**, 92 (2017)
203. S. Shiki, G. Fujii, S. Tomita, K. Sasa, Nucl. Instrum. Methods in Phys. Res. B **554**, 165451 (2024)
204. M. Ohkubo, Y. Shigeri, T. Kinumi, N. Saito, M. Ukibe, Y. E. Chen, A. Kushino, A. Kurokawa, H. Sato, S. Ichimura, Nucl. Instrum. Methods in Phys. Res. A **559**, 779 (2006)
205. M. Ohkubo, M. Ukibe, Y. Chen, S. Shiki, Y. Sato, S. Tomita, S. Hayakawa, J. Low Temp. Phys. **151**, 760 (2008)
206. K. Suzuki, M. Ohkubo, M. Ukibe, K. Chiba-Kamoshida, S. Shiki, S. Miki, Z. Wang, Rapid Commun. Mass Spectrom. **24**, 3290 (2010)
207. M. Ohkubo, N. Zen, T. Kitazume, M. Ukibe, S. Shiki, M. Koike, Prog. Supercond. **14**, 77 (2012)
208. S. McHugh, B.A. Mazin, B.S. Serfass, S. Meeker, K. O'Brien, R. Duan, R. Raffanti, D. Werthimer, Rev. Sci. Instrum. **83**, 044702 (2012)
209. M.J. Prest et al., Solid-State Electron. **103**, 15 (2015)
210. J. Li, P.S. Barry, T. Cecil, C.L. Chang, K. Dibert, R. Gualtieri, M. Lisovenko, Z. Pan, V. Yefremenko, G. Wang, J. Zhang, IEEE Trans. Appl. Super. **33**, 2400105 (2023)
211. A. Miceli, T.W. Cecil, L. Gades, O. Quaranta, J. Low Temp. Phys. **176**, 497 (2014)
212. B.A. Mazin, B. Bumble, P. Day, M.E. Eckart, S. Golwala, J. Zmuidzinas, F.A. Harrison, Appl. Phys. Lett. **89**, 222507 (2006)
213. B.A. Mazin, B. Bumble, S.R. Meeker, K. O'Brien, S. HcHugh, E. Langman, Otpic Express **20**, 1503 (2012)
214. P. Szypryt et al., Opt. Express **25**, 25894 (2017)

215. T. Matsumura et al., J. Low Temp. Phys. **176**, 733 (2014)
216. J. Hubmayr et al., J. Low Temp. Phys. **193**, 633 (2018)
217. E.G.P. O'Connor, A. Shearer, K. O'Brien, New Astro. Rev. **87**, 101526 (2019)
218. J. Baselmans, J. Low Temp. Phys. **167**, 292 (2012)
219. J. Ullom, J. Austermann, K. Koehler, J. Hubmayr (eds.), Special Issue: Low Temperature Detectors LTD19. J. Low Temp. Phys. **209**(3–6) (2022)
220. Y.H. Kim, W. Cui, H-S. Jo, L. You (eds.), Special Issue: Low Temperature Detectors LTD20. J. Low Temp. Phys. **214**(3–4) (2024)

Chapter 4
Thermal Detection

Abstract Chapter 4 for thermal detection describes the fundamental statistical physics of energy resolution, and the operation of practical devices such as SPTD, TESD, MMCD, and SSPD. A historical overview highlights the works of pioneers. The SSPD is categorized as thermal detection because the normal-domain thermal-expansion induced by Joule sef-heating governs signal generation and pulse recovery is facilitated by electrothermal feedback (ETF) that is common to other thermal detectors. As observed in quantum detectors, spatial inhomogeneity occurs due to the sharp superconducting transition that functions as an extremely sensitive thermometer. The best superconductor thermal detectors record a high energy resolution of 1.25–1.58 eV at 5.9 keV, which enables X-ray line-shape analysis that is X-ray emission spectroscopy (XES).

Keywords Thermal detection · Canonical ensemble · Electrothermal feedback · Transition edge · SQUID · Nanostrip · Spatial inhomogeneity

4.1 Energy Resolution of Thermal Detectors

4.1.1 Statistical Model of Canonical Ensemble

Figure 4.1 shows the schematic structure and thermal model of thermal detection. Thermal detectors are frequently used for measuring incoming power of radiation. Typical radiation thermometers measure a temperature increase due to infrared light from objects, covering a temperature range such as − 50 to 3500 °C. The radiation thermometers cannot distinguish individual photons, but measure nE_0 or $\int_{E_1}^{E_2} n(E_0)E_0 \mathrm{d}E_0$, where E_0 is the photon energy, n is the photon flux, E_1 and E_2 are the lower limit and the upper limit of a radiation spectrum. Thermal detectors with a superconductor realize single particle detection even one photon of infrared light, which has an energy of ~ 10^{-19} J or ~ 1 eV. The essence of the single particle thermal detectors is a combination between thermometer temperature sensitivity dR/dT and absorber heat capacity C. In a cryogenic temperature range, C becomes

M. Ohkubo, *Superconductor Detectors*, https://doi.org/10.1007/978-3-031-84193-4_4

small and thus a large temperature rise is expected; however, conventional thermometers are generally inadequate for measuring single particles with regard to temperature sensitivity. A thermometer that is more sensitive than semiconductor thermistor thermometers is required. Transition between a normal-conducting state and a superconducting state is extremely steep as a function of temperature; therefore, the transition can be a superior thermometer for a cryogenic use. When we select a proper detector design, an energy resolution of superconductor thermal detectors for single particles is in general better than that of superconductor quantum detectors. Almost all the particle kinetic energy is finally transferred to a thermal system in contrast to the unavoidable energy loss below an energy gap in quantum detection. In thermal detection, there is no lower detection limit for photon energy or wavelength. On the other hand, in quantum detection, absorption events or excitations lower than an energy gap are lost and cannot contribute to electric signal generation. Nevertheless, superconductor quantum detectors operates at a higher temperature than superconductor thermal detectors, and have a high particle count rate and a high tolerance to base temperature fluctuation due to the energy gap.

The thermal system in Fig. 4.1 a is translated into Fig. 4.1b, which highlights the relation between the absorber and the heat bath to investigate an energy fluctuation of the absorber subsystem [1]. The entire system is isolated from the external environment. There is no exchange of atoms between the subsystem and the heat bath,

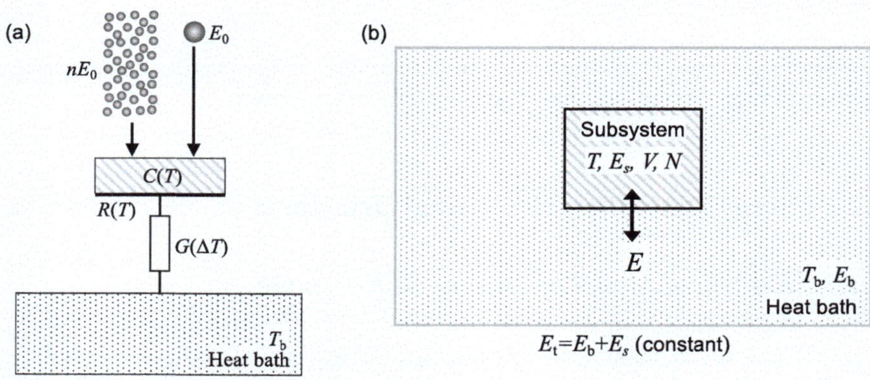

Fig. 4.1 Thermal detection: (**a**) schematic thermal structure and (**b**) canonical ensemble model. In (**a**), the heat conductance G, which is a function of $\Delta T(= T - T_b)$, connects the absorber with heat capacity C at temperature T and the heat bath at T_b. The thermometer with a resistance value of $R(T)$ is denoted by the black bar attached to the absorber. The incoming particles in (**a**) represent two measurement schemes: detection of single particles with a kinetic energy E_0 in J or eV and detection of particle flux nE_0 in W, where n is the number of particles per second. Either single particle detection or flux detection is possible, depending on a relation between detector's time-response and particle incident frequency. In (**b**), the thermal subsystem at T is surrounded by the large heat bath with energy E_b at T_b. The subsystem has an energy of E_s, a constant volume V, and a fixed number of atoms N. No exchange of atoms between the subsystem and the heat bath is allowed, but energy exchange can occur. This is called the canonical ensemble. The total energy of the entire system is constant at E_t

but energy exchange is possible through the thermal connection. The heat bath size is extremely larger than that of the subsystem, so that the bath temperature T_b is practically constant. The temperature of the subsystem is equal to the bath temperature while allowing energy variation. This system is called the canonical ensemble. We assume that the subsystem has discrete microstates of E_1, E_2, E_3..., and E_s, while holding the constant energy E_t of the entire system. The probability p_s for the subsystem to occupy a state E_s is proportional to the number of states of the bath with $\Omega(E_b)$, where $E_b = E_t - E_s$. Considering the normalization factor $\sum_s p_s = 1$, p_s is expressed by

$$p_s = \frac{\Omega(E_t - E_s)}{\sum_s \Omega(E_t - E_s)}. \tag{4.1}$$

Considering entropy $S(E_t - E_s) = k_B \ln\Omega(E_t - E_s)$ and inverse temperature $1/T_b = \partial S(E_t)/\partial E_b$, Eq. (4.1) is transformed to Boltzmann distribution $p_s = \frac{1}{Z}e^{-E_s/k_B T_b}$, where $Z = \sum_s e^{-E_s/k_B T_b}$. The mean energy of the subsystem is given by $\overline{E} = \sum_s p_s E_s$. The energy fluctuation variance expressed by $\sigma^2 = \overline{(E - \overline{E})^2} = \overline{E^2} - \overline{E}^2$ yields the energy fluctuation in FWHM:

$$\Delta E = 2.355\sigma = 2.355\sqrt{k_B T^2 C}, \tag{4.2}$$

considering a thermal equilibrium of $T = T_b$. The details of the mathematical derivation appear in Appendix A.3. It should be noted that $\sqrt{k_B T^2 C}$ is occasionally referred to as root mean square (RMS) fluctuation; however, it is the standard deviation of the subsystem's energy-fluctuation, which is independent of the average energy $\overline{E}$. Only when $\overline{E} = 0$, the RMS value is equal to the standard deviation $\sqrt{k_B T^2 C}$. The factor 2.355 converts the standard deviation to FWHM, providing that the fluctuation is well approximated by a Gaussian distribution in the same way as quantum detection. Energetic particle incidence ($\overline{E} = E_0$) produces a temperature difference between T and T_b; nevertheless, the effect of the difference on ΔE is negligibly small in a condition of $\Delta T = (T - T_b) \ll T_b$.

4.1.2 Lessons from Statistical Physics of Thermal Detection

Comparing $\Delta E = 2.355\sqrt{F E_0 \varepsilon}$ in Eq. (3.4) and $\Delta E = 2.355\sqrt{k_B T^2 C}$ in Eq. (4.2), the large difference is noticeable. Thermal detectors have ΔE that is independent of the incoming particle energy E_0 in contrast to the $\sqrt{E_0}$ dependence of quantum detectors. Practically, ΔE has a small E_0 dependence, when the condition for Eq. (4.2) is broken in actual thermal systems. Quantum detectors have a certain lower limit of ΔE because F cannot be close to zero and ε is nonzero due to energy gaps. On the other hand, ΔE of thermal detectors can asymptotically approach zero when either

T or C approaches zero, although the demand for cooling a thermal system to a temperature of ~ 0 K is impractically heavy.

Equation (4.2) is the theoretical limit of energy resolution integrating a whole frequency range of detector output signals. Power spectrum of fluctuation is spread over a wide range of frequencies. Output signal pulses in single particle detection usually have a shape of exponential decay similar to that of quantum detectors. When we use a narrow readout bandwidth, it is possible to realize a ΔE value smaller than Eq. (4.2). Therefore, Eq. (4.2) is not the practical limit in modern thermal detectors; the signal processing method is called optimal filtering. Nonetheless, keeping in mind that Eq. (4.2) is a guidance on designing thermal detectors such that an energy resolution becomes better (or smaller) as C or T decreases. Thermal detectors have energy resolution independent of E_0; therefore, they cover a wide range of E_0 from infrared rays to γ-rays at a constant ΔE. Practically, C must be adjusted to a proper value depending on particle energy, which limits an achievable ΔE. This is in contrast to the case of quantum detectors, in which ΔE is proportional to $\sqrt{E_0}$. In general, thermal detectors can cover a wide energy range, especially holding a high energy resolution in a high energy range, whereas quantum detectors are suitable for a low energy range. The advantage of quantum detectors is that energy gaps ensure a stable operation in a higher temperature range than that for thermal detectors: no precise temperature regulation is necessary.

In conventional non-superconductor thermal detectors, the steady-state absorber temperature is the same as the heat bath at T_b. On the other hand, superconductor thermal detectors utilize temperature self-regulation using electrothermal feedback (ETF), as described later. The steady state T of the subsystem is slightly raised from T_b by self-heating of a superconductor thermometer, which is held in the middle of superconducting transition near T_c [2, 3]. For example, an Ir/Au-bilayer transition-edge-sensor detector (TESD) has a T_c value of 110 mK [4]. When a cryostat with $T_b = 50$ mK is used, the energy flow from the subsystem at 110 mK to the heat bath at 50 mK occurs continuously. This energy flow is inconsistent with the condition for the canonical ensemble model in Fig. 4.1b and thereby adds a statistical fluctuation. Nevertheless, the additional fluctuation is negligibly smaller than other fluctuation sources that are described in the subsequent sections.

For spectroscopic applications, it is necessary to consider a dynamic range (covering energy range) of superconductor thermometers such that they stay within a superconducting transition width against temperature rise due to particle incidence. The dynamic range design affects energy resolution and count rate.

4.2 Superconductor Thermal Detectors

4.2.1 A Historical Overview

A thermal detector is equipped with a thermometer to measure a temperature rise of an absorber. Bolometers and calorimeters are typical thermal detectors. In addition, we categorize superconductor stirp particle detectors (SSPDs) as the thermal type, although they have no thermometers and electronic excitation plays an important role in signal generation at the initial stage. The operation of the superconductor strip (SS) type is based on superconductor-normal (S–N) transition, followed by normal domain expansion and contraction induced by electrothermal feedback (ETF) after particle incidence breaks Cooper pairs or induce magnetic flux movement [5–7]. The ETF is the central function of the transition edge sensor (TES) type.

The development of thermal detector development for energy-dispersive spectroscopy (EDS) is to realize a smaller C value and a higher $\mathrm{d}R/\mathrm{d}T$ value for detecting single particles and improving energy resolution in calorimetry. For reviews on the TES and SS types, we can refer to Refs. [8–10]. For terminology, we use bolometer for power measurement and calorimeter for single particle detection, as shown in Fig. 4.1a. This section briefly describes a history of thermal detectors.

For single-particle detection, calorimeters must have an absorber exhibiting reasonably large temperature rise to be measured with a thermometer. Solids have heat capacity that decreases with temperature. Therefore, cryogenic temperature is effective to obtain a small C value as well as a low thermal noise. In the 1930s, however, no proper thermometers exist in a cryogenic temperature range near 4.2 K. A possible use of sharp S–N transition as a thermometer was proposed in 1939–1941 [11, 12], after the discovery of superconductivity in mercury in 1911 [13]. The first superconductor thermal detector was a bolometer fabricated by Andrews et al.; the bolometer had the form of a Pb ribbon or strip that was 1 mm wide, 10 μm thick, and 1.5 cm long, evaporated onto a mirror-like glass surface [12]. The Pb strip functioning as both absorber and thermometer was held at a temperature within the Pb S–N transition at 7.2 K by controlling base temperature. In the 1930s, SQUIDs were unavailable; therefore, a galvanometer detected a voltage change due to a 35-K black body radiation. However, a large thermal fluctuation prohibited precise radiation power measurement.

In 1942, Andrews reported the first successful bolometer with a superconductor [2], which is called the superconductor phase transition (SPT) bolometer or TES later. The SPT bolometer was also the first functioning superconductor device in all the application areas of superconductivity. A superconductor electromagnet appeared 13 years later in 1955 [14]. The SPT thermometer using a Ta wire coil was attached to an Al sheet absorber with a thickness of 80 μm and an area of $8 \times 10\ \mathrm{mm}^2$. Silk threads supported the Al absorber sheet, isolating the Al absorber from the heat bath. The detector response was consistent with a blackbody radiation change between 24 and 55 K, based on the Stefan-Boltzmann law. Figure 4.2a shows the structure of the Ta-SPT bolometer in a cryostat with the radiation source [2]. The Ta wire coil exhibited the superconducting transition in a temperature range between 3.22 and 3.23 K in Fig. 4.3a. The equivalent circuit in Fig. 4.2b has the biasing circuit with the shunt resistor R_s in parallel with the Ta coil. This is the first self-bias circuit for holding temperature automatically in the middle of narrow superconducting transition with ETF. The IEC graphical symbol for the normal-superconductor (N-S) boundary has a pair of dots and a right-angle wedge [15, 16]. The temperature-dependent resistor sandwiched by the two N-S boundary symbols denotes the superconductor Ta thermometer. The Ta coil biased in the middle of the superconducting transition is not completely in a superconducting state. Strictly speaking, the term of "superconductor detector" is suitable for the SPT bolometer rather than "superconducting detector." Other elements in the equivalent circuit are located at room temperature. Andrew contrived the ETF concept such that the heat escape to the heat bath balanced with the Joule self-heating in the Ta coil thermometer [2], although he did not use the term ETF.

While the switch in Fig. 4.2b is open, the Ta wire conveys an effectively constant current expressed by

$$I_1 = I_b = \frac{U}{R + 2R_w + R_b} \approx \frac{U}{R_b}, \tag{4.3}$$

provided that the bias resistance is extremely larger than the sum of the Ta resistance and the wiring lead resistance, $R_b \gg R + 2R_w$. We postulate that the temperature dependence of R is linear in the middle of superconducting transition. A typical superconducting transition of Ta is shown in Fig. 4.3a [2]. In the ETF operation principle depicted in Fig. 4.3b, the black $I_1{}^2R$ line shows self-heating power of the Ta coil. The $I_1{}^2R$ curve crosses the heat-escape light-blue line indicated as $G\Delta T$ at a temperature shift of ΔT_b, where G is the heat conductance between the Al absorber and the heat bath. This cross point is unstable because the gradient of the black $I_1{}^2R$ line is steeper than that of the light-blue $G\Delta T$ line. A stable balance is obtained by decreasing I_1 or increasing G; however, both approaches significantly degrade the sensitivity of the thermometer [2].

The solution by Andrew was to add the shunt resistor R_s as shown in Fig. 4.2b. Engaging the switch, the current flowing the Ta coil is approximately expressed by

$$I_2 \approx \frac{I_1}{1 + \frac{R+2R_w}{R_s}}. \tag{4.4}$$

This leads to a gradual slope depicted by the green line $I_2{}^2R$ in Fig. 4.3b when we specify a suitable R_s value such that the blue $G\Delta T$ line is steeper than the green $I_2{}^2R$ line. Although it is not stated in Ref. [2], the typical R_s values should be 3–25 Ω for $R_w = 1$–10Ω. The temperature of the Ta coil is automatically self-regulated at the crossing point that is in the middle of the superconducting transition. The temperature of the Ta coil is increased by ΔT_b from the base temperature T_b. This thermodynamic system is also stable against radiation heat input J, which raises the Ta coil temperature by ΔT_J and shifts the SPT operation from the green line to

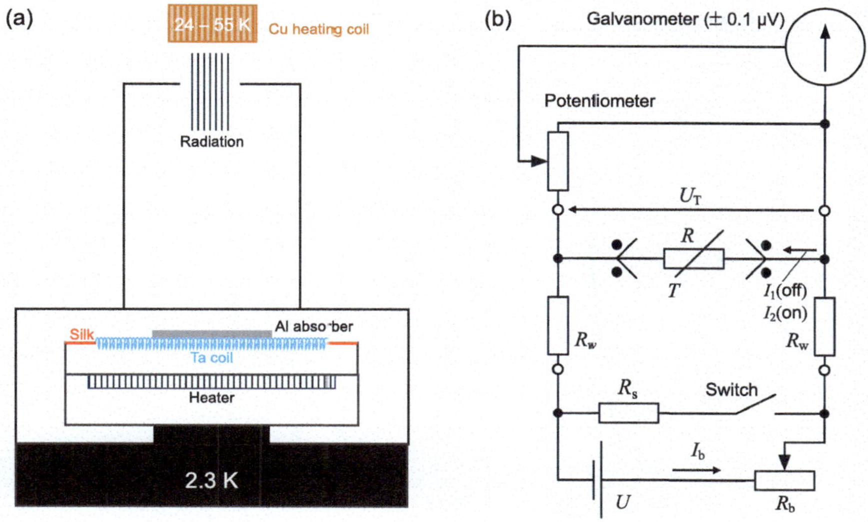

Fig. 4.2 First SPT bolometer in 1942 [2]: (**a**) schematic drawing of bolometer structure and radiation source, (**b**) equivalent circuit for bias and readout, based on the IEC 60617 database for graphical symbols [15, 16]. In (**a**), the Ta coil thermometer is suspended by the silk threads. The cobalt-black-coated Al absorber receives the black body radiation from the source. Radiation shields surround the SPT structure. In (**b**), the two IEC graphical symbols with a pair of dots and a right wedge denote the boundaries between the normal-conducting and superconducting regions. The right wedge side is in a superconducting state while the opposite side with two dots is in a normal-conducting state. The superconductor Ta coil is denoted by the resistor symbol with temperature dependence because it is not in a superconducting state at a midpoint of superconducting transition under operating condition. In (**b**), R_w is the wiring lead resistance, R_b is the bias resistance that is exceedingly larger than $R + 2R_w$, and R_s is the shunt resistor. Firstly, the temperature of the SPT block in (**a**) is raised to ~ 3.3 K considerably higher than T_c while turning on the switch and applying I_2 to the Ta coil. Secondly, the temperature is reduced to T_b such that the temperature of the Ta coil is self-regulated in the middle of the superconducting transition. In 1942, no SQUIDs were available, so that constant voltage biasing was impossible; the combination of a potentiometer and a galvanometer had an accuracy of ± 0.1 μV in potential measurement using a needle deflection

the red line by $\Delta H + J$, where ΔH is the additional Ta-coil heating with $(\Delta I_2)^2 R$. Accordingly, for the heat input, I_2 decreases following Eq. (4.4). This provokes a weak negative ETF.

The combination of the potentiometer and the galvanometer requires a voltage change, so that the negative ETF is weak. On the other hand, modern SQUID current–voltage conversion amplifiers enable the measurement of a current change under a constant bias-voltage U_{TES} in a condition of $R_s \ll R$ and $R_w = 0$ under a superconducting state. The self-heating power U_{TES}^2/R in a TES thermometer decreases rapidly when heat input raises R of the superconductor thermometer. The constant-voltage biasing scheme rotates the green $I_2^2 R$ line clockwise to the dashed orange U_{TES}^2/R line in Fig. 4.3b. This induces the strong negative ETF due to a negative slope of the U_{TES}^2/R line, devised by Irwin in 1995 [3]. It is notable that

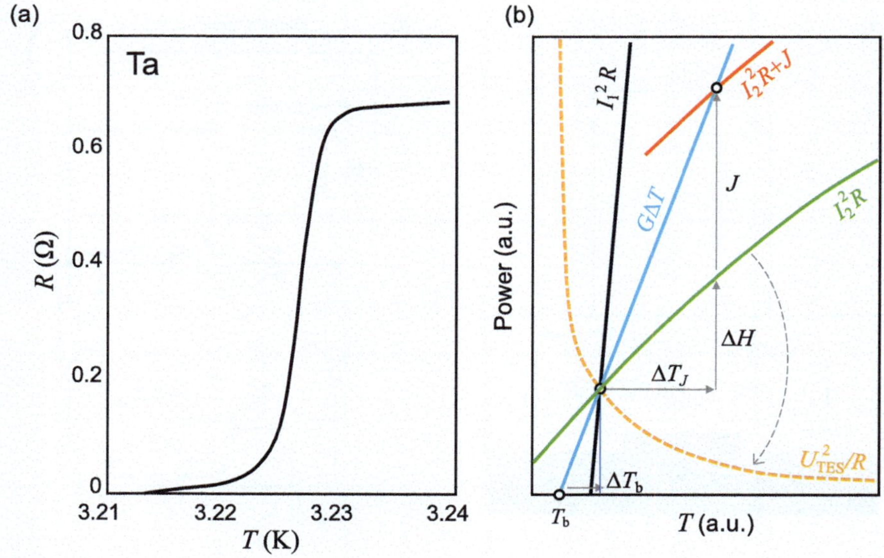

Fig. 4.3 Electrothermal feedback (ETF) operation using superconductor Ta thermometer in1942 [2]: (**a**) typical superconducting transition of Ta coil and (**b**) heat balance between Joule self-heating of the Ta coil and heat escape to the heat bath. In (**b**), T_b is the base temperature of the cryostat in Fig. 4.2a, G is the heat conductance due to the silk threads and wiring leads. The details of the heat balance are described in the text. Under the stable self-biasing in the middle of the superconducting transition at the cross point between the light-blue $G\Delta T$ curve and the green I_2^2R curve, the galvanometer reads a black-body-radiation induced voltage-change across the Ta coil. Instead of the voltage change, modern SQUID amplifiers with zero input resistance enables current change measurement under a constant voltage bias at U_T. The modern TES type with SQUID induces clockwise rotation from the green I_2^2R line to the orange dashed U_{TES}^2/R line by Joule self-heating of a TES thermometer, resulting in the strong ETF [3]

the first superconductor bolometer in 1942 already had the self-regulation function to automatically hold a temperature within a superconducting transition width. The temperature self-regulation and ETF in 1942 was ahead of its time. The modern SQUID amplifiers based on a superconductor galvanometer (superconductor low-inductance undulatory galvanometer: SLUG) [17] conceived in 1966 by Pippard and Clarke realize high sensitive thermometers and the strong negative ETF.

Nineteen years later after the Andrews's work, a semiconductor Ge resistor appeared as a thermometer compatible with cryogenic operation in 1961 [18]. It is intriguing that the superconductor thermometer was invented before the semiconductor thermometer. A semiconductor X-ray imaging calorimeter for an energy range between 0.3 and 12 keV was implemented on the astronomical satellite, Hitomi X-ray Observatory launched in 2016 [19]. The thermometer for the Hitomi calorimeter was an ion-implanted Si thermistor operating at 50 mK, which recorded an energy resolution of 4.9 eV for 5.9-keV X-rays after integrating pulses from all the 36 pixels. Currently, the TES type with a better energy resolution is a suitable selection for future X-ray satellite missions. For the pioneering works on calorimeters for single X-ray photon detection, we can refer to Refs. [20, 21] by Moseley, McCammon et al.

A variety of superconductor thermometers such as Al films, Josephson junctions, and normal-insulator-superconductor (NIS) junctions were utilized by Clarke et al. in 1977 [22]. The superconductor thermometers were attached to a dielectric absorber with a low heat capacity. The Clarke's paper is a guidance in designing modern bolometers. In addition, there are many excellent reviews on bolometers, for example, Ref. [23] by Richards in 1994, which includes the reason why superconductivity plays an important role in thermal detectors. In 1995, Irwin proposed a new superconductor thermometer with a strong negative ETF operation, which is known as the TES type [3]. A Mo/Au-TESD recorded an energy resolution of 1.58 eV for 5.9-keV X-ray photons with a fall time of ~ 1 ms at NASA in 2012 [24, 25], which is significantly better than that of the Hitomi calorimeter with the semiconductor thermometers. Besides an X-ray range, TESDs excellently operate for single photons in an infrared-to-optical range, similar to the performance of quantum superconductor detectors such as STJDs. A W-TESD achieved an energy resolution of 0.15 eV and a fall time of 60 μs in a photon energy range as low as 0.3 eV in 1998 [26]. The fall time of output pulses was improved to 200 ns in 2009 by Fukuda et al. [27]. Furthermore, an Au/Ti TESD by Hattori et al. recorded an energy resolution of 0.067 eV and a fall time of 4.6 μs in this photon energy range in 2022 [28].

Another important thermal detector type is magnetic calorimetry. The first magnetic calorimeter had a Er-doped garnet thermometer, proposed by Bühler and Umlauf in 1988 [29, 30]. The detector operation was demonstrated for 5.5 MeV α-particles. Magnetic calorimetry using metallic hosts, which exhibit a stronger spin-phonon interaction than dielectric garnet hosts, was proposed by Bandler et al. in 1993 [31]. The demonstration of X-ray and γ-ray detection was performed by Enss et al. in 2000 [32] The metallic magnetic calorimetric detectors (MMCDs) recorded an energy resolution of 1.58 eV and a fall time of ~ 1 ms for 5.9-keV X-ray photons in 2018 [33], which was improved to 1.25 eV in 2024 [127].

As categorized in Chap. 2, Sect. 2.3, the SS type known as SSPD also operates under ETF. The SS type has a simple form of superconductor strips in a nanoscale: the TES type usually has a square superconductor thin film form. It is often called the superconductor nanowire sigle photon detector (SNSPD). However, the IEC/ISO terminology recommends superconductor strip photon (or particle) detector (SSPD) [34, 35], because operation principle is in 2D nature of superconductor strips with a width (~ 100 nm to ~ 5 μm) that is wider than ξ in most cases. The term of nanowire is used for longitudinal 1D phenomena in superconductors with a diameter less than ξ excluding the transverse extent, which conflicts with the 2D physics on SSPDs with thin film strips. Note that there are cases in which the term "nanowire" is correct. A resistive change due to particle incidence onto a superconducting strip generates a voltage pulses due to a superconducting-normal (S-N) transition induced by Joule self-heating under a constant bias current. The SS type also uses a negative ETF mode with a shunt resistor to recover a superconducting state, as explained later. The first SSPD was developed 38 years after the discovery of superconductivity, although the superconductor's dimensions are not in a nanoscale. A 6-μm thick,

0.4-mm wide, and 3.5-mm long NbN superconductor strip successfully detected α-particles in 1949 [36], which was reported by Andrews, who was also the pioneer of the first superconductor bolometer [2]. The NbN-SSPD was directly mounted on a Cu base, which is different from calorimeters that are thermally isolated from a cold stage. When the NbN-SSPD was held in the middle of superconducting transition just like SPT or TES, the counting performance was the best. This was the first SSPD for single particle counting, although they used the term "bolometer" [36].

Instead of holding SSPDs in the middle of superconducting transition, a different operation mode was proposed many years later by Sherman in 1962 [37]. He proposed applying a bias current approximately half of I_c, while holding a Sn-SSPD at a temperature significantly lower than T_c. The presumed detection mechanism for 5-MeV α-particles or 60-MeV fission fragments was the S-N transition of the 100-nm thick and 10-μm wide Sn strip. The simulation anticipated no normal region growth due to Joule self-heating, which is the operation principle of modern SSPDs assisted with the negative ETF. After the Sherman's proposal, the experimental paper on an 0.1-μm thick and 34-μm wide In-SSPD and a Sn-SSPD was reported in 1965; Spiel et al. observed voltage pulses for 5.3-MeV α-particles [38]. The SS type was also capable of detecting electrons from a beta-decay source with a NbN strip that is 0.36-μm thick, 4-μm wide, and 2-mm long in 1990 [39]. Moreover, a 0.24-μm thick, 1.8-μm wide, and 0.55-mm long W strip detected 6-keV X-ray photons in 1992 [40]. The W strip operated in either a high bias current region for hot spot formation and subsequent Joule self-heating or a low bias region for direct normal-region formation bridging the strip width without Joule self-heating. Recovery times were in a range of 10–50 ns, which is ~ 10 times longer than modern SSPDs. Nonetheless, the operation mode for 6-keV photons is exactly the same as modern SSPDs for infrared photons.

The infrared single photon detection with a 5-nm thick, 0.2-μm wide, and 1-μm long NbN strip was realized by Gol'tsman et al. in 2001 [41]. Device miniaturization to a nanoscale with a bias current close to I_c enabled the detection of ~ 1-eV photons, the energy of which is three to six orders of magnitude smaller than those in the early reports. The proposed operation mode is the same as the W strip, that is the normal band formation bridging the strip width and the following normal domain expansion induced by Joule self-heating [40]. Due to the negative ETF with shunt resistance of readout circuitry, the normal region contracts. SSPDs normally generate pulses with a constant height independent of particle type and energy. The initial voltage drop or hot spot formation is also induced by vortex motion in superconductor strips. The photon detection in a telecommunication wavelength region with SSPDs covers applications such as astronomy, quantum cryptography, quantum computing, and TOF ranging using single photon detection capability [42]. These applications require no energy measurement. Furthermore, a possible application of SSPD to ion detection in TOF–MS was suggested by Frank et al. in a review paper on superconductor detectors for MS in 1999 [43]. Another operation mode for MS was reported by Pagano in 2005; it is kinetic inductance operation [44]. Experimentally, the first successful acquisition of MS spectra with an NbN-SSPD was reported by Suzuki et al., who detected a biomolecule of bovine serum albumin (BSA) in 2008 [45]. Neutrons were also detected with a ^{10}B-enriched MgB_2 strip by Ishida [46].

It is notable that the pioneers obtained the ideas using superconductivity for thermal detection in the 1940s. Especially, we have a deep reverence to Andrews, who is the pioneer of both TES type in 1942 and the SS type in 1949 [2, 36]. The modern TESDs and SSPDs exhibit a significantly better performance and a wider particle-energy coverage compared with those in the early years. Superconductor thermal detectors are also installed into analytical instruments for synchrotron radiation X-ray analysis, mass spectrometry, infrared spectroscopy, and visible-infrared spectroscopy [47–51].

4.2.2 Biasing into Superconducting Transition Edge

Bias conditions are mentioned in the previous section for different superconductor thermal detectors. Briefly, superconductor thin film thermometers (TESs) are biased just in the middle of superconducting transition to use a high dR/dT sensitivity, whereas superconductor strip photon counters cooled below T_c are biased close to the superconducting transition by a bias current close to I_c. It is worth summarizing here to clarify similarities and dissimilarities in biasing into superconducting transition edges. In early years of the SPTbolometers, they were biased with an almost constant current to measure a voltage change. Radiation absorption raises the temperature of a superconductor sensor attached to an absorber, the temperature of which is controlled within a superconducting transition width by a heater. The almost constant bias current induces a Joule heating power of $P = I^2R$, which is the positive ETF that causes instability. By adding a shunt resistor, the current diverted to the parallel resistor reduces the Joule heating power in the superconductor. This negative ETF stabilizes the SPT bias point in the middle of S-N transition and also against radiation power input. This weak negative ETF is also applicable to SSPDs; a similar ETF recovers a superconducting state or a zero voltage state.

The self-biasing concept of modern TESDs with shunt resistance is similar to the SPT bolometers; the difference is that the shunt circuit of TESDs satisfies $R_s \ll R$ to maintain a constant voltage across the TES thermometer. Joule heating power is expressed by $P = U_{\mathrm{TES}}^2/R$ at a constant voltage. Radiation or single particle incidence raises R, which rapidly reduces P. This strong negative ETF provides high stability and superior energy resolution. The T_b is lower than T_c; for example, $T_b = 60$ mK and $T_c = 107$ mK for Mo/Cu TESD with a 124 × 124 × 2.8-μm^3 Bi absorber [47]. We can control T_c of superconductor thermometers by attaching a normal metal layer, which is the proximity effect in the same way as controlling Δ_g in STJDs. A temperature rise due to particle incidence is usually designed to be ~ 1 mK.

SSPDs are also biased close to the superconducting transition by a constant bias current. Inevitably, the SS type is shunted by an input resistor of a wideband amplifier, which is normally 50 Ω. Therefore, the equivalent circuit is the same as that of the first Ta SPT bolometer, stabilizing SSPDs with the weak negative ETF against latching in a normal state. In contrast to the TES type, the SS type is cooled below T_c. The SS type utilizes current biasing close to I_c while holding a superconducting state.

As an example of biasing, a 10-nm thick, 800-nm wide, and 40-mm long NbN for molecule detection operate at $T_c = 13$ K, $I_c = 590$ μA, and $I_b = 355–545$ μA at 4.2 K [52]. The SS type operates near a transition edge; therefore, it can also be categorized as the TES type.

4.2.3 Transition Edge Sensor (TES) Type

4.2.3.1 Structure of TES Type

The TES type for X-rays generally consists of an absorber and a superconductor bilayer thermometer. For infrared-optical photons, superconductor thermometers also function as an absorber. Figure 4.4a shows a typical structure of superconductor thermometers, omitting an absorber for X-rays. The thermometer has a proximity bilayer structure of Ir(100 nm)/Au(25 nm) with an area of 200 × 200 μm^2 on a 300-nm thick SiN_x membrane [4]. The Ir/Au bilayer film has T_c values of 110 mK at a low bias current of 1 μA and 107 mK at a high bias current of 45 μA. The T_c values are lower than 140 mK of pure Ir due to the proximity effect of the 20-nm thick Au layer. Figure 4.4b depicts the thermal system: R_{TES} and C are the resistance value in the middle of superconducting transition and the heat capacity for the Ir/Au thermometer, respectively, and G is the heat conductance due to the SiN_x membrane and the Nb leads. The Si substrate functions as the heat bath with a base temperature T_b of 50 mK.

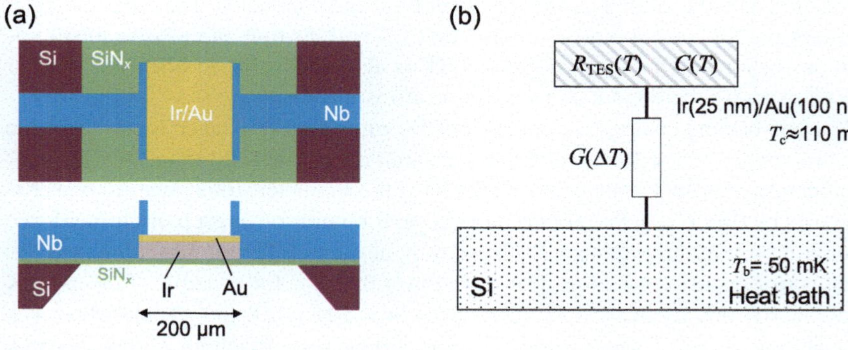

Fig. 4.4 Structure of Ir/Au TESD: (**a**) schematic drawing of top and side views of the superconductor thermometer and (**b**) thermal system. In (**a**), the thermometer has a proximity bilayer structure of Ir(100 nm)/Au(25 nm) with an area of 200 × 200 μm^2 on the 300-nm thick SiN_x membrane [4]. For soft X-rays, the thermometer also functions as an absorber. For high-energy X-rays or γ-rays, an absorber is added onto the thermometer. In (**b**), the Ir/Au TESD connected to the Si heat bath with heat conductance G. The Ir/Au film has a T_c value of 110 mK and a transition width of ~ 2 mK at a low current measurement. This Ir/Au TESD produced the experimental data in Figs. 4.10 and 4.12

The thermal conductance is determined by the Nb leads and the SiN_x membrane, which is fabricated by etching the entire Si wafer underneath the TESD from the back side. The membrane structure in Fig. 4.4a is fragile because the backside is open. By adopting a silicon-on-insulator (SOI) wafer, it is possible to leave the Si substrate underneath the SiN_x membrane by etching the Si layer between the membrane and the insulator layer [53]. The entire structure of the SOI-TES detector chip is robust because the thick Si substrate reinforces the SiN_x structure.

4.2.3.2 ETF Operation

Figure 4.5 depicts an equivalent circuit for the bias and readout circuit for TESDs in (a) and *I*–*U* and *P*–*U* characteristics in (b). The temperature-dependent resistor with R_{TES} denotes a TES thermometer. It is noticeable that, compared with the SPT thermometer in Fig. 4.2b, the shunt resistor R_s is repositioned to a place nearby the TES thermometer within a cryogenic superconducting circuit. The R_s value is selected such that a relation of $R_s \ll R_{\mathrm{TES}}(T)$ is held for an operating temperature range for particle incidence. A constant bias current generates a constant bias voltage U_{TES} across the TES thermometer. The constant bias voltage engages the strong negative ETF with a Joule self-heating power of $P = U_{\mathrm{TES}}^2/R_{\mathrm{TES}}$, which balances with $G(\Delta T)$.

Figure 4.5b shows an I_{TES}–U_{TES} curve in green, which explains the ETF operation in a bias range below 1.2 μV within a superconducting transition width, for the Ir/Au TES in Fig. 4.4. Above 1.2 μV, the Ir/Au film is in a fully normal state at 154 mΩ, which is almost constant near T_c . The ETF self-regulation of temperature shifts the TES temperature by ~ 60 mK from the heat bath temperature T_b of 50 mK. This temperature rise occurs under the relation of

$$\frac{U_{\mathrm{TES}}^2}{R_{\mathrm{TES}}(T)} = U_{\mathrm{TES}} I_{\mathrm{TES}} = G\Delta T. \tag{4.5}$$

Equation (4.5) yields a $G\Delta T$ value of 9.35 pW at 1.2 μV. In the ETF operation below 1.2 μV, I_{TES} changes such that a relation of $U_{\mathrm{TES}} I_{\mathrm{TES}} = 9.35$ pW is satisfied, holding the almost constant heating power, as shown by the red curve. Therefore, when U_{TES} decreases below 1.2 μV, the green curve deviates from the dashed black line of 154 mΩ. $R_{\mathrm{TES}}(T)$ decreases rapidly below the onset temperature of the superconducting transition, following $U_{\mathrm{TES}}{}^2/9.35$ pW. As a result, the temperature of the Ir/Au TES is self-regulated within the superconducting transition width of a few mK in a voltage bias range below 1.2 μV.

Another noticeable change from the first SPT bolometer in Fig. 4.2a is that the SQUID current–voltage conversion amplifier replaces the potentiometer-galvanometer combination that measures a voltage change of the Ta coil at room temperature. The two IEC symbols for Josephson junctions [15] form a superconducting ring, which denotes a DC SQUID. The cross symbol placed at the center of the SQUID ring is the attribute to magnetic dependence or magnetic field effects [15].

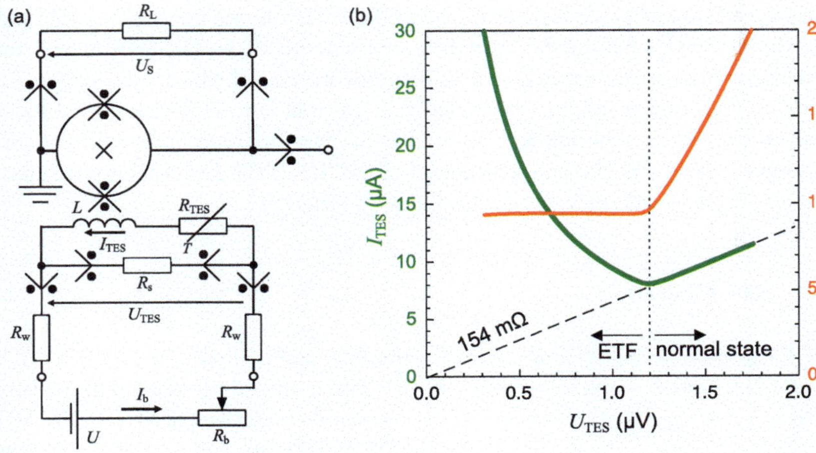

Fig. 4.5 Operation principle of strong negative ETF in TESDs: (**a**) equivalent circuit, (**b**) experimental I–U curve in green and heating power curve in red for the Ir/Au TESD [4]. In (**a**), the IEC graphical symbols with a cross and a pair of dots denote Josephson junctions [15]. The superconducting ring with two Josephson junctions symbolizes a DC SQUID. The IEC cross symbol at the center of the SQUID ring denotes magnetic field dependence [15]: this cross is usually omitted. The other graphical symbols follow the notation in Fig. 4.2b. The magnetic field due to I_{TES} is transmitted to the SQUID ring by the mutual inductance between the coil L and the SQUID ring, which enables current-voltage conversion at zero input resistance. The Ir/Au film is in a normal-conducting state in $U_{TES} > 1.2\ \mu V$, whereas it is temperature-biased within a superconducting transition width in $U_{TES} < 1.2\ \mu V$. The SQUID readout circuit in (**a**) is simplified from actual amplifiers

The cross symbol explicitly expresses the magnetic-field dependent oscillation of the voltage across the SQUID ring: it is usually omitted from electric circuit diagrams in the literature.

While holding the TES temperature within a superconducting transition width, a single particle absorption event raises the TES temperature slightly by an order of μK. The small temperature increase induces a large change $R_{TES} + \Delta R$ and a current change of $\Delta I_{TES} = -\frac{U_{TES}}{R_{TES}}\left(\frac{\Delta R}{R_{TES}+\Delta R}\right)$. This current reduction is converted to a change of external magnetic field for the SQUID ring through the mutual inductance that couples the coil L in series to the TES and the SQUID ring. The SQUID current–voltage conversion amplifier generates a voltage change U_S. In Fig. 4.5a, one DC SQUID is present together with a simplified auxiliary circuit. Practically, a variety of SQUID arrangements are possible to obtain a sufficiently high output voltage for signal processing with semiconductor circuits at room temperature; for example, a first single SQUID stage followed by a second SQUID array. Each particle event produces a negative current pulse. A SQUID array amplifier converts each current pulse to a voltage pulse.

In addition to a high temperature-sensitivity with superconductivity and self-regulation of temperature, the strong ETF contributes to shortening the fall time of output pulses. The intrinsic recovery time τ_0 to a base temperature is equal to *C/G*. Providing that a feedback band width of ETF is wide enough to hold U_{TES} constant throughout the duration of each particle absorption event, the effective fall time τ_{eff} of output pulses is expressed by

$$\tau_{\mathrm{eff}} = \frac{\tau_0}{1 + \frac{\alpha}{n}}, \tag{4.6}$$

where α is the unitless parameter of superconducting transition, $\alpha = (T/R)(\mathrm{d}R/\mathrm{d}T)$, and n is the parameter ranging from 4 to 6 depending on thermal impedance between a superconductor film and a heat bath [3]. For a TESD with realistic values of $\alpha = 80$ and $\tau_0 = 500$ μs, the pulse fall time can be shortened to 24 μs form ~ 1 ms. Even before the TES invention[3], it was pointed out that strong negative ETF would facilitate the shortening of pulse duration and reducing noise [20]. Irwin explains the details of the thermal response time of TESDs in Ref. [54]. Equation (4.6) is valid only when the constant-voltage bias circuit responds very quickly. However, bias circuitry has a typical input coil inductance of ~ 150 nH and a TES resistance of ~ 1 mΩ; the time constant is 150 μs, which is slower than τ_{eff}. Furthermore, multiplex readout for TES array detectors accompanies a Nyquist inductance of a few 100 nH [55]. The circuit cannot hold the bias voltage constant. Therefore, an actual time constant is normally between 200 μs and 1 ms for X-ray TESDs with a superior energy resolution [47].

The strong ETF transforms Eq. (4.2) to

$$\Delta E = 2.355\sqrt{\frac{k_{\mathrm{B}} T^2 C}{\alpha}}. \tag{4.7}$$

The ETF improves the energy resolution by a factor of $\alpha^{0.5}$, exceeding the limit of conventional calorimeters. TESDs optimized for a soft X-ray range record a ΔE value of 1.1–1.5 eV for 500-eV X-ray photons [47, 48], which is a few times better than those of STJDs. A Mo/Au TESD with an Au or Bi/Au absorber recorded a ΔE value of 1.58 eV for 6-keV photons [24, 25]. This ΔE value considerably exceeds an STJD best energy resolution of 29 eV at 6 keV, obtained by a Nb/Al-STJD with a SQUID amplifier [56]. Furthermore, the TES type covers a wide range from infrared to γ rays, designing suitable thickness and heat capacity for an absorber. On the other hand, STJDs normally cover a range of infrared to soft X-rays up to ~ 2 keV. Such superior energy resolution of TESDs is normally achievable at a pulse fall time of ~ 1 ms, which implies that the photon count rate is ~ 100 times lower than that of STJDs. Nevertheless, the superior energy resolution of TESDs have pioneered numerous analytical applications at a low count rate [9, 47].

4.2.3.3 Fundamental Physical Issues in TESDs

The performance of TESDs depends on the form of superconducting transition curves. TESDs operate in an intermediate state between normal and superconducting states. What is the intermediate state? There are a few models for the intermediate state, based on two-fluid Ginzburg-Landau model [57] and Josephson weak link (or resistively shunted junction) model [58, 59]. Superconductor films for TES have macroscopic dimensions beyond ξ and λ, so that spatil inhomogeneity can occur in an intermediate state. One extreme case was observed in a pure-Ir TESD, which exhibited a clear phase separation into the superconducting region and normal region along the bias-current flow direction due to self-heating [64]. Self-heating sustains stable phase separation in current-carrying superconductors with small transverse dimensions, such as wires or strips [60]. This phase separation in TESDs was visualized by LTSSM in the next section.

One of the recent topics on TESDs is the long-range proximity effect as discussed in Chap. 3, Sect. 3.2.4.4 [58, 61–63]. Both TES and STJ types operate by means of the proximity effect: superconductor/normal-metal bilayers for controlling T_c in TESDs and superconductor/superconductor bilayers for reducing Δ_g in STJDs. The proximity effect is often treated within uniform superconducting properties in bilayers thinner than ξ. The STJDs and TESDs shed light on the spatial extent of the proximity effect by using different device sizes. The experimental results indicate that the proximity effect in thin film bilayers extends over several tens of μm [63]. The spatial extent of the proximity effect is a few orders of magnitude longer than a common belief of a few times ξ, which is typically 50–500 nm depending on superconductor materials. Although the horizonal extent is emphasized in STJDs, the long-range proximity effect can also be valid along the direction perpenducular to film surfaces. A possible explanation is that Δ_g is constant throughout bilayer films, although DOS rapidly decreases with increasing the distance from a bilayer interface or a material boundary. When a perturbation due to particle incidence or bias current is small, the proximity superconductivity extends over a few μm, holding a constant Δ_g. The long-range proximity effect is an interesting future topic of unsolved detector physics.

4.2.3.4 Spatial Inhomogeneity of TESD

TESDs often have a detector size larger than STJDs. A TES size of several hundreds of μm exceeds ξ, λ, and even the extent of the long-range proximity effect, so that spatial inhomogeneity can be an issue to achieve a superior ΔE value. In the worst case, a TESD exhibits no sharp full-energy absorption peak for a ^{55}Fe X-ray source, as shown in Fig. 4.6. Moreover, the response function drastically changes with bias voltage, similar to the STJDs with Al layers thinner than 50 nm as shown in Fig. 3.22. The appearance of the TESD is shown in in Fig. 4.7a, which exhibits a pure Ir film with a thickness of 50-nm thick and an area of $500 \times 500\ \mu m^2$ on a SiN_x membrane without a discrete absorber and Nb leads. Figure 4.7 reveals spatial inhomogeneity of the Ir-TESD by using LTSSM with a 3-keV X-ray microbeam, which provides guidance on designing TESDs appropriately. The LTSSM images visualize spatial profiles of pulse height, rise time τ_{rise}, and fall time τ_{fall} of output pulses in Fig. 4.7b–d [64]. The 500-μm square Ir film has a T_c value of 134.5 mK, a superconducting transition width of 1.3 mK, and a normal state resistance of 2.15 Ω just above T_c. While biasing the Ir-TES at 3.61 μV within the ETF region, a 3-keV X-ray microbeam with a diameter of 20 μm scanned the TESD surface. The red open circle in the ETF region on the $I_{\text{TES}}-U_{\text{TES}}$ curve in Fig. 4.7f indicates the bias point. Typical current pulses in (e), inverted from negative current pulses, at two locations indicated by the red and light green dots in (b) possess totally different pulse parameters: the red and light green numbers denote the rise time and fall time values. The extremely large spatial inhomogeneity to the 3-keV X-ray photons provides important implication of detector physics and energy resolution.

The large inhomogeneity in Fig. 4.7 stems from a clear phase separation into the superconducting region on the left side and the normal-conducting region on the right side, the boundary of which is depicted by the dashed black line in Fig. 4.7a at the bias point indicated by the red open circle in (f). The TES temperature sensitivity is the highest in the middle of superconducting transition; therefore, it is natural that pulse heights become maximum at the N-S boundary, where the Ir film has the maximum temperature sensitivity. The maximum pulse height locations in (b) indicates that the N-S boundary is slightly curved as depicted in (a). Figure 4.7e shows that the red pulse near the N-S boundary has a higher pulse height and shorter τ_{rise} and τ_{fall}

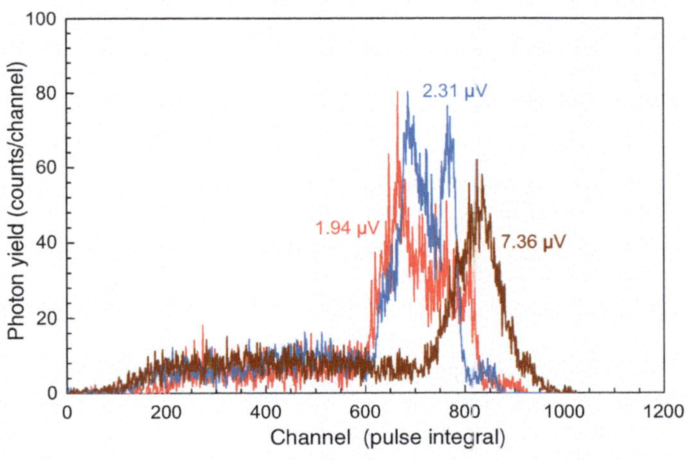

Fig. 4.6 Bias voltage dependence of full illumination pulse integral spectra of 500-μm square Ir-TESD. The appearance of the Ir-TESD is shown in Fig. 4.7**a**. The pulse integral corresponds to the charge of each pulse of SQUID output. The plot of pulse peak (pulse height) values results in a similar spectrum. Note that pulse height values in semiconductor detectors represent output charge values because a charge-sensitive preamplifier integrates each output-current pulse: see Fig. 3.1. X-rays from a ^{55}Fe source irradiated the whole area of the TESD biased at 1.94, 2.31, and 7.36 μV. The spectral shapes are called "response function." The incident X-rays can be considered as monochromatic X-rays at 5.9 keV, which is the main line of the ^{55}Fe X-ray source

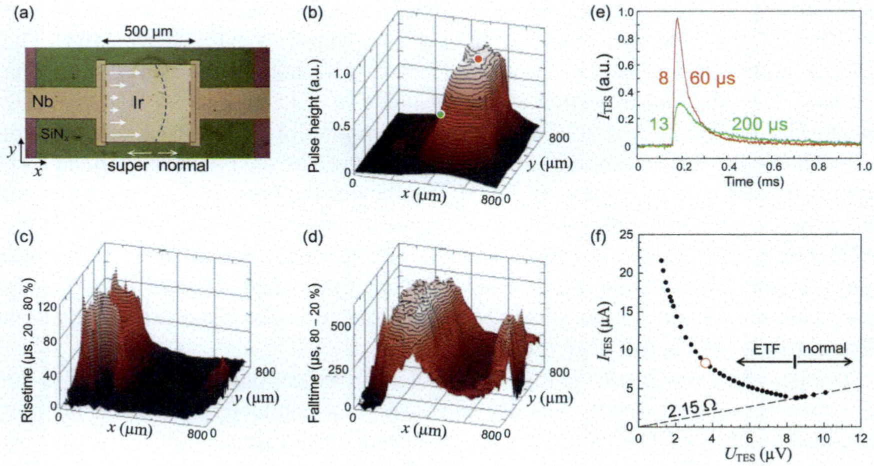

Fig. 4.7 Spatial inhomogeneity of Ir-TESD output-pulse parameters measured with LTSSM [64, 65]: (**a**) optical image of pure-Ir TESD, (**b**) pulse height (current peak) image, (**c**) rise time image, (**d**) fall time image, (**e**) typical output current pulses at two locations indicated by the red and light green dots in (**b**), and (**f**) 3.61-μV bias position (red open circle) on I_{TES}–U_{TES} curve. The 3-keV X-ray microbeam with a diameter of 20-μm scanned the surface of the Ir-TESD, which is 50-nm thick and 500-μm square, on a 300-nm thick SiN_x membrane. The white arrows in (**a**) denote the current density distribution expected from spike pulses in Fig. 4.9. The pulse parameter values in (**b–d**) are represented by a color scale with black for the lowest count, red for the middle, and white for the highest. The dotted back line in (**a**) indicates the boundary between normal and superconducting regions. The red and light-green numbers in (**e**) indicate the τ_{rise} and τ_{fall} values. The Ir-TESD has a normal resistance of 2.15 Ω, a T_c value of 134.5 mK, and a transition width of 1.3 mK in low current measurement. This imaging analysis was performed at a bath temperature of 55 mK

values than the light-green pulse inside the superconducting region. The τ_{fall} values in (d) range from 60 μs at the N-S boundary to 500 μs inside the superconducting region far from the boundary. This phenomenon indicates that signals inside the superconducting region are due to heat diffusion from absorption locations to the N-S boundary. Consequently, the TES operation is effective exclusively at the N-S boundary.

Figure 4.8 shows the bias dependence of LTSSM line scans along the x direction across the center of the Ir-TESD in Fig. 4.7a [64]. The current pulse heights at different bias positions are plotted as a function of x. As U_{TES} increases, the maximum pulse height decreases and simultaneously the maximum positions indicated by the open arrows shifts from right to left in Fig. 4.8a. The inset in (a) plots the measured Ir-TES resistance values at the different bias positions as a function of the distance between the film right edge and the maximum pulse-height positions. The straight line demonstrates that that the N-S boundary moves from right to left as the bias voltage increases. At 8.5 μV, the whole area of the Ir film is in a normal-conducting state, as expected from Fig. 4.7f. The bias-current induced self-heating maintans the normal-conducting region. As the normal region expands, the maximum pulse height value decreases and the τ_{fall} (τ_{eff}) value increases, which represents a reduction of α in Eqs. (4.6) and (4.7). According to Eq. (4.5), the Ir-TES temperature is maintained close to T_c within a variation of several mK, which exceeds a superconducting transition width of 1.3 mK. Figure 4.8b shows expected temperature variation curves inside the Ir film, providing a thermal conductance of 1.5 nW/K and a thermal healing length of ~ 700 μm. The temperature variation is ± 8 mK, which is significantly larger than a transition width of 1.3 mK, so that the occurrence of the superconducting-normal phase separation is reasonable. The observed distinct phase separation during the ETF operation stems from the low thermal conductivity of the pure Ir film, similar to the phase separation in current-carrying superconductors with negligibly small transverse extent (one-dimensional model) [60]. It is intriguingly that the quasi-one-dimensional behavior in Figs. 4.7 and 4.8 occurs in the 500-μm square film. Precisely, the phase separation has the two-dimensional character in the square Ir film, similar to a two-dimensional model for single-photon detection in SSPDs.

Additionally, a single photon absorption event interchanges the superconducting and normal-conducting regions. The dashed red line at 3.3 μV in Fig. 4.8a is an example of the position interchange; the normal region is situated on the left side instead of the right side. The normal region always sticks to either end of the Ir

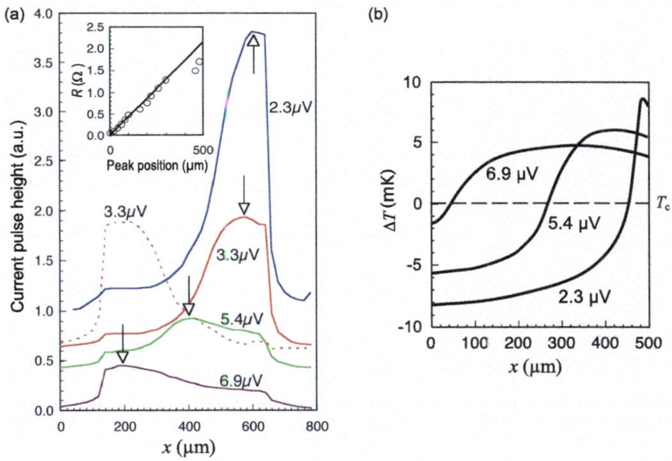

Fig. 4.8 Normal-superconducting (N-S) phase separation in pure-Ir TESD [64]: (**a**) LTSSM line scan profiles of current pulse height at different U_{TES} values and (**b**) estimated temperature variation profiles. The 20-μm-diameter 3-keV X-ray microbeam scanned the Ir-TESD along the x-direction across the detector center along the bias current flow direction. The open arrows in (**a**) locate the maximum pulse height positions. The inset in (**a**) plots the measured TES resistance values against the distance between the right edge and the black open arrows. This confirms the expansion of the normal-conducting region from the right edge toward the left edge. One single photon event interchanges the normal-conducting and superconducting regions, as indicated by the dashed red curve at 3.3 μV

film. This phenomenon is independent of bias current direction. Unlike in long superconductor wires, a central normal region sandwiched between two superconducting regions was never observed in the 500-μm square Ir film.

4.2.3.5 SSPD-Like Spike Pulses in TESD

The τ_{rise} image in Fig. 4.7c shows a hole-like feature near the lower left corner inside the superconducting region, while holding the N-S boundary near the right edge at 3.61 μV, at which the LTSSM line scan is substantially the same as that at 3.3 μV in Fig. 4.8a. The hole indicates that the τ_{rise} values are extremely short: the upper left corner has the same hole. Figure 4.9a displays the output current pulse shapes at the different y locations relative to the TES center at $x = 180$ μm in (b). Figure 4.9b shows the top view of the τ_{rise} image, which is represented by a color scale with black for the shortest τ_{rise} and yellow for the longest τ_{rise}. The sharp spikes with $\tau_{\mathrm{rise}} < 1$ μs precede the normal TES pulses with a τ_{rise} value of ~ 10 μs [65]. Due to the preceding sharp spikes, a pulse-parameter computation routine yields the two black holes at both corners near the left edge in Fig. 4.9b, whose 3D plot is shown in Fig. 4.7c. Even near the film center at 10 and − 30 μm relative to the center along the y-axis, small spikes are recognizable. The spike height is maximum at $y = 150$ μm. These observations imply that the spike amplitude is proportional to the local supercurrent density, considering electron-beam-induced signal images of inhomogeneous superconductor films by using low temperature scanning electron microscope (LTSEM) [66]. The supercurrent densities in wide superconductor strips are essentially inhomogeneous and related to the magnetic penetration length: the current densities near both side-edges of wide strips are higher than that at the center. By using an approximation for current distribution in wide superconductor films based on the Ginzburg–Landau (GL) theory [67], the perpendicular magnetic penetration depth $\lambda_{\perp}$ is evaluated at ~ 100 μm by fitting the model to the pulse height data in Fig. 4.9c [65]. The white arrows in Fig. 4.9b denote an expected supercurrent density distribution. The slightly reduced pulse height just on the film edge at 250 μm is due to the absence of superconductor film beyond the edge, which limits current flow. The spike pulse generation is explainable on the basis of the hot spot model for SSPDs biased below I_c, as discussed in Sect. 4.2.5. The pure-Ir TESD is a rare case that both TESD-ETF and SSPD-ETF operations coexist due to the N-S phase separation. This behavior confirms that both TES and SS types operate near a superconducting transition edge, reinforcing that SSPDs should be considered within the thermal detection framework together with the TES type.

4.2.3.6 Optimum TES Design

We learned that the spatial inhomogeneity severely degrades energy resolution both in superconductor thermal and quantum detectors; no clear full-energy absorption peaks appear. Such detectors cannot be applied to analytical science. For STJDs, an

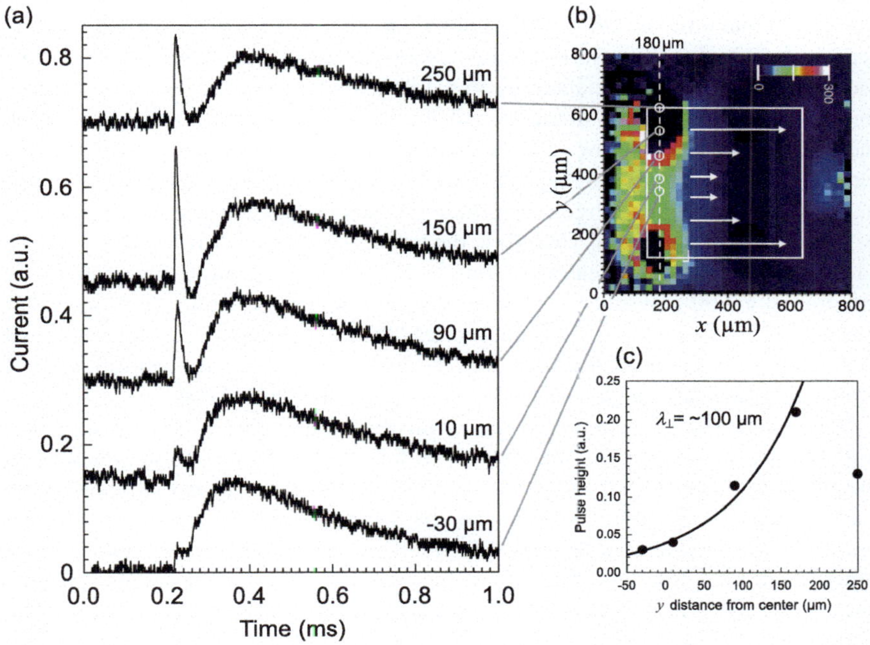

Fig. 4.9 Sub-μs short spikes observed in pure-Ir TESD [65]: (**a**) y-position dependence of output pulses preceding normal TES pulses at $x = 180$ μm, (**b**) top image of τ_{rise}, equivalent to the 3D plot in Fig. 4.7c, and (**c**) spike pulse height as a function of the distance from the center along the y-axis. The numbers in μm in (**a**) are relative to the film center along the y direction. The bias voltage was 3.61 μV. The spikes appear in the superconducting region of the left side of the Ir film. The white arrows in (**b**) denote an expected supercurrent density distribution. The spike amplitudes, which should be proportional to the supercurrent density, are plotted against the distance from the center along the y direction in (**c**). The perpendicular magnetic penetration depth $\lambda_{\perp}$ is evaluated at ~ 100 μm for the Ir film with a thickness of 50 nm and an area of 500×500 μm^2

appropriate detector design involves reducing Δ_g and increasing DOS by thickening Al layers, based on the gap-reduction model, and thereby enhancing the proximity effect. This design improves spatial homogeneity. Intriguingly, proximity bilayers also improve spatial homogeneity of TESDs, improving thermal conductivity as described below. There is common detector physics of the long-range proximity effect, on both TESDs and STJDs.

In the previous section, the 500-μm square Ir superconductor film exhibits a temperature variation of $\pm$ 8 mK, which is considerably larger than a transition width of 1.3 mK, under the bias condition where the negative ETF functions. A low thermal conductance of 1.5 nW/K, which corresponds to a thermal healing length of 700 μm, results in the N-S phase separation. In $T < T_c$, the thermal conductivity of superconductors drops due to a less contribution of normal electrons; for example, ~

70% at 7 K and ~ 0% at 1 K in Nb. Although no large reduction of thermal conductivity occurs just at T_c, a form of superconductor/normal-metal bilayers is effective to increase thermal conductivity in a cryogenic temperature range. In normal-conducting metals, phonon thermal conductivity is proportional to T^3 and electron one is proportional to T below ~ 2 K. Nevertheless, proximity bilayers possess thermal conductivity significantly higher than that of superconductors alone. Additionally, the proximity effect in superconductor/normal-metal bilayers is useful to control T_c by varying film thicknesses. Therefore, the proximity bilayer structure is a common selection for TES thermometers to avoid the N-S phase separation. The temperature variation of a 200-μm square Ir(100 nm)/Au(25 nm) bilayer film, which is smaller than the Ir TESD in Fig. 4.7, should be less than 1.3 mK from expected thermal parameters: a thermal conductance of 7.6 nW/K and a thermal healing length of 1 mm. No phase separation is expected to occur for the Ir/Au TESD.

Figure 4.10 shows an experimental verification of the above-mentioned prediction. The gray-scale image of the number of 6-keV photon absorption events in (a) visualizes the Ir/Au TESD and the Nb leads on both sides. The event numbers are scaled from black to white. The number of the X-ray absorption events on the 300-nm thick SiN_x membrane area is negligibly small, so that the membrane area is close to black. Figure 4.10b plots the LTSSM line scan profiles of pulse height along the black arrow in (a) at the different bias voltages. Compared with Fig. 4.8a, the line profiles are flat inside the Ir/Au film region, ensuring that no N-S phase separation occurs at any bias points. The absorption events in the Nb leads also generate measurable output pulses due to thermal diffusion, similar to the quasiparticle diffusion in STJDs in Fig. 3.21a. The full-illumination energy resolution was 9.4 eV for the 5.9-keV X-rays from a ^{55}Fe source. This energy resolution exceeds ~ 120 eV of SDDs by a factor of more than 10.

Although no meaningful spatial inhomogeneity in pulse height is observed, the τ_{rise} image between 5 and 6.4 μs in Fig. 4.10c still exhibits small inhomogeneity that probably represents an inhomogeneous bias-current distribution, similar to the white arrows in Fig. 4.9b. The horizontal long-range proximity effect [58, 62, 63] due to the Nb leads also affects this τ_{rise} inhomogeneity. In the middle of superconducting transition at T_c, inhomogeneity is possible in both two-fluid model and Josephson weak link models. An unexpected excess noise [68] observed in simple square bilayer films can stem from spatial inhomogeneity, although the excess noise was explained by the RSJ model for Josephson weak links [59, 69]. Experimentally, the excess noise can be reduced by adding a Cu metal bar feature on a superconductor bilayer or applying a perpendicular magnetic field [24]. The normal metal feature is effective to suppress bias current inhomogeneity. A Mo/Au TESD with a normal metal feature and an Au absorber recorded 1.58 eV at 5.9 keV, as shown in Fig. 4.11 [24, 25]. Note that another thermal detector type of SSPDs has a narrow strip form, the width of which is normally less than 1 μm, so that spatial inhomogeneity is less important. Moreover, the SS type only has a particle counting function due to a type of avalanche effect, not measuring particle energy. Spatial inhomogeneity in SSPDs can affect output pulse timing that is called jitter.

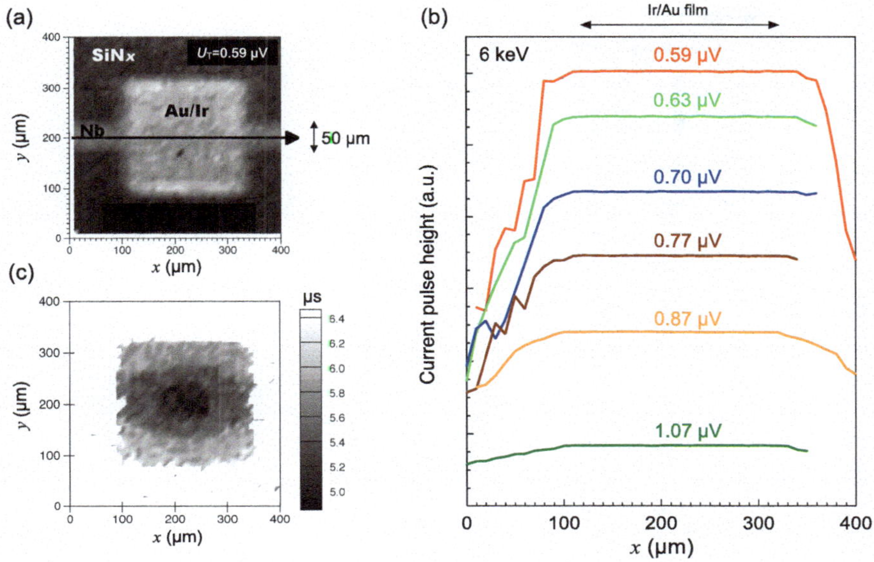

Fig. 4.10 Improvement of spatial homogeneity in Ir/Au-TESD: (**a**) LTSSM image of the number of detected 6-keV photon events for 200-μm square Ir(100)/Au(25) proximity bilayer on 300-nm thick SiN_x membrane, (**b**) LTSSM line scan profiles of pulse height along the black arrow in (**a**) at the different bias points, and (**c**) image of rise time τ_{rise} between 5 and 6.4 μs. The Ir/Au bilayer has a T_c value of 110 mK. The pulse height curves along the arrow in (**a**) are almost flat inside the Ir/Au area at any bias points, in contrast to the N-S phase separation in Figs. 4.7 and 4.8. The energy resolution for ^{55}Fe-source full-illumination was 9.4 eV at 5.9 keV. In (**c**), the τ_{rise} image still exhibits small inhomogeneity, which is attributed to bias current density distribution similar to the white arrows in Fig. 4.9b. One of the latest TESDs in the literature, an NASA's Mo/Au TESD has an energy resolution of 1.58 eV at 5.9 keV [24, 25]. Reproduced with permission from Ohkubo [8].

4.2.3.7 Spectroscopy with Ir/Au TESD

The Ir/Au TESD was utilized for X-ray fluorescence analysis of the NIST standard reference material SRM 1297 (stainless steel) at a KEK-PF synchrotron radiation beam line. Figure 4.12 shows the comparison between the Ir/Au TESD and a Si detector mounted on a commercial table-top X-ray fluorescence spectrometer. The 31-eV energy resolution for the Mn K-$L_{2,3}$ (Mn K$\alpha_{1,2}$) line at 5.9 keV is worse than the best value of 9.4 eV because large electronic noise at the beam line. Nonetheless, the Mn K-$L_{2,3}$ line and the Cr K-$M_{2,3}$ line from the stainless steel are distinguishable on the pulse height spectrum of the Ir/Au TESD in contrast to the broad dashed line of the Si detector.

In addition to X-ray spectroscopy, a Ti/Au TESD has been utilized for confocal microscopy bioimaging by Niwa et al. [70, 71]. Covering a wave-length range of 400–2800 nm, the Ti/Au TESD realized single photon color imaging at a 100 times lower illumination than that with complementary metal oxide semiconductor (CMOS)

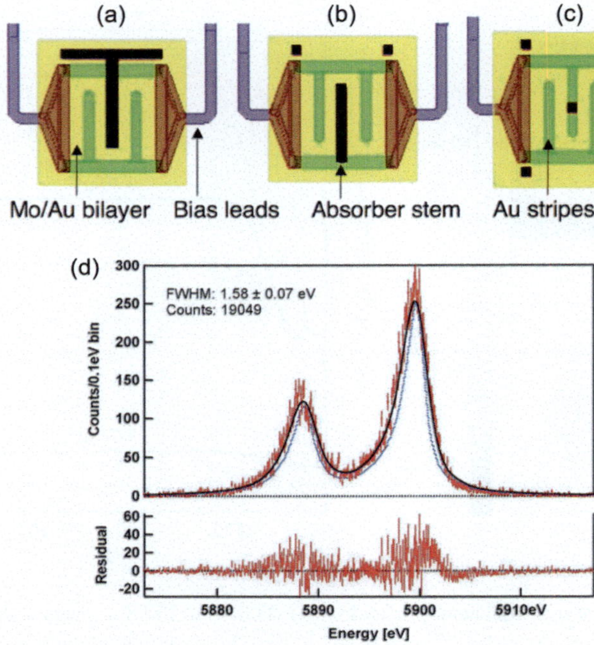

Fig. 4.11 Mo/Au TESD [24]: (**a**–**c**) different designs of black stems to connect Mo/Au bilayer TES and absorber with bias leads in purple, and Au stripes in green on TES; and (**d**) X-ray spectrum of the design (**c**) for ^{55}Fe source. On the 35-μm-square yellow Mo(62)/Au(180) bilayer in nm, the black stems support an unshown Au absorber with an area of 64×64 μm^2 and a thickness of 4.5 μm. The energy resolution of the design **c** is 1.58 eV in FWHM at 5.9 keV by subtracting a width of natural line (blue) of ^{55}Fe X-rays from a width of the right black peak fitted to the experimental spectrum in red. Reproduced with permission from Smith et al. [24]. Copyright 2012 Springer Nature

imagers. In addition, the instrument was applied to obtain single photon images of a fluorescence-labeled cell sample stained with three different dyes. TESDs prevent sample damage and decolorization of the dyes.

4.2.4 Metallic Magnetic Calorimetric (MMC) Type

4.2.4.1 Concept of MMCD

The MMC type introduces another important thermal detection scheme. It also requires SQUID readout. Details of the operation and performance appear, for example, in Ref. [33]. The MMC type consists of a paramagnetic temperature sensor attached to an absorber as shown in Fig. 4.13. The magnetic susceptibility of paramagnetic materials strongly depends on temperature. A SQUID magnetometer measures

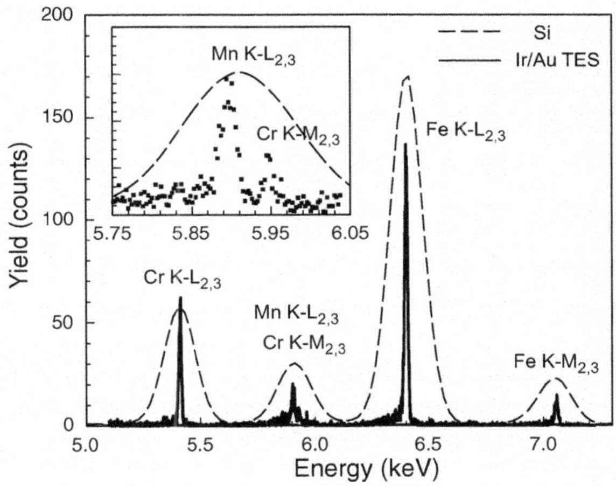

Fig. 4.12 Comparison between Ir/Au TESD and Si detector in X-ray fluorescence analysis for NIST standard reference material SRM 1297 (stainless steel). Electronic noise at a synchrotron radiation beam line considerably degrades the energy resolution from 9.4 to 31 eV at 5.9 keV. The ΔE value is still better than 183 eV of the Si detector mounted on a table-top X-ray fluorescence spectrometer. Even at the degraded energy resolution of the Ir/Au TESD, it is possible to distinguish the Mn K-$L_{2,3}$ (5.9 keV) and Cr K-$M_{2,3}$ (5.947 keV)

a change of magnetization $M(T)$ due to particle incidence. A high dM/dT coefficient is realized by diluted alloys. An example of the diluted alloys is Au-Er with an Er concentration of ~ 300 ppm [32]. The absorber is often Au. A weak link with small heat conductance G connects the absorber-sensor block and the heat bath, similar to other calorimetric or bolometric types. When an energetic particle transfers its energy to the absorber, $M(T)$ of the paramagnetic temperature sensor attached to the absorber changes as indicated by the arrow in (b). The magnetic flux in the paramagnetic temperature sensor changes under a bias magnetic field H. The pickup and coupling coils transmit the flux change to the SQUID ring, which generates a voltage pulse. Actual SQUID readout circuits are more complex than the circuit in (a). Figure 4.13 shows a simplified conceptual structure; several detector designs are possible [33].

The time response or τ_{fall} is simply given by C/G, similar to conventional calorimeters. Achievable energy resolution exceeds the thermostatistical limit of Eq. (4.2) by using optimal filtering as performed in TESDs. When the heat capacity C_{s} of the sensor equals to C_{a} of the absorber ($C = C_{\text{s}} = C_{\text{a}}$), the energy resolution at the optimal filtering limit is expressed by

$$\Delta E = 2.355\sqrt{4k_{\text{B}}T^2C}(4\tau_{\text{rise}}/\tau_{\text{fall}})^{1/4}, \tag{4.8}$$

where τ_{rise} and τ_{fall} are the rise time and the fall time of output pulses [33]. The expected energy resolution at 18 mK is 0.64 eV for 5.9 keV [72]. Experimentally,

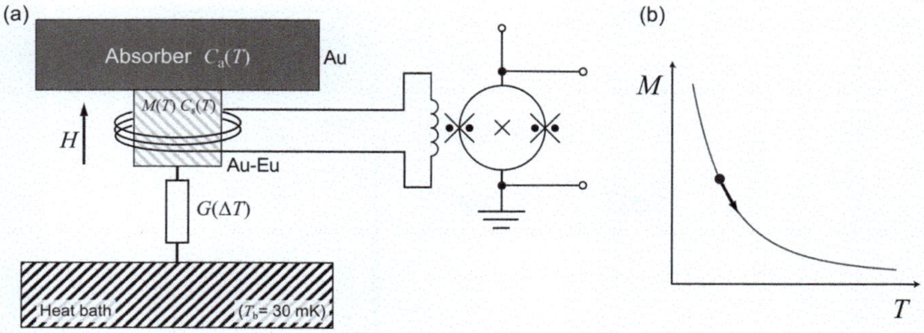

Fig. 4.13 Structure and operation of MMCD [33]: (**a**) schematic structure with SQUID readout and (**b**) temperature dependence of magnetization $M(T)$ of a paramagnetic material under a magnetic field H. The paramagnetic temperature sensor consists of an Au-Er alloy with heat capacity $C_s(T)$. The particle incidence raises the absorber temperature with heat capacity $C_a(T)$, which often consists of Au. A temperature rise of the paramagnetic temperature sensor leads to a reduction in M. The reduction in M is transduced to a magnetic flux change by the pickup coil and the coupling coil, and then the SQUID ring generates a voltage change. Two IEC graphical symbols with a cross and a pair of dots that symbolizes a Cooper pair denote the Josephson junctions [15]. The small cross symbol at the center of the DC SQUID ring with two Josephson junctions is the attribute of magnetic field dependence [15]. The single DC SQUID is shown in (**a**) for the sake of simplicity. Reproduced with permission from Ohkubo [8]. Copyright 2023 AIP Publishing LLC

the MMCD with a 50-μm square Ag-Er(260 ppm) paramagnetic thermometer and an Au absorber with an area of 150×150 μm^2 and a thickness of 3 μm records an energy resolution of 1.25 eV at 5.9 keV [72].

4.2.4.2 Practical MMCD

Figure 4.14 shows an SEM image of the MMCD with a Ag-Eu paramagnetic temperature sensor. Figure 4.15 is a response function or X-ray spectrum for a ^{55}Fe X-ray source: the energy resolution is 1.25 eV in FWHM. Typical resolving powers $E_0/\Delta E$ of MMCDs are ~ 3000 for X-rays and ~ 6000 for γ-rays, which correspond to ~ 10 eV in a wide energy range up to 60 keV [73]. The MMC type is utilized for a wide range of applications: X-ray spectroscopy, γ-ray spectroscopy, neutrinoless double beta decay for the Majorana particle nature of neutrinos [74], neutrino mass measurement with electron capture in Ho [75], MS for neutral fragments in an ion storage ring [76, 77], and dark matter search [78]. Although the MMC type has a rather long fall time of output pulses, similar to conventional calorimeters with a C/G time constant, the strong spin-phonon interaction in diluted alloy thermometers realizes the fastest rise time of 70 ns among calorimeters [79]. This is advantageous to the timing measurement in imaging MS in a low particle flux condition [76, 77].

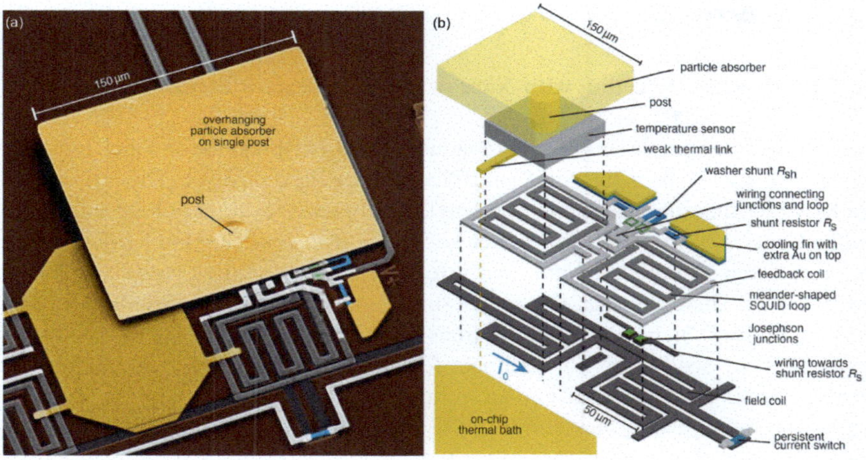

Fig. 4.14 MMCD with Ag-Er paramagnetic thermometer and Au absorber [72]: (**a**) SEM image and (**b**) layer structure. The orange plate represents the 150-μm square Au absorber with a thickness of 3 μm. The temperature sensor is a diluted alloy of Ag-Er(260–450 ppm). The structure underneath the temperature sensor consists of pickup and coupling coils, DC-SQUID, and a field coil to apply a magnetic field to the temperature sensor. Reproduced from Krantz et al. [72]. AIP Publishing LLC; licensed under CC BY 4.0: https://creativecommons.org/licenses/by/4.0/

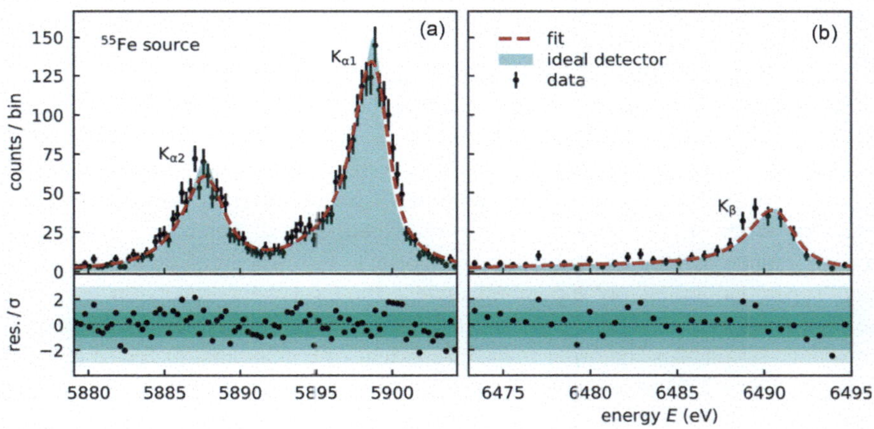

Fig. 4.15 Pulse height spectrum or response function of MMCD for ^{55}Fe X-ray source emitting three lines of Mn K-L$_3$ (Mn Kα_1) at 5.8988 keV, Mn K-L$_2$ (Mn Kα_2) at 5.8877 keV, and Mn K-M$_{2,3}$ (Mn K$_{\beta1,3}$) at 6.4906 keV. The red dashed line is fitted to the experimental data denoted by closed circles with vertical bars for uncertainty. After subtracting the natural line width of the Mn K-L$_3$ line from the measured data, the energy resolution is 1.25(18) eV. The light green area indicates a response function of an ideal detector with a ΔE value of zero; the MMCD performance is close to that of the ideal detector. Reproduced from Krantz et al. [72]. AIP Publishing LLC; licensed under CC BY 4.0: http://creativecommons.org/licenses/by/4.0/

4.2.5 *Superconductor Strip (SS) Type*

4.2.5.1 Particle Detection with Narrow Superconductor Strips

The SS type consists of a simple superconductor strip, the thickness of which typically falls in the nanoscale range of 1 − 100 nm in modern designs. The strip width can be wider than the nanoscale: the length is normally extremely longer than the width. Generally, strips have meander geometry with a thickness of a few nm, a width of a few hundreds of nm, and a length of a few mm to cover a practical sensitive area. Therefore, the 2D extent of superconductor thin-film strips is in conflict with the term of "nanowire" without transverse extent, as explained below and in Appendix C.4. SSPDs are not adequate for high resolution spectroscopy but fast particle counting in a temporal domain, although the discrimination of particle energy is possible by adjusting bias current [49], and the measurement of single-photon energy is achievable in a limited energy range between 0.4 and 2.0 eV [80]. SSPDs normally generate identical pulses with the same pulse height, independent of particle type and particle energy, similar to Geiger-Müller (GM) counters and avalanche photodiodes (APDs). Although the SS type outperforms semiconductor based fast photon counters, the nanoscale is one of the disadvantages. For a practically large sensitive area, the meander geometry is one solution; however, it is still insufficient. Other solutions include parallel connection of multiple nanostrips [81] and thermally coupled 2D layout [82]. For additional references to achieve a large sensitive area, we should refer to the section "2. Operation in parallel configuration" on p. 43 and the section "5. Spatial resolution and arraying" on p. 38 in Ref. [8].

The term of "superconducting nanowire single photon (particle) detector (SNSPD)" is widely disseminated. However, as described in Appendix C.4, IEC standards recommend "superconductor strip photon detector (SSPD)" owing to two reasons. Firstly, the SS type becomes a normal state after particle incidence; superconductor strips are not always in a superconducting state. Therefore, strictly, the term of "superconducting nanowire" is incorrect, and "superconductor" is better than "superconducting" or "superconductive." The term "superconductor" resolves the confusion between these two adjective terms. Secondly, the term "nanowire" represents 1D physics, ignoring transverse extent of conductors. Detector physics treats SSPDs as 2D superconductor strips [83] because strip thickness t_{f} is smaller than ξ, and width w is wider than ξ or even comparable to λ, that is $t_{\mathrm{f}} \ll \xi \ll w \lesssim \lambda$. Signal generation mechanisms of the hot-spot and vortex-motion models described in this section are valid for 2D superconductor films. Due to the 2D nature of SSPDs, this book follows the term "strip" used by the pioneer in the early years [36] and the IEC terminology [34, 35]; possible variations are "nanostrip," or "micro-strip" instead of "nanowire" or "microwire." Note that "microstrip" is the term for transmission lines consisting of a conducting strip and a parallel extended conducting surface bonded to opposite sides of a thin dielectric substrate under the IEC glossary [84].

The nanoscale dimensions of superconductors enable detection of single photons in an infrared-optical range near ~ 1 eV, which is 10^6 times smaller than a MeV range of α-particles for the microscale SSPDs in the early years [36]. Besides photons near 1 eV, SSPDs can detect electrons [39], X-ray photons [40, 85, 86], and macro-molecules that generate phonons on a superconductor surface [45, 52, 81, 87]. The detection of these particles normally requires strips narrower than 1 μm. On a certain operation condition, several-μm ultrawide strips can detect ~ 1-eV photons [88, 89]. In this section, we describe detection principles for a variety of strip dimensions.

Initial perturbation models for a photon energy range between ~ 0.5 and ~ 3 eV including a telecommunication wavelength of 1.55 μm (0.8 eV) are still controversial. The operation of SSPDs resembles that of classical current-carrying superconductors [60]. Models for output signal generation include normal-region bridging assisted by bias current (hot spot), initially proposed in 2001 [41, 90], single vortex crossing (SVC) [91, 92], vortex-antivortex pair (VAP) deparing [93, 94], and phase slip center (PSC) formation [83, 95]. The hot spot model is based on normal core formation with a temperature of higher than T_c or hot-electron core formation at a local spot of photon absorption. Bias supercurrent detours around the hot spot region, so that a current density beside the hot spot region exceeds a critical current density. This leads to a normal band bridging across the strip width [40, 96]. Constant current biasing causes a voltage pulse. On the SVC model, a single vortex crosses a strip, which generates a voltage drop. On the VAP model, Cooper pair breaking causes vortex-antivortex pair generation, depairing, and subsequent movement to counter directions due to Lorenz force. In general, superconductors used for SSPDs belong to type-II, allowing vortex nucleation. Niobium is intermediate between type-I and type-II: polycrystalline Nb films are normally categorized as type-II due to coherence length shortening. The vortex motion induced by Lorentz force causes a voltage drop, the origin of which is different from that on the hot spot model. The initial response to a single photon below 1 eV can be either SVC or VAP model. The PSC model for one-dimensional conductors is inappropriate for SSPDs in most cases, although it is suggested that a relatively narrow NbN strip with a width of 100 nm exhibited a behavior of thermally activated phase slipping [97]. The SVC model is plausible in a strip-width range significantly smaller than the Pearl length in Table 1.1 [91, 98]. The Pearl length expressed by $\Lambda = 2\lambda^2/t_f$ is the magnetic penetration depth for films, whose surfaces are perpendicular to a magnetic field [99]. The SVC model is valid when $\xi \ll w \ll \Lambda$. An appropriate detection model varies among the hot spot, SVC, and VAP, depending on superconductor materials and strip dimensions.

In a low energy range that is insufficient to induce a hot spot, the initial stimulation is governed by one of the SVC , VAP, or SVC-VAP mixture models, depending on the superconductor material properties listed in Table 1.1. The SVC-VAP mixture means that either mechanism can randomly occur with equal probability or that a temporal shift occurs between SVC and VAP. Note that VAP can be induced by a hot spot [96]. When assuming typical strip dimensions of $t_f = \xi$ and $w =$ 200 nm, the probable models for the initial response to ~ 1-eV photons are SVC for NbN and YBCO, VAP for Nb, and SVC-VAP mixture for MgB_2. However, none of the detection models currently explain all observations completely. The different mechanisms are engaged depending on particle energy, strip dimensions, the time elapsed after particle incidence, the distance from a particle incidence location, and operation condition. Further theoretical and experimental studies are required [100–102].

We categorize the SS type as thermal detection. More precisely, the SS type stands in an intermediate detection regime between quantum and thermal categories: quantum phenomena such as hot spot formation due to Cooper pair breaking, SVC, and VAP precede thermal phenomenon of normal region expansion due to Joule self-heating. The self-heating determines the output pulse height. This intermediate regime probably occurs in both normal-conducting films and superconducting films [66, 93, 96]. Here, we postulate the hot spot formation for keV ion detection [52] and the vortex motion due to either SVC or VAP for photon detection near and below 1 eV. The photon energy boundary between the hot spot formation and the vortex motion is ~ 1 eV for the 4-μm square NbN SSPD with $t_f = 4.5$ nm and $w = 100$ nm [80]. Given the ongoing controversy over the detection mechanisms, it is worthwhile to begin the following section with a discussion of basic perturbations using electrons that are spatially and temporally controllable, in order to understand SSPD operation.

4.2.5.2 Perturbation in Superconductor Thin Films and Output Signals

Electron beam stimulation experiments using LTSEM [66, 103] provide valuable insights into the response of superconductor films or strips to perturbations. LTSEM imaging exerts scanning stimulation due to an electron beam on current-carrying superconductor strips. Figure 4.16 illustrates an ultrawide inhomogeneous strip with $t_f \leq \xi, w \gg \lambda$. The bias current I_b is applied to the blue strip close to I_c along the y-axis. Current flow directions are not necessarily in parallel to the y-axis because w is considerably larger than ξ and λ. In addition, local current densities j_c can vary depending on locations. This recalls the pure-Ir film used for the Ir-TESD in Figs. 4.8 and 4.9. The white x_r line in Fig. 4.16a is perpendicular to the supposed current flow directions indicated by the red arrows. We assume a j_c profile along the x_r line in Fig. 4.16b. The integration of j_c equals to the critical current I_c of the ultrawide film: $I_c = \int j_c \, dx_r$. The x_r line traces the locations of the weakest superconductivity. The electron beam stimulation typically at a few keV induces a local current reduction δj_c in (b). When we bias the strip such that the relation $I_b \geq I_c - \delta j_c$ holds, a resistive band runs across the strip width along the x_r line [38]. This normal band produces a voltage drop under the constant bias current. This picture of the signal generation is equivalent to the hot spot model in modern SSPDs. When increasing I_b to I_c, the entire x_r line appears on the LTSEM electron beam induced voltage (EBIV) image. The LTSEM voltage amplitude is proportional to the local j_c values as shown in Fig. 4.16b, when we postulate a constant j_c reduction rate under the current bias condition $I_b = I_c$, in which everywhere along the x_r line carries j_c [66]. At the same bias condition, the first macroscale NbN-SSPD achieved the best counting performance for α-particles [36]. A further I_b increase prompts a second x_r line with a higher I_c elsewhere in the strip. The LTSEM employs frequency modulation and lock-in techniques to improve an S/N ratio, which allows a small beam current with a spatial resolution of a few μm. It is important to understand the signal generation in inhomogeneous film strips before discussing SSPD signals, which have almost constant pulse height due to negligibly small inhomogeneity of j_c in narrow strips in most cases.

The extent of the δj_c reduction area in Fig. 4.16b is illustrated in Fig. 4.17 with two diffusion models: thermal treatment in (a) and nonthermal treatment in (b) [66]. A superconductor strip on a substrate runs along the direction perpendicular to the sheet; the bias current also flows along that direction. When phonon escape into the substrate is negligible (thermal treatment), a thermal healing length, which is governed by thermal diffusion, determines the perturbation area expansion [104]. On the other hand, when phonon escape is significant (nonthermal treatment), a quasiparticle diffusion length determines the perturbation area expansion [104]. These two pictures also depend on perturbation magnitude. Large stimulation due to ~ keV molecules is appropriate to the thermal treatment owing to a probable temperature rise, whereas small stimulation due to ~ 1-eV photon is appropriate to the nonthermal treatment with electronic excitation. The quasiparticle dynamics on diffusion and recombination inside a hot spot was theoretically studied in detail [105]. Physical parameters to be considered are T, C, k_t for the thermal treatment in (a) and N_q, D_q, τ_{eff} for the nonthermal treatment in (b), where T, C, and k_t are the temperature, heat capacity, and thermal conductivity; and N_q, D_q, and τ_{eff} are the quasiparticle density, quasiparticle diffusion constant, and quasiparticle effective recombination

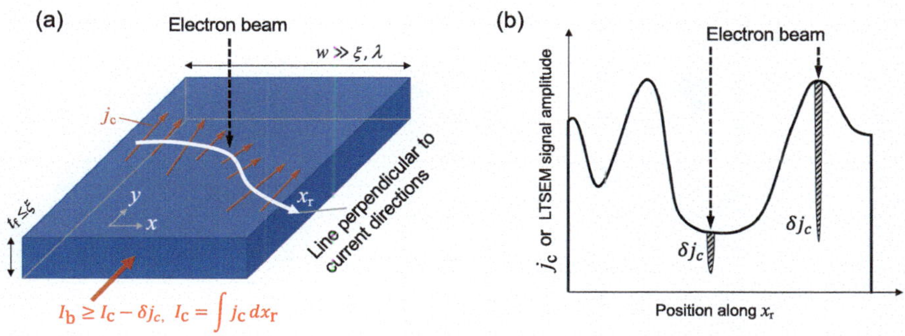

Fig. 4.16 LTSEM electron beam induced voltage (EBIV) generation [66]: (**a**) electron beam stimulation in current-carrying ultrawide inhomogeneous superconductor strip and (**b**) hypothetical profile of local critical current density j_c. or LTSEM signal amplitude in bias condition $I_b = I_c$. The typical electron beam with a current range of 2–8 pA has an energy of 10 keV, a pulse width of 25 μs, and a modulation frequency of 10 kHz. The inhomogeneous superconductor strip with $w \gg \lambda$ in (**a**) necessarily has neither uniform j_c distribution nor identical current flow directions in parallel with the y-axis. The perpendicular line to the assumed current flow directions is depicted by the white x_r line in (**a**). In (**b**), an assumed j_c distribution along the x_r line is depicted by the black solid line with current reduction δj_c due to the electron-beam local-stimulation. The extent of perturbed regions is shown in Fig. 4.17. The inhomogeneous bias current distribution resembles the pure-Ir TESD in Fig. 4.9. This scenario of signal generation corresponds to the hot spot model in modern SSPDs

time. The molecule in (a), as an example of keV particles, is immunoglobulin G (IgG) with a size of 15 nm. The size of IgG is enlarged relative to a hot-spot diameter d of ~ 100 nm. On the other hand, a 1-eV photon induces almost no temperature rise, which can be treated by quasiparticle diffusion in (b). In either case, a typical time scale for the initial hot-spot core formation in red is normally shorter than 100 ps, which corresponds to the relaxation time of hot electrons or quasiparticles down to the order of Δ_g [66]. For 10-keV electrons, the hot-spot diameter d is between 1 and 10 μm in most film-substrate combinations. For 1-keV Ar ions, d is evaluated at 110 nm for NbN [52].

The perturbed region for the internal photoelectric effect at ~ 1-eV is smaller than that for the phonon bubble generation of molecule incident at ~ 10 keV. It is reasonable that, in the photon detection, a thinner film thickness is better for a higher detection efficiency. Superconductor strips for photon detection generally possess a thickness of < 10 nm and a width of a few hundreds of nm. For energetic molecule detection, on the other hand, it is possible to utilize 10 times thicker and wider strips. This is due to the 10^4 times higher energy even though it is phonon-mediated excitation at a few tens of meV as shown in Fig. 3.4a. The maximum strip width for keV molecule detection is empirically ~ 1 μm within the framework of the hot spot model [52], which leads to a voltage drop due to a normal band bridging the strip width. Another initial model, vortex motion, generates electromotive force and a voltage drop without breaking the long-range coherency of superconductivity.

Initial perturbations relax at a certain time scale. An interesting experiment was reported on this matter [106]. The relaxation time for the 1.17-eV photon perturbation was measured by using a two-photon detection technique based on the pump-probe

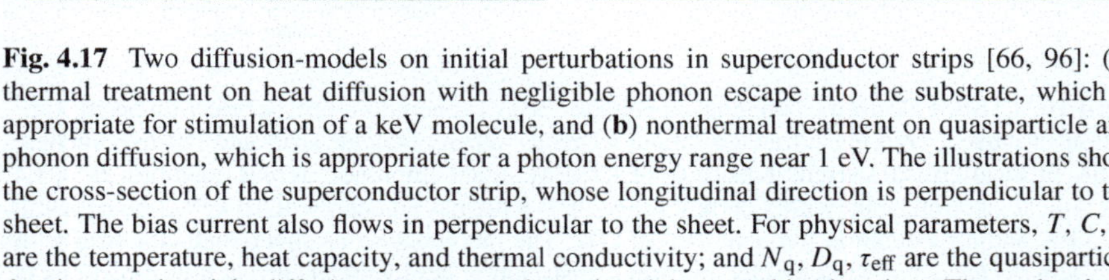

Fig. 4.17 Two diffusion-models on initial perturbations in superconductor strips [66, 96]: (**a**) thermal treatment on heat diffusion with negligible phonon escape into the substrate, which is appropriate for stimulation of a keV molecule, and (**b**) nonthermal treatment on quasiparticle and phonon diffusion, which is appropriate for a photon energy range near 1 eV. The illustrations show the cross-section of the superconductor strip, whose longitudinal direction is perpendicular to the sheet. The bias current also flows in perpendicular to the sheet. For physical parameters, T, C, k_t are the temperature, heat capacity, and thermal conductivity; and N_q, D_q, τ_{eff} are the quasiparticle density, quasiparticle diffusion constant, and quasiparticle recombination time. The molecule in (**a**) is immunoglobulin G (IgG) with a size of 15 nm, as an example of particle impact. The IgG size in (**a**) is enlarged relative to a typical hot-spot diameter d of ~ 100 nm [52]: Aactually, the molecule size is ~ 10% of the hot spot size. The 1-eV photon absorption in **b** induces electronic excitation, which leads to Cooper pair breaking and generates quasiparticles that diffuse in a circle around the initial hot spot

method in an NbN stirp with $t_f = \sim 4$ nm, $w = 100$ nm, covering $2 \times 8\ \mu m^2$ on a sapphire substrate. The measured relaxation time was ~ 15 ps. The dependence of the relaxation time on substrate material for 0.80-eV photons was also measured in NbN strips with $t_f = 4.5$–7.5 nm and $w = 90$–130 nm, covering an area of 15-μm in diameter. The measured relaxation time ranged from 12 to 35 ps, depending on substrates such as MgO, MgF_2, Al_2O_3, SiO_2, and Si. The photon energy of 0.80 eV falls into a boundary range between the hot spot and the vortex motion mechanisms. The observed fast relaxation times probably stem from the vortex motion instead of the hot spot formation.

4.2.5.3 SSPD Structure and Output Pulse

We need a reasonable detector size for analytical science; a straightforward approach involves fabricating a long meandering superconductor line. However, the inductance of long lines degrades temporal response. Therefore, we face conflict between a large sensitive area and a fast response time. Figure 4.18a shows an example of simple meander SSPDs. This SSPD has a single-crystalline NbN strip with $t_f = 10$ nm, $w = 800$ nm, and $l = 40$ mm, forming a meander geometry with a line spacing of 200 nm on a MgO substrate [107]. The detector area is 200-μm square with a filling factor of 80%. This NbN strip has a T_c value of 14 K, so that it is operative at 4.2 K. Figure 4.18b shows a typical example of output pulses from an analogous 50-μm square NbN SSPD with a meander nanostrip with $t_f = 7$ nm, $w = 300$ nm, $l = 4.15$ mm, and a line space of 300 nm. A biomolecule of angiotensin-I accelerated at 17.5 kV impacted the nanostrip carrying a bias current of 170 μA that is 94% of I_c. The angiotensin-I is a peptide biomolecule with a molecular weight of 296. The temporal response to the molecule impact has a rise time τ_{rise} of 360 ps and a fall time τ_{fall} of 9 ns. The output pulse shape is independent of molecular weight and molecular types such as 66.4-kDa BSA and 146-kDa IgG. This NbN SSPD also enables detection of ~ 1-eVphotons [107]. The pulse height and shape for the 1-eV photons are exactly the same as those for the macromolecules with an energy of 17.5 keV. The NbN strip always generates the identical pulses regardless of the 17,500-fold difference in particle energy and the mass range difference from zero mass of photons to146-kDa of IgG. The 146 kDa is equal to the mass of 12,170 atoms of ^{12}C. This phenomenon implies that the initial perturbations are merely a trigger, not affecting output pulses: the reason is explained in the next section. This invariance in pulse height is in contrast to the absorption-position-dependent spike pulses observed inside the superconducting region in the 500-μm square Ir-TESD in Fig. 4.9 and the LTSEM signal generation in the extremely wide superconductor strip in Fig. 4.16.

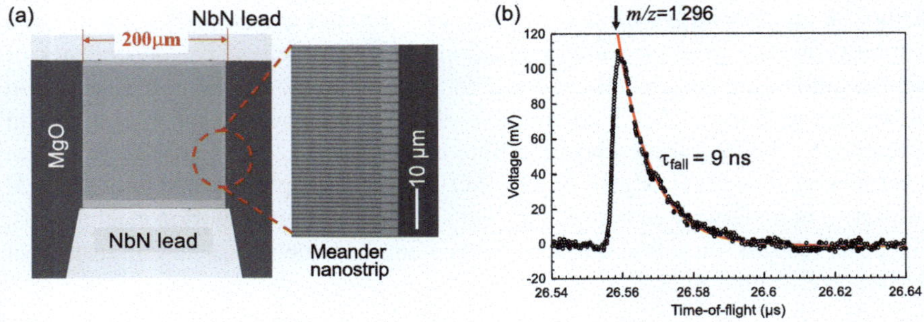

Fig. 4.18 NbN-SSPD and output pulse [87]: (**a**) laser scanning microscope image of 200-μm square meander SSPD and (**b**) output pulse of analogous 50-μm square NbN-SSPD for an angiotensin-I ion at a bias current of $0.94I_c$ and at 4.2 K. In (**a**), the 200-μm square SSPD has $t_f = 10$ nm, $w = 800$ nm, $l = 40$ mm, and a line spacing of 200 nm. The 200-μm square SSPD recorded $\tau_{rise} = 900$ ps and $\tau_{fall} = 27$ ns. The pulse in (**b**) is generated by the 50-μm square SSPD with $t_f = 7$ nm, $w = 300$ nm, $l = 4.15$ mm, and a line spacing of 300 nm:$\tau_{rise} = 360$ ps and $\tau_{fall} = 9$ ns. The NbN strips have a T_c value of 14 K. The angiotensin-I is a 1296-Da peptide biomolecule. The biomolecules were ionized by matrix-assisted laser desorption/ionization (MALDI) and accelerated at 17.5 kV. The red dashed exponential decay curve in (**b**) depicts a fitted curve with $\tau_{fall} = 9$ ns. The rise time measured with a wide band RF amplifier was 360 ps. The time of the x-axis represents a time-of-flight (TOF) for a flight pass length of 1.37 m. The TOF value of 26.56 μs corresponds to 1296 Da for the singly-charged angiotensin-I. Photons with an energy of ~ 1 eV generate the identical pulse height and shape. The NbN SSPD was also applied to charge state discrimination in MS [49]. Reproduced with permission from Ohkubo [8]. Copyright 2023 AIP Publishing LLC

4.2.5.4 ETF in SSPD

We empirically know that SSPDs generate a constant pulse height regardless of particle types and particle energies. Compared with the extremely wide inhomogeneous strip in Fig. 4.16, modern SSPDs consist of a high-quality uniform narrow strip, the width of which is normally less than 1 μm. Therefore, both the bias supercurrent distribution and the j_c distribution along the transverse direction are uniform. Furthermore, the formation of a resistive region is dissimilar from that of ultrawide strips in Fig. 4.16a. In narrow strips, a normal-conducting resistive region easily expands across the strip width. Therefore, the identical output pulses are understandable as quasi-1D behavior with respect to pulse height, contrary to the output voltage proportional to j_c in Fig. 4.16b. Nevertheless, more precisely, signal generation in narrow strips is 2D behavior. An ingenious method of quantum detector tomography achieved a surprisingly high spatial resolution of 10 nm, which enabled the analysis of position dependence of both local detection efficiency and detection-threshold bias current for ~ 1-eV photons [111]. For an NbN strip measuring 150-nm wide, 100-nm long, and 5-nm thick, the local distribution of the threshold bias current is inhomogeneous: 27.5 μA at the center and 25.3 μA near the transverse edges of the strip. This behavior is explained by the SVC model [111]. The SVC and VAP models require 2D superconductor films. In addition to the 2D behavior of the threshold bias current, the conventional hot spot formation and succeeding bias current detouring also require transverse extent. Consequently, modern SSPDs exhibit the 2D nature even in a strip-width range of a few hundreds of nm.

Both initial resistive band formation in the hot spot model and voltage drop due to vortex motion is followed by Joule self-heating ($P = I_b^2 R = I_b U$) inside current-carrying superconductor strips, provoking subsequent resistive domain expansion. The resistive domain expansion and succeeding contraction described below are determined by circuitry regardless of the initial response types such as hot-spot formation, SVC, and VAP. When I_b is constant for the pulse duration, the normal domain expands throughout the entire strip due to positive ETF with increasing R. This is called normal-state latching or thermal runaway, which prohibits continuous photon counting. Practical applications require the negative ETF that reduces a bias current such that subsequent particles are detectable. Other thermal detector types such as the SPT and TES have a shunt resistor to stabilize detector operation with the negative ETF.

Figure 4.19 illustrates an SSPD operation sequence with the negative ETF in the case of the hot spot model, providing a homogeneous initial bias-current distribution. Figure 4.19a shows an IgG molecule and a created hot spot. A photon can replace the macromolecule; a voltage drop due to vortex motion replaces the hot spot when SVC or VAP occurs. SSPDs are high-speed counters; therefore, we use a wide-band amplifier, which normally has a 50-Ω input resistor. Intentionally or unintentionally, the input resistor shunts the SSPD as shown in (b). The shunt resistor or load resistor R_L causes a decrease of the current flowing the strip after a particle induces a normal-conducting domain, which restores a superconducting state due to negative ETF.

The time scale in Fig. 4.19 refers to the experimental data on 17.5-keV BSA ion detection by using the 50-μm square NbN SSPD with $t_f = 7$ nm, $w = 200$ nm, $l =$ 6.25 mm, and $L_k = 1.1$ μH [87]. The time scale for 17.5-keV molecule detection is exactly the same as that for 1-eV photon detection. The hot spot in Fig. 4.19a leads to the supercurrent detour around the hot spot, which increases current density outside the hot spot [40, 96]. When the blue j_b curve exceeds the dashed black j_c line for the critical current density outside the hot spot, a normal-conducting resistive band bridges the strip width. This condition is expressed by [38, 112]

$$I_b \geq I_c - \delta I \text{ or } I_b \geq I_c\left(1 - \frac{d}{w}\right), \qquad (4.9)$$

where d is the diameter of the hot spot , $I_b = \int j_b \mathrm{d}x = \int j_s \mathrm{d}x$, $I_c = \int j_c \mathrm{d}x$, and $\delta I = \int^d j_s \mathrm{d}x$: black j_c is the critical current density in (a), red j_s is the supercurrent density before events, and blue j_b is the bias current density after the hot spot formation. Equation (4.9) is equivalent to the bias condition in Fig. 4.16a; LTSEM scanning measures the j_c distribution at $I_b = I_c$, as shown in Fig. 4.16b. The normal band in (b) subsequently induces resistive domain expansion due to Joule self-heating until the resistance R_k increases enough such that the current I_k decreases due to the diverted current on R_L of the amplifier. Therefore, the power of Joule self-heating decreases, removing the heat by an escape into the substrate at $t < 100$ ps. Although the strip

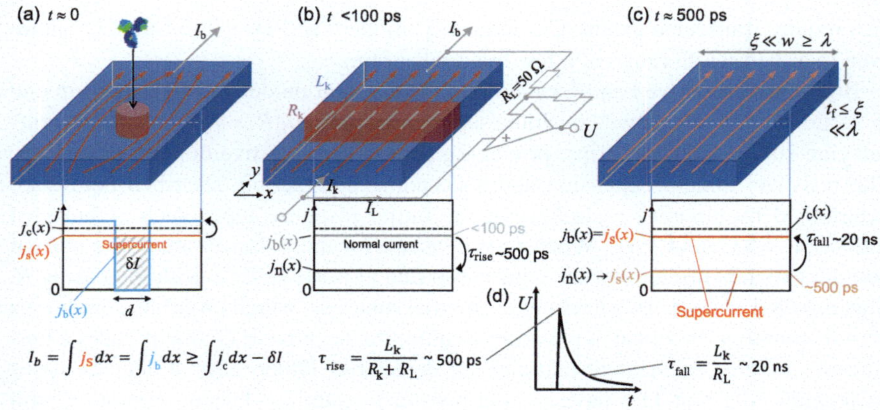

Fig. 4.19 Negative electrothermal feedback (ETF) against Joule self-heating after initial stimulation: (**a**) initial stimulation with molecule impact at $t = 0$, (**b**) normal band formation in $t < 100$ ps and readout circuit, (**c**) recovery of supercurrent at $t = \sim 500$ ps, and (**d**) output voltage pulse. The time scale is based on the experimental observation of the 50-μm square NbN SSPD with $t_f =$ 7 nm, $w = 200$ nm, $l = 6.25$ mm, and $L_k = 1.1$ μH. The 17.5-keV BSA ions and 1-eV photons generate the identical output pulses in the same time scale. Photons induce an initial voltage drop due to either hot spot or vortex motion depending on photon energy. The bias current I_b should meet the condition $I_b = \int j_s(x)dx = \int j_b dx \geq \int j_c dx - \delta I$ to detect molecules or photons. In (**a**), the black $j_c(x)$ is the critical current density, the red $j_s(x)$ is the bias supercurrent density before the event, the blue $j_b(x)$ is the current density distribution after the keV molecule event. In (**b**), the gray $j_b(x)$ is the current density after the normal band is formed, the black $j_n(x)$ is the normal current density after forming the normal band with a resistance value of R_k, which determines τ_{rise}. In (**c**), the light red $j_s(x)$ supercurrent returns to the initial red $j_s(x)$ at a readout-circuit time constant of ~ 20 ns. The Joule self-heating in the strip is suppressed by the negative ETF with the 50-Ω shunt or load resistor R_L in (**b**). The part of I_b is diverted to the shunt resistor. This negative ETF recovers the superconducting state without latching when suitable circuitry parameters are selected. The amplifier load resistor ($R_L = 50$ Ω) plays the same role as the shunt resistor for the first SPT thermometer in Figs. 4.2 and 4.3 [2]. After restoring the superconducting state in (**c**), the output pulse decays with a time constant τ_{fall} fixed by the readout circuit, as shown in (**d**). Reproduced with permission from Ohkubo [8]. Copyright 2023 AIP Publishing LLC

resistance varies in the course of the resistive domain expansion and contraction, the constant resistance R_k represents an effective value. Based on Eq. (4.9), the hot spot formation in ion bombardment in a keV range is confirmed by measuring the threshold bias current for ion detection as a function of ion kinetic energy [8, 52, 113] The estimated hot spot diameter for 1-keV Ar ions is 110 nm [52].

The diverted current flowing R_L generates an output pulse with a rise time constant expressed by

$$\tau_{rise} = \frac{L_k}{R_k + R_L}, \tag{4.10}$$

where L_k is the kinetic inductance of the strip [114]. The τ_{rise} value in the experiment for the BSA ion impact is 640 ps; it is approximated by ~ 500 ps in Fig. 4.19c.

The R_k is evaluated at 1.7 kΩ, which corresponds to a normal-domain expansion length of 1.5 μm [87], by using a L_k value of 1.1 μH that is evaluated from the observed τ_{fall} later. The output voltage becomes maximum just after the maximum expansion of the normal domain [115]. Due to I_L flowing R_L, the current flowing the strip is reduced from $j_b(x)$ to $j_n(x)$ in (b), also reducing the self-heating power. This is the negative ETF. The temperature of the normal domain rapidly decreases by thermal diffusion into the substrate because the strip has a good thermal contact with the substrate unlike TESDs. When the strip temperature decreases below T_c with a typical time scale of ~ 500 ps, the normal current j_n is transformed to the supercurrent j_s in light red in Fig. 4.19c.

The temporal expansion of a resistive domain was measured using parallel NbN strips with a thickness of 9 nm and a width of 100 nm [115]. The NbN SSPDs with one of the parallel configurations of 4, 8, 12, and 24 is connected to a series bias inductance of 470 nH to induce normal domains inside all the parallel strips. The resistive domain exponentially expands for ~ 6 ns, which is slower than a resistive-domain contraction time of ~ 1 ns. The maximum output signal is delayed by ~ 1 ns from the occasion of the maximum expansion of the normal domains. The expansion lengths are 0.59 μm for the 4 parallel strips and 3.5 μm for the 24 parallel strips. These lengths are consistent with the effective maximum length of 1.5 μm in the meander NbN SSPD, although we cannot directly compare these lengths due to the different configurations.

After recovering the superconducting state at ~ 500 ps in Fig. 4.19c, the output pulse starts to relax at a circuit time constant. The supercurrent returns to the value of the steady-state bias current j_b or deep-red j_s in (c) with a time constant expressed by

$$\tau_{fall} = \frac{L_k}{R_L}. \tag{4.11}$$

Correspondingly, the current flowing R_L returns to zero, which is the initial condition before the particle event. The measured τ_{fall} value of the meander NbN SSPD is 22 ns; L_K is evaluated at 1.1 μH. Both 22 ns of the 200-nm wide SSPD and 9 ns of the 800-nm wide SSPD in Fig. 4.18 are comparable with those of avalanche photodiodes (APDs). However, APDs have a rather long dead time or recovery time due to a afterpulse phenomenon. SSPDs exhibit a shorter dead time without afterpulse, which is important for quantum photonics [116].

The total inductance of a strip is the sum of kinetic inductance L_K and magnetic inductance. Because superconductor strips have an extremely small cross-section, the magnetic inductance is normally negligible [114]. The kinetic inductance L_k of a superconductor strip is given by [117]

$$L_k = \frac{\mu_0 \lambda^2 l}{w t_f}, \tag{4.12}$$

where μ_0 is the vacuum permeability. The λ(4.2 K) of the NbN film is evaluated at 440 nm from L_k and the strip dimensions. This value is not inconsistent with 194 nm at 0 K for a single-crystalline NbN film in Table 1.1, considering that λ depends on film quality and measurement method. Equation (4.12) is valid for straight strips. A universal formula of L_k is derived for a variety of meander patterns [118].

For stable operation without thermal runaway, it is important to consider the cooling time of the normal domain. The cooling time is primarily governed by the inelastic electron–phonon scattering time τ_{ep} in Table 1.1. When the duration of an I_L pulse flowing R_L is longer than the normal domain cooling time, I_K flowing the strip rapidly decreases [5, 7]. Therefore, with this negative ETF, the strip restores the superconducting state without latching. If it is opposite, the normal region expands to the whole stirp. We also need to select a correct kinetic inductance [86]. When the cooling time of the normal domain is sufficiently fast, the load resistor R_L functions as a stabilizer. This condition is exactly the same as the shunt resistor for the first Ta-SPT bolometer in 1942 in Fig. 4.2 [2], and resembles the operation in the strong negative ETF in TESDs.

We must select a correct R_L value for the stable operation without latching. If the negative ETF with the shut resistor is insufficient, the entire strip region remains latched in the normal state after molecule or photon incidence. No latching occurs when the ratio R_L/R_K is significantly lower than ~ 10. On the other hand, if the equivalent circuit in Fig. 4.19b has a fall time faster than the cooling time of the normal domain in R_L/R_K > ~ 10, the Joule self-heating continues on and the resistive domain expands to the entire strip length [5–7]. This is the positive ETF. Therefore, a small R_L is better to avoid latching; however, this makes both τ_{rise} and τ_{fall} long according to Eqs. (4.10) and (4.11). The R_L value should be selected such that continuous particle counting is as fast as possible without latching, similar to the conflict between a kinetic inductance and a sensitive area. The typical ratios of R_L to the effective resistance R_k are diverse: ~ 10 mΩ/1 Ω = 0.01 for the Ir TESD, ~ 50 Ω/1.7 kΩ = 0.03 for the NbN SSPD, and ~ 4 Ω/0.4 Ω = 10 for the first Ta-SPT bolometer. The TES type can operate at a negligibly small L_k value; L_k is not a factor limiting temporal response. This allows extremely small R_L. The Ta-SPT bolometer is close to the border of R_L/R_K = ~ 10, which is due to the available readout instruments such as galvanometers and potentiometers. The R_L/R_k ratios determine the strength order of the negative ETF: TESD > SSPD > SPT.

The invariable output pulse shape and height are secured by the negative ETF with Joule self-heating, independent of particle types, particle energies, and initial particle stimulation modes such as the hot spot and the vortex motion. This resembles the operation of GM counters on the Townsend avalanche mode to obtain a high quantum efficiency. A high voltage applied to a GM tube beyond a proportional region induces avalanche multiplication of electron–ion pairs. GM tubes can achieve a high quantum efficiency but lose energy resolution. APDs are a solid-state version of the GM counters. Unlike proportional counters, SSPDs cannot essentially measure particle energy even when bias current is reduced, although the detection threshold level can be adjusted.

It is worth considering superconductor materials for fast response. The τ_{ep} values listed in Table 1.1 regulate the count-rate upper limit for high frequency applications such as quantum communication. Aluminum is an essential material for the STJ type, but the extremely long τ_{ep}, 395 μs, is unsuitable for fast response. Niobium with 330 ps is acceptable, and NbN with 88 ps is favorable. The SS type simply utilizes the S-N transition; high temperature superconductors (HTSs) are promising for operation at a liquid nitrogen temperature regardless of the *d*-wave symmetry or multiple superconducting energy gaps. The compounds MgB_2 and YBCO with a τ_{ep} value of a few ps are attractive for ultra-high-speed applications. Photon detection with MgB_2-SSPDs is successful and reviewed [10]. We face a difficulty in preparing uniform YBCO films. A meander YBCO-SSPD with a thickness of 200 nm, a width of 1.4 μm, and a T_c value of 77 K was unable to detect 17.5-keV molecules in 2009 because film inhomogeneity led to a poor *I–U* characteristics [119]. No single photon detection has been realized with YBCO-SSPDs, although it is close to single photon detection [120, 121]. A 5-μm square BSCCO-SSPD is able to detect single photons at temperatures up to 25 K. The BSCCO thin film was mechanically exfoliated from a single crystal at a van-der-Waals gap in addition to a LSCO/LCO bilayer SSPD fabricated by atomic layer deposition [122]. Furthermore, a BSCCO nanobridge with a width of 250 nm and a length of ~ 1-μm is able to detect single photons at temperatures up to 20 K [123]. These high-T_c SSPDs are extremely small for practical applications. Further development of HTS SSPDs is needed to enlarge their sensitive area and realize operation at 77 K.

4.2.5.5 Co-existence of Hot Spot and Vortex Motion

Mathematical simulation is one approach to more precisely investigate the ETF signal generation processes in Fig. 4.19; there are some details that cannot be observed experimentally. Figure 4.20 shows a calculation result of the spatial–temporal evolution of the normal domain, based on a 2D time-dependent Ginzburg–Landau (TDGL) equation coupled with a thermal diffusion equation for the MgB_2-SSPD with t_f = 10 nm, w = 250 nm, and l = 200 μm [124]. The result exhibits a temporal expansion and contraction process of the normal domain in (a), a change of the superfluid electron density in (b) at the different locations marked by A, B, C, and D in (a), and an output voltage pulse appeared on $R_L(= 50\ \Omega)$ in (c). The power input Q_{ion} depicted by the deep-red dashed dotted line in (b) simulates a molecule impact event at the longitudinal center to create the initial hot spot. The hot-spot-diameter dependence of subsequent processes is negligible between 50 and 200 nm. The normal domain expands up to ~ 1 μm at an elapsed time of 16 ps due to self-heating; the temperature rises to 48 K at the incident location. The strip entirely recovers the superconducting state at 400 ps, when the output pulse reaches the maximum voltage in (d). The pulse decay follows a circuit time constant. The calculation agrees well with a typical experimental output pulse of the 10-μm square MgB_2-SSPD for a 20-keV lysozyme ion in (c).

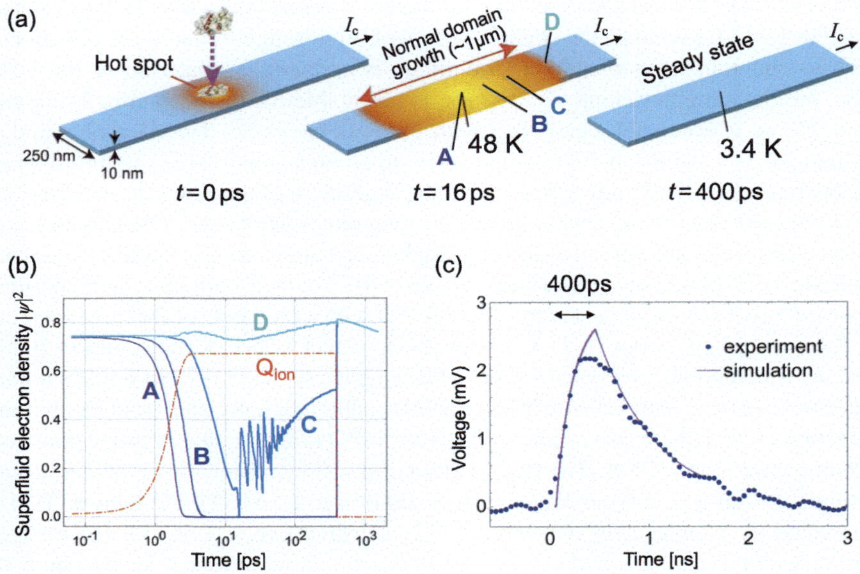

Fig. 4.20 Simulation with 2D time-dependent Ginzburg–Landau (GL) equation coupled with thermal diffusion equation for MgB_2 SSPD with $t_f = 10$ nm, $w = 250$ nm, and $l = 200$ μm [124]: (**a**) temporal scenario deduced from the simulation, (**b**) position dependence of superfluid electron density at locations marked A, B, C, and D in (**a**), and (**c**) comparison between the simulation and an experimental output voltage pulse on R_L. The MgB_2 SSPD with $L_k = 30$ nH is connected to a load resistor R_L (50 Ω) in parallel. The molecule impact is simulated by a power input Q_{ion} in the middle of the MgB_2 strip. The used simulation parameters are a bath temperature of 3.4 K, $I_b/I_c = 0.98$, and $\lambda = 550$ nm that is estimated from the fall time of the output pulse. The magnetic penetration depth 550 nm is considerably longer than 40–100 nm of high quality MgB_2 bulks and close to the Pearl length of 650 nm listed in Table 1.1. Reproduced with permission from Ohkubo [8]. Copyright 2023 AIP Publishing LLC

In Fig. 4.20b, the superfluid density curve A decreases to zero after a few ps at the center of the MgB_2 strip. The curves A and B near the molecule impact location shows a smooth temporal change without oscillation, indicating that the normal-core hot-spot is formed. The curve D exhibits a small variation at the N-S boundary, which is the forefront of the normal domain expansion at an elapsed time of 16 ps when the normal domain size reaches its maximum. Intriguingly, the curve C oscillates at a location halfway between the center and the N-S boundary within a time span from ~ 10 ps to 100 ps for the contracting period of the normal domain. The oscillation lasting for ~ 100 ps is attributed to vortex generation and motion. The VAP can coexist with the primary hot spot formation for the certain elapsed-time interval after restoring an intermediate state. This coexistence of the hot-spot and vortex motion can presumably resolve the controversy on detection mechanisms for the photon energy range $E_0 \leq 1$ eV.

The above-mentioned coexistence of the different phenomena can also explain the experimental result on particle-energy sensitivity in the SS type. The energy sensitivity is in contrast with the common belief that SSPDs are categorized as the photon counting detector type that generates a constant pulse height independent of energy. There is an interesting report on photon energy measurement by using a 4-μm square NbN SSPD measuring 4.5-nm thick and 100-nm wide [80]. Above ~ 2 eV, no pulse height variation is observed, which symbolizes the hot spot formation with ETF. Below ~ 0.3 eV, dark counts appear with a constant pulse heigh, which symbolizes the vortex motion without ETF. In a photon-energy range between ~ 0.3 and ~ 2 eV, the pulse amplitude almost linearly decreases with increasing the photon energy. The best energy resolution was 0.55 eV in a range of 0.8–1.2 eV. The energy sensitivity is attributed to a detection mechanism change from the vortex motion to the hot spot formation [80, 93, 125]. A photon energy of 1 eV stands on the boundary between the two mechanisms. An activated detection mechanism depends on nanostrip dimensions and superconductor material as well as photon energy. The pulse height reduction with increasing photon energy is puzzling when we assume that the hot spot with ETF in a high photon-energy range produces a higher pulse height due to normal domain expansion. Alternatively, the VAP or SVC process without ETF likely generates short pulses, which are expected to be higher than the long pulses of the hot spot process with ETF. Further studies are required.

A simulation similar to the MgB_2 SSPD was performed for a Nb SSPD. The material parameters of Nb are significantly different from those of MgB_2 as listed in Table 1.1,; nevertheless, the simulation result resembles Fig. 4.20. This suggests that the high energy of keV particles obscures the material parameter difference. In other word, the time evolution is predominantly governed by the normal domain expansion and contraction due to the negative ETF. Initial stimulation has a negligible effect on the pulse amplitude and shape, analogous to the behavior observed in GM counters and APDs.

4.2.5.6 Photon Detection with Ultra-Wide SSPD

Photon counting in a telecommunication wavelength normally requires a strip width of a few hundreds of nm and a thickness of several nm. Even the detection of a keV ion requires a strip width less than 1 μm. However, there are reports that NbN SSPDs with a thickness of 5.8 nm and a width up to 5.15 μm successfully detect single photons [88, 89]. The factors to be considered are the heat capacity ratio of electrons to phonons and critical current. For single photon detection, the heat capacity ratio should be larger than 1, and the critical current should be larger than 70% of depairing current (critical pair-breaking current) [88]. In addition to NbN, the number of superconductor materials used for ultrawide SSPDs is increasing; for example, amorphous superconductor $Nb_{0.15}Re_{0.85}$ with a thickness of 4 nm and a width of up to 2.5 μm, which enables detection of photons at 0.8 eV at a low bias current of 21% of the departing current [126]. Ultrawide strips solve the problem of the conflict between a high detection efficiency and a practically large sensitive area.

Furthermore, fast response is possible because L_k is inversely proportional to strip cross-section following Eq. (4.12).

References

1. F. Mandl, *Statistical Physics*, 2nd edn. (Wiley, 1988), p. 52
2. D.H. Andrews, W.F. Brucksch, W.T. Ziegler, E.R. Blanchard, Rev. Sci. Instrum. **13**, 281 (1942)
3. K. Irwin, Appl. Phys. Lett. **66**, 1998 (1995)
4. R.M.T. Damayanthi, Y. Kunieda, N. Zen, F. Mori, K. Fujita, H. Takahashi, M. Nakazawa, F. Fukuda, M. Ohkubo, Jpn. J. Appl. Phys. **45**, 6259 (2006)
5. J.K.W. Yang, A.J. Kerman, E.A. Dauler, V. Anant, K.M. Rosfjord, K.K. Berggren, IEEE Trans. Appl. Super. **17**, 581 (2007)
6. A.J. Kerman, J.K.W. Yang, R.J. Molnar, E.A. Dauler, K.K. Berggren, Phys. Rev. B **79**, 100509(R) (2009)
7. A. Annunziata, O. Quaranta, D.F. Santavicca, A. Casaburi, L. Frunzio, M. Ejrnaes, M.J. Rooks, R. Cristiano, S. Pagano, A. Frydman, D.E. Prober, J. Appl. Phys. **108**, 084507 (2010)
8. M. Ohkubo, J. Appl. Phys. **134**, 081101 (2023)
9. J. Ullom, D. Bennett, Supercond. Sci. Technol. **28**, 084003 (2015)
10. H. Shibata, IEICE Trans. Electron. **E104-C**, 429 (2021)
11. A. Goetz, Phys. Rev. **55**, 1270 (1939)
12. D.H. Andrews, W.F. Brucksch, W.T. Ziegler, E.R. Blanchard, Phys. Rev. **59**, 1045 (1941)
13. H. Kamerlingh-Onnes, Akad. van Wetenschappen, Amsterdam **14**, 818 (1911)
14. G.B. Yntema, Phys. Rev. 98, **1144** (1955), *Minutes of the 1955 Annual Meeting*, p. 1197. https://doi.org/10.1103/PhysRev.98.1144
15. IEC 60617—Graphical Symbols for Diagrams. http://std.iec.ch/iec60617
16. IEC 60617 S01925: Normal-superconducting boundary. https://products.iec.ch/view/grs/8257?q=eyJxdWVyeSI6Ik5vcm1hbC1zdXBlcmNvbmR1Y3RpbmcgYm91bmRhcnkiLCJtb2RlIjoiR1JBUEhJQ0FMLVNZTUJPTCIsInNvcnRCeSI6InJlZmVyZW5jZS0tYXNjIiwibGFuZ3VhZ2UiOiJlbiJ9
17. J. Clarke, Phil. Mag. **13**, 115 (1966)
18. F. Low, J. Opt. Soc. Am. **51**, 1300 (1961)
19. Hitomi collaboration, Nature **535**, 11 (2016)
20. S.H. Moseley, J.C. Mather, D. McCammon, J. Appl. Phys. **56**, 1257 (1984)
21. D. McCammon, S.H. Moseley, J.C. Mather, R.F. Mushotzky, J. Appl. Phys. **56**, 1263 (1984)
22. J. Clarke, G.I. Hoffer, P.L. Richards, N.H. Yeh, J. Appl. Phys. **48**, 4865 (1977)
23. P.L. Richareds, J. Appl. Phys. **76**, 1 (1994)
24. S.J. Smith, J.S. Adams, C.N. Bailey, S.R. Bandler, J.A. Chervenak, M.E. Eckart, F.M. Finkbeiner, R.L. Kelley, C.A. Kilbourne, F.S. Porter, J.E. Sadleir, J. Low Temp. Phys. **167**, 168 (2012)
25. A.R. Miniussi et al., J. Low Temp. Phys. **193**, 337 (2018)
26. B. Cabrera, R.M. Clarke, P. Colling, A.J. Miller, S. Nam, R.W. Romani, Appl. Phys. Lett. **73**, 735 (1998)
27. D. Fukuda, G. Fujii, T. Numata, A. Yoshizawa, H. Tsuchida, H. Fujino, H. Ishii, T. Itatani, S. Inoue, T. Zama, Metrologia **46**, S288 (2009)
28. K. Hattori, T. Konno, Y. Miura, S. Takasu, D. Fukuda, Supercond. Sci. Technol. **35**, 095002 (2022)
29. M. Bühler, E. Umlauf, Europhys. Lett. **5**, 297 (1988)
30. E. Umlauf, M. Bühler, *Low Temperature Detectors for Neutrinos and Dark Matter*, ed. by N.E. Booth, G.L. Salmon (Editions Frontieres, Gif-sur-Yvette, 1991), p. 229
31. S.R. Bandler, C. Enss, R.E. Lanou, H.J. Maris, T. More, S.F. Porter, G.M. Seidel, J. Low Temp. Phys. **93**, 709 (1993)

32. C. Enss, A. Fleischmann, K. Horst, J. Schoönefeld, J. Sollner, J.S. Adams, Y.H. Huang, Y.H. Kim, G.M. Seidel, J. Low Temp. Phys. **121**, 137 (2000)
33. S. Kempf, A. Fleischmann, L. Gastaldo, C. Enss, J. Low Temp. Phys. **193**, 365 (2018)
34. IEC 61788-22-1:2017, Superconductivity—Part 22-1: Superconducting electronic devices - Generic specification for sensors and detectors. https://webstore.iec.ch/publication/26674
35. IEC 61788-22-3: Superconductivity—Part 22-3: Superconducting strip photon detector—Dark count rate. https://webstore.iec.ch/publication/63471
36. D.H. Andrews, R.D. Fowler, M.C. Williams, Phys. Rev. **76**, 154 (1949)
37. N.K. Sherman, Phys. Rev. Lett. **8**, 438 (1962)
38. D.E. Spiel, R.W. Boom, E.C. Crittenden, Appl. Phys. Lett. **7**, 292 (1965)
39. M. Caria, T.O. Niinikoski, A. Rijllart, B. Strehl, A. Nigro, R. Vaglio, IEEE Trans. Nucl. Sci. **37**, 1127 (1990)
40. A. Gabutti, K.E. Gray, G.M. Pugh, R. Tiberio, Nucl. Instrum. Methods in Phys. Res. A **312**, 475 (1992)
41. G.N. Gol'tsman, O. Okunev, G. Chulkova, A. Lipatov, A. Semenov, K. Smirnov, B. Voronov, A. Dzardanov, C. Williams, R. Sobolewski, Appl. Phys. Lett. **79**, 705 (2001)
42. D.V. Morozov, A. Casaburi, and R.H. Hadfield, Contemp. Phys. **62**, 69 (2021). https://doi.org/10.1080/00107514.2022.2043596.
43. M. Frank, S. Labov, G. Westmacott, W. Benner, Mass Spectrom. Rev. **18**, 155 (1999)
44. S. Pagano, E. Esposito, M. Ejrnaes, C. Nappi, R. Cristiano, I.E.E.E. Appl, Super. **15**, 940 (2005)
45. K. Suzuki, S. Miki, Z. Wang, Y. Kobayashi, S. Shiki, M. Ohkubo, J. Low Temp. Phys. **151**, 766 (2008)
46. T. Ishida, IEICE Trans. Electron. **E103** (2020)
47. W.B. Doriese et al., Rev. Sci. Instrum. **88**, 053108 (2017)
48. S.J. Lee et al., Rev. Sci. Instrum. **90**, 113101 (2019)
49. K. Suzuki et al., Rapid Commun. Mass Spectrom. **24**, 3290 (2010)
50. L. Chen, D. Schwarzer, V.B. Verma, M.J. Stevens, F. Marsili, R.P. Mirin, S.W. Nam, A.M. Wodtke, Acc. Chem. Res. **50**, 1400 (2017)
51. R. Cheng, C.L. Zou, X. Guo, S. Wang, X. Han, H.X. Tang, Nat. Commun. **10**, 4104 (2019)
52. K. Suzuki, S. Shiki, M. Ukibe, M. Koike, S. Miki, Z. Wang, M. Ohkubo, Appl. Phys. Express **4**, 083101 (2011)
53. K. Tanaka, T. Morooka, K. Chinone, M. Ukibe, F. Hirayama, M. Ohkubo, M. Koyanagi, Appl. Phys. Lett. **77**, 4196 (2000)
54. K.D. Irwin, G.C. Hilton, D.A. Wollman, J.M. Martinis, J. Appl. Phys. **83**, 3978 (1998)
55. J.A. Chervenak, K.D. Irwin, E.N. Grossman, J.M. Martinis, C.D. Reintsema, Appl. Phys. Lett. **74**, 4043 (1999)
56. C.A. Mears, S.E. Labov, M. Frank, M.A. Lindeman, L.J. Hiller, H. Netel, A.T. Barfknect, Nucl. Instrum. Methods Phys. Res. A **370**, 53 (1996)
57. D.A. Bennett, D.S. Swetz, R.D. Horansky, D.R. Schmidt, J.N. Ullom, J. Low Temp. Phys. **167**, 102 (2012)
58. J.E. Sadleir, S.J. Smith, S.R. Bandler, J.A. Chervenak, J.R. Clem, Phys. Rev. Lett. **104**, 047003 (2010)
59. A. Kozorezov, A. Golubov, D. Martin, P. de Korte, M. Lindeman, R. Hijmering, J. van der Kuur, H. Hoevers, L. Gottardi, M. Kupriyanov, J. Wigmore, J. Low Temp. Phys. Phys. **167**, 108 (2012)
60. A. VI. Gurevich, R.G. Mints, Rev. Mod. Phys. **59**, 941 (1987)
61. R. den Hartog, A. Golubov, D. Martin, P. Verhoeve, A. Poelaert, A. Peacock, M. Krumrey, Nucl. Instrum. Methods Phys. Res. A **444**, 28 (2000)
62. R. den Hartog, A. Golubov, P. Verhoeve, A. Poelaert, D. Matin, A. Peacock, A. van Dordrecht, D.J. Goldie, Phys. Rev. B **63**, 214507 (2001)
63. J.E. Sadleir, S.J. Smith, I.K. Robinson, F.M. Finkbeiner, J.A. Chervenak, S.R. Bandler, M.E. Eckart, C. Kilbourne, Phys. Rev. B **84**, 184502 (2011)

64. H. Pressler, M. Koike, M. Ohkubo, D. Fukuda, Y. Noguchi, M. Ohno, H. Takahashi, M. Nakazawa, Appl. Phys. Lett. **81**, 331 (2002)
65. M. Ohkubo, H. Pressler, D. Fukuda, T. Inou, H. Takahashi, M. Nakazawa, IEEE Trans. Appl. Super. **13**, 634 (2003)
66. R. Gross, D. Koelle, Rep. Prog. Phys. **57**, 651 (1994)
67. W.J. Skocpol, Phys. Rev. B **14**, 1045 (1976)
68. J.N. Ullom, W.B. Doriese, G.C. Hilton, J.A. Beall, S. Deiker, W.D. Duncan, L. Ferreira, K.D. Irwin, C.D. Reintsema, L.R. Vale, Appl. Phys. Lett. **84**, 4206 (2004)
69. L. Gottardi, M. de Wit, E. Taralli, K. Nagayashi, A. Kozorezov, Phys. Rev. Lett. **126**, 217001 (2021)
70. K. Niwa, T. Numata, K. Hattori, D. Fukuda, Sci. Rep. **7**, 45660 (2017)
71. K. Niwa, K. Hattori, D. Fukuda, Front. Bioeng. Biotechnol. **9**, 789709 (2021)
72. M. Krantz, F. Toschi, B. Maier, G. Heine, C. Enss, S. Kempf, Appl. Phys. Lett. **124**, 032601 (2024)
73. T. Sikorsky et al., Phys. Rev. Lett. **125**, 142503 (2020)
74. V. Alenkov et al., Eur. Phys. J. C **79**, 791 (2019)
75. L. Gastaldo et al., Eur. Phys. J. Special Topics **226**, 1623 (2017)
76. O. Novotny, S. Allgeier, C. Enss, A. Fleischmann, L. Gamer D. Hengstler, S. Kempf, C. Krantz, A. Pabinger, C. Pies, D.W. Savin, D. Schwalm, A. Wolf, J. Appl. Phys. **118**, 104503 (2015)
77. L. Gamer, D. Schulz, C. Enss, A. Freischmann, L. Gastaldo, S. Kempf, C. Krantz, O. Novotny, D. Schwalm, A. Wolf, J. Low Temp. Phys. **184**, 839 (2016)
78. Y.H. Kim, S.-J. Lee, B. Yang, Supercond. Sci. Technol. **35**, 063001 (2022)
79. H. Rotzinger, J. Adams, S.R. Bandler, J. Beyer, H. Eguchi, E. Figueroa-Feliciano, W. Hsieh, G.M. Seidel, T. Stevenson, J. Low Temp. Phys. **151**, 351 (2008)
80. A.D. Semenov, P. Haas, B. Günther, H-W. Hübers, K. Il'in, M. Siegel, A. Kirste, J. Beyer, D. Drung, T. Schurig, A. Smirnov, Supercond. Sci. Technol. **20**, 919 (2007)
81. R. Cristiano, M. Ejrnaes, A. Casaburi, N. Zen, M. Ohkubo, Supercond. Sci. Technol. **28**, 124004 (2015)
82. B.G. Oripov, D.S. Rampini, J. Allmaras, M.D. Shaw, S.W. Nam, B. Korzh, A.N. McCaughan, Nature **622**, 730 (2023)
83. A. Murphy, A. Semenov, A. Korneev, Y. Korneeva, G. Gol'tsman, and A. Bezryadin, Sci. Rep. **5**, 10174 (2015)
84. IEC Products & Services Portal. https://products.iec.ch/
85. L. Parlato, G. Peluso, G. Pepe, R. Vaglio, C. Attanasio, A. Ruosi, S. Barbanera, M. Cirillo, R. Leoni, Nucl. Instrum. Methos Phys. Res. A **348**, 127 (1994)
86. K. Inderbitzin, A. Engel, A. Schilling, K. Il'in, M. Siegel, Appl. Phys. Lett. **101**, 162601 (2012)
87. K. Suzuki, S. Miki, S. Shiki, Z. Wang, M. Ohkubo, Appl. Phys. Express **1**, 031702 (2008)
88. D.Y. Vodolazov, Phys. Rev, Appl. **7**, 034014 (2017)
89. Y.P. Korneeva, D.Y. Vodolazov, A.V. Semenov, I.N. Florya, N. Simonov, E. Baeva, A.A. Korneev, G.N. Gol'tsman, T.M. Klapwijk, Phys. Rev. Appl. **9**, 064037 (2018)
90. A. Verevkin et al., Appl. Phys. Lett. **80**, 4687 (2002)
91. L.N. Bulaevskii, M.J. Graf, C.D. Batista, V.G. Kogan, Phys. Rev. B **83**, 144526 (2011)
92. A. Engel, A. Schilling, J. Appl. Phys. **114**, 214501 (2013)
93. A.M. Kadin, M. Leung, A.D. Smith, Phys. Rev. Lett. **65**, 3193 (1990)
94. A.N. Zotova, D.Y. Vodolazov, Phys. Rev. B **85**, 024509 (2012)
95. M. Tinkham, Rev. Mod. Phys. **46**, 587 (1974)
96. A.M. Kadin, M.W. Johnson, Appl. Phys. Lett. **69**, 3938 (1996)
97. M. Bell, A. Sergeev, V. Mitin, J. Bird, A. Verevkin, G. Gol'tsman, Phys. Rev. B **76**, 094521 (2007)
98. L.N. Bulaevskii, M.J. Graf, V.G. Kogan, Phys. Rev. B **85**, 014505 (2012)
99. J. Pearl, Appl. Phys. Lett. **5**, 65 (1964)
100. C.M. Natarajan, M.G. Tanner, R.H. Hadfield, Supercond. Sci. Technol. **25**, 063001 (2012)

101. A. Engel, J. J. Renema, K. Il'in, A. Semenov, Supercond. Sci. Technol. **28**, 114003 (2015)
102. A. Semenov, Supercond. Sci. Technol. **34**, 054002 (2021)
103. R.P. Huebener, Rep. Prog. Phys. **47**, 175 (1984)
104. R. Gross, M. Koyanagi, J. Low Temp. Phys. **60**, 277 (1985)
105. A.G. Kozorezov et al., Phys. Rev. B **92**, 064504 (2015)
106. L. Zhang, L. You, X. Yang, J. Wu, C. Lv, Q. Guo, W. Zhang, H. Li, W. Peng, Z. Wang, X. Xie, Sci. Rep. **8**, 1486 (2018)
107. S. Miki, M. Fujiwara, M. Sasaki, B. Baek, A.J. Miller, R.H. Hadfield, S.W. Nam, Z. Wang, Appl. Phys. Lett. **92**, 061116 (2008)
108. A. Casaburi, M. Ejrnaes, N. Zen, M. Ohkubo, S. Pagano, R. Cristiano, Appl. Phys. Lett. **98**, 023702 (2011)
109. S.B. Kaplan, C.C. Chi, D.N. Langenberg, J.J. Chang, S. Jafarey, D.J. Scalapino, Phys. Rev. B **14**, 4854 (1976)
110. S.B. Kaplan, J. Low Temp. Phys. **37**, 343 (1979)
111. J.J. Renema, Q. Wang, R. Gaudio, I. Komen, K. Hoog, D. Sahin, A. Schilling, M.P. van Exter, A. Fiore, A. Engel, M.J.A. de Dood, Nano Lett. **15**, 4541 (2015)
112. L. Maingault, M. Tarkhov, I. Florya, A. Semenov, R.E. de Lamaëstre, P. Cavalier, G. Gol'tsman, J. -P. Poizat, J. -C. Villégier, J. Appl. Phys. **107**, 116103 (2010)
113. M. Sclafani, M. Marksteiner, F.M. Keir, A. Divochiy, A. Korneev, A. Semenov, G. Gol'tsman, M. Arndt, Nanotech. **23**, 065501 (2012)
114. A.J. Kerman, E.A. Dauler, W.E. Keichr, J.K.W. Yang, K. Berggren, G. Gol'tsman, B. Voronov, Appl. Phys. Lett. **88**, 111116 (2006)
115. M. Ejrnaes, A. Casaburi, F. Mattioli, R. Leoni, S. Pagano, R. Cristiano, Phys. Rev. B **81**, 132503 (2010)
116. C. Autebert, G. Gras, E. Amri, M. Perrenoud, M. Caloz, H. Zbinden, F. Bussières, J. Appl. Phys. **128**, 074504 (2020)
117. R. Meservey, P.M. Tedrow, J. Appl. Phys. **40**, 2028 (1969)
118. Y. Mawatari, J. Appl. Phys. **133**, 173902 (2023)
119. N. Zen, Y. Chen, K. Suzuki, M. Ohkubo, S. Miki, Z. Wang, IEEE Trans Appl. Super. **19**, 354 (2009)
120. M. Ejrnaes, L. Parlato, R. Arpaia, T. Bauch, F. Lombardi, R. Cristiano,F. Tafuri, G.P. Pepe, Supercond. Sci. Technol. **30**, 12LT02 (2017)
121. P. Amari et al., Phys. Rev. Appl. **20**, 044025 (2023)
122. I. Charaev et al., Nat. Nano. **18**, 343 (2023)
123. R.L. Merino, P. Seifert, J.D. Retamal, R.K. Mech, T. Taniguchi, K. Watanabe, K. Kadowaki, R.H. Hadfield, D.K. Efetov, 2D Mater. **10**, 021001 (2023)
124. N. Zen, H. Shibata, Y. Mawatari, M. Koike, M. Ohkubo, Appl. Phys. Lett. **106**, 222601 (2015)
125. M. Hofherr, D. Rall, K.S. Ilin, A. Semenov, N. Gippius, H.-W. Hübers, M. Siegel, J. Phys. Conf. Ser. **234**, 012017 (2010)
126. M. Ejrnaes, C. Cirillo, D. Salvoni, F. Chianese, C. Bruscino, P. Ercolano, A. Cassinese, C. Attanasio, G.P. Pepe, L. Parlato, Appl. Phys. Lett. **121**, 262601 (2022)
127. M. Krantz et al, Appl. Phys. Lett. **124**, 032601 (2024)

Chapter 5
Performance Comparison for Analytical Science

Abstract Chapter 5 compares the performance of different superconductor detector types with respect to energy resolution, response time, energy coverage, spatial resolution, and operating temperature to support informed detector selection for targeted analytical applications. Particles to be detected are photons in an energy range from ~ 0.3 eV to a few hundreds of keV, ions or molecules in a mass range from 1 to a few MDa, and electrons, including both zero-mass particles and nonzero-mass particles such as ions and neutral molecules. Typical analytical approaches are X-ray spectroscopy and mass spectrometry (MS).

Keyword Performance comparison · Detector selection

5.1 Two Detector Research Communities

The detector technology of all the superconductor detector types is maturing, although there are still engineering barriers and unsolved/open questions on detector physics. Their performance exceeds that of non-superconductor detectors such as semiconductor detectors. Especially, the energy resolution of superconductor detectors is at least 10 times better. They are ready for practical use to address fundamental scientific problems, which require the unconventional detector performance. Typical scientific problems are the rare event search of neutrinoless double beta decay and dark matter [1, 2]. The important factor in these fields is low background or background rejection; detector development is the central part of research projects. In addition, a large area coverage is also important. For example, a Si wafer with a surface area of 45.6 cm^2 should be covered by 1031 TES-pixels with an energy resolution of ~ 4 eV at 20 eV and a timing resolution of 2.3 μs [1]. The rare event search does not require a high count rate; for example, a detector test is performed with 5.9-keV X-rays at ~ 5 cps and thus a τ_{fall} value of 1 ms is sufficient for most rare event search programs. The count rate is ~ 5 orders of magnitude lower than that required for analytical science.

M. Ohkubo, *Superconductor Detectors*, https://doi.org/10.1007/978-3-031-84193-4_5

The applications of superconductor detectors to analytical science treat not only pure scientific samples such as asteroid rocks and biomolecules but also industrial samples such as advanced functional and structural materials. Analytical science requires a wide range of techniques from hardware to software. At present, machine learning and even emerging quantum computing are beginning to replace man-powered data analysis. With this diversity of analytical processes, detectors are merely located at the ends of hardware instruments to generate electric signals. In contrast to the rare event search, the importance of detectors is often underestimated. The detector research is frequently neglected in multidisciplinary areas such as proteomics, even though the central analytical method of mass spectrometry (MS) for proteomics requires an ion detector [3, 4]. Analytical scientists focus on data analysis acquired by automated analytical instruments. However, the detector performance affects the quality of data that are created by converting the physical quantities of waves or particles to electric signals. Therefore, detectors are a potential bottleneck in data quality [3, 4]: both references mention superconductor MS detectors. This fact symbolizes the undoubtful importance of detectors in analytical science.

The detector research community is divided into two groups: (1) those whose primary interest is the detectors and the measurement and (2) those whose primary interest is the science enabled by the detectors [3]. This book stands in the first group while stepping into the second group.

The detector size is typically ~ cm for analytical instruments. When a detector operates on the pulse mode described in Chap. 2, Sect. 2.1, a count rate required for high throughput analysis reaches several hundreds of k cps; a τ_{fall} value of output pulses should be less than 10 μs. The weaknesses of superconductor detectors are the small sensitive area and the low count rate even in superconductor quantum detectors, compared with non-superconductor detectors. SSPDs have a high count-rate, but they essentially have no sensitivity to particle energy. Nevertheless, by conceiving a proper application for each superconductor detector type and utilizing auxiliary devices, we can overcome the fundamental limits of analytical instruments equipped with conventional non-superconductor detectors, revolutionizing both science and industry.

5.2 Performance Figures

Table 5.1 lists the important performance figures to clarify strengths and weaknesses of the different detector types: the performance figures include energy resolution, response time or count rate, energy coverage range, spatial resolution, and operating temperature. Achieving energy resolution beyond the limit of conventional semiconductor detectors is essential for analytical science; concurrently, analytical instruments frequently require a high count-rate, for example, to scan sample surfaces within a tolerable measurement time. Direct comparison of energy resolution among different detector types is practically meaningless because energy resolution depends on particle energy and each detector type specializes in a specific energy range. In consideration of this difficulty in comparison, Table 5.1 lists ΔE values in FWHM followed by photon energy after "@." In addition, the performance figures focus on X-ray photons, although the superconductor detectors are utilized for detecting diverse particle types such as electrons, ions, and molecules. Analytical requirements add the response time to Table 5.1 as the second important figure. The response time corresponds to pulse fall time τ_{fall}, which is also equivalent to photon count rate. Naturally, a shorter τ_{fall} is better for a high count-rate. Additionally, pulse rise time τ_{rise} is also important for timing measurement. When a particle arrives in the period of τ_{fall} after a preceding particle, it is called tail pile-up that results in energy resolution degradation on pulse height spectra. The second type of pile-up is called peak pile-up that produces sum peaks of two events within τ_{rise}. The probability of the peak pile-up also depends on τ_{fall} because a readout circuit time constant is generally adjusted to τ_{fall} to reduce electronic noise. We can conceive an application suitable for each detector type by considering spatial resolution and operating temperature as well as ΔE and τ_{fall}, comparing with semiconductor detectors such as SDD, CMOS, and CCD. The advantages and disadvantages of the different superconductor detector types are described in detail in Ref. [5].

The STJ quantum type can count particles up to ~ 50 kcps per pixel, covering an energy range between ~ 0.3 eV and ~ 2 keV at an energy resolution of ~ 10 times better than that of Si detectors. It is also possible to cover up to ~ 10 keV or more by adding an absorber, although improvement of energy resolution is required to outperform semiconductors [16, 17]. The number of STJ pixels connected to FET amplifiers is practically limited to 100–200 because the STJ type is incompatible with multiplex readout due to a fast fall time of 2–10 μs. Note that there is a proposal that radio-frequency single-electron transistor readout can be employed for frequency multiplexing for STJDs, although it does not target single particle detection [18, 19]. An STJ array detector with 100 pixels records an integrated energy resolution of 6 eV including electronic noise for 400-eV X-rays up to ~ 100 kcps, which is limited by parallel readout circuits for the STJ pixels. The detection efficiency above ~ 2 keV is rather poor because the thin film layer structure has a low absorption efficiency for X-rays. The energy resolution 6 eV is smaller than natural linewidths of characteristic X-rays such as the K lines of light elements and the L lines of heavy elements when atoms are embedded in a crystal or a matrix substance. Elemental mapping requires no further ΔE improvement. Consequently, one of the suitable applications of the STJ type is elemental mapping on SEM or TEM at a high count-rate up to ~ 200 kcps below ~ 2 keV. Another application is florescence-yield X-ray absorption spectroscopy for light elements in synchrotron radiation facilities. X-ray emission spectroscopy (XES) is also possible with EDS; characteristic X-ray line-shape analysis enables the different oxidation states of atoms. However, a better ΔE value is desirable for EDS-XES. Detection of phonon at a few tens of meV is also possible, although there are no reports on applications of phonon spectroscopy at present. This has the potential to open new analytical science with phonons for unsolved solid state physics.

Table 5.1 Superconductor detector types and important performance figures

Detection category[a]	Detector name[b]	Acronym[b]	ΔE (eV)[c]	Fall time τ_{fall} (μs)	Coverage	Spatial resolution[d] (μm)	T (K)
Quantum	Silicon drift detector (10 mm^2)	SDD	42 @ 277 124 @ 5900	0.01–0.1	X-ray (100 eV–100 keV)	N/A	240–300
Quantum	Superconductor tunnel junction detector	STJD	0.1 @ 1 2.5 @ 250 13@ 5900	2–10	Infrared—X-ray	~ 100	~ 0.3
Quantum	Microwave kinetic inductance detector	MKID	0.3 @ 3 62@ 5900	2–10	Infrared—X-ray	~ 100	~ 0.3
Thermal	Transition edge sensor detector	TESD	0.07@ 0.8 1.5 @ 500 1.58 @ 5900	0.2 100 –1000	Infrared—γ-ray	~ 100	~ 0.05
Thermal	Metallic magnetic calorimetric detector	MMCD	1.25 @ 5900	~ 1000	Infrared—γ-ray	~ 100	~ 0.05

(continued)

Table 5.1 (continued)

Detection category[a]	Detector name[b]	Acronym[b]	ΔE (eV)[c]	Fall time τ_{fall} (μs)	Coverage	Spatial resolution[d] (μm)	T (K)
Thermal[e]	Superconductor strip particle detector	SSPD	N/A[f] (0.55@0.8–1.2)	~ 0.01	Infrared—X-ray	5[g] (< 1)	~ 4 (13–25)[h]

The detection category column indicates the basic signal generation principles: quantum detection and thermal detection. The performance figures for photons include energy resolution ΔE at a photon energy after "@," typical fall time τ_{fall} of output pulses, energy covering range, typical spatial resolution, and typical operating temperature T

[a] We follow the widely accepted common definitions in radiation detection; there are basically two primary detection modes of quantum detection and thermal detection[6]. The former directly produces an electric signal due to electronic excitation. The latter measures a temperature rise. However, superconductor thermal detectors also employ quantum phenomena of superconductivity; for example, superconductor thermometers and SQUID readout circuits. Therefore, in this perspective, the superconductor thermal detection can be also classified into quantum detection

[b] For consistency, all detector names end with "detector" according to the IEC standard for detector nomenclature [7]. In the text, the acronyms such as STJ, TES, and MMC are also used in the forms such as STJ pixel, TES pulse, and MMC type. The term "superconducting" appears in the current IEC standard documents. The correction of "superconducting" to "superconductor" in the standard documents is under discussion; opinions from the community are welcome. The reason is mentioned in Appendix C.3.

[c] These are the best values as of 2024, referring to the references in this article. For comparison, the latest Si drift detector (SDD) with a sensitive are of 10 mm^2 has ΔE values of 38 eV for B K-line (183.3 eV), 42 eV for C K-line (277 eV), 48 eV for O K-line (524.9 eV), and 124 eV for Mn K-line (5.9 keV) [8, 9]. SDDs with a larger sensitive areas of 30–80 mm^2 are also available on the market

[d] These are typical pixel sizes. The sizes of 10 μm or smaller are possible. The imaging detector type with a superconductor long strip, whose each transverse edge is attached with an STJD or an MKID, has a better spatial resolution[10]

[e] Joule self-heating with ETF governs output signal generation, although initial interaction with a particle and a superconductor has the quantum nature. Therefore, the SSPD is classified as the thermal detection type. As described in the preceding Chap. 4, the SSPD operation including a readout circuit resembles those of the first SPT bolometer and TESDs with a shunt resistor for ETF

[f] There is a report that an SSPD has an energy resolution in parentheses. A mechanism is the transition between the hot spot formation and vortex motion[11]. However, unlike other types, SSPDs are impractical for energy-dispersive spectroscopy (EDS)

[g] A 2D-thermally-coupled SSPD imager has 400,000 pixels with an area of 5×5 μm^2 [12]. The sub-μm resolution in parentheses can be achieved by a delay-line format [13]

[h] The keV-molecule detection with a MgB_2-SSPD was achieved up to 13 K. The single photon detection near 1 eV with BSCCO-SSPDs was achieved up to 20–25 K [14, 15]

Nonzero-mass particles require an open space in front of STJDs; therefore, we cannot avoid the illumination of black body radiation from a sample stage or an analyte. Nonetheless, the superconducting energy gap ensures the proper operation of STJDs even when temperature fluctuates. Therefore, STJDs are advantageous to MS compared with thermal detectors. A filter that allows massive particles to pass through and reflects IR radiation prevents a temperature increase beyond an operating temperature [20]. Ions or molecules cannot deeply penetrate into superconductors; they even completely stop on junction surfaces. Nevertheless, a particle-impact-induced phonon bubble enables kinetic energy measurement of massive particles [21]. STJDs cover an energy range up to ~ 50 keV at a unity quantum efficiency for nonzero-mass particles, although an energy nonlinearity is not negligible above ~ 20 keV [5].

The MKI quantum type is probably the most suitable for imaging by covering a large sensitive area because thousands of pixels can be read out by a couple of coaxial cables. The MKI type demonstrates superior performance in a low photon-energy range below ~ 10 eV. Although X-ray detection is possible, the energy resolution performance is not suitable for analytical science compared with other superconductor X-ray detectors. This is understandable from the experimental facts that an STJD exhibits a significant ΔE degradation in an energy range above ~ 300 eV in Fig. 3.20 and an abnormally large spatial inhomogeneity for 6-keV photons in Fig. 3.23. This suggests that the number of excited quasiparticles is beyond a range of linear detector-operation. Consequently, one of the suitable applications of the MKI type is imaging for infrared-visible photons in the fields of astronomy and spectroscopic analysis. An MKID photon-counting camera for a range of 800–1400 nm is one of the visiting instruments for the Subaru telescope [22]. Currently, however, there is no report in analytical science. Nevertheless, the MKI type as well as the STJ type shows promise for future analytical instruments covering infrared-visible photons.

The TES and MMC thermal types hold the best energy resolution among the superconductor detectors: 1.25 eV with an MMCD and 1.58 eV with a TESD for 5.9-keV X-rays. They also function efficiently in a wide range from infrared photons to γ-rays by designing a proper absorber. The ΔE values are far smaller than the characteristic X-ray linewidths below ~ 5 keV; therefore, they even enable EDS-XES for chemical state analysis. However, they have a fall time of ~ 1 ms at the best energy resolution for X-rays; the photon count rate is limited to ~ 100 cps per pixel. Currently, the TES and MMC types require a long measurement time, which is often comparable with that of WDS that is configured for a similar energy resolution for material analysis. Nevertheless, development of fast thermal detectors is promising. Florescence-yield X-ray absorption spectroscopy in synchrotron radiation facilities and spectroscopy covering infrared-visible photons are the active application areas of the thermal detectors.

In exchange for the slow response in thermal detection, the long response-time allows multiplex readout for a large array format. Several space mission programs are underway, building TES X-ray imagers with pixels of ~ 2400 to ~ 1 million for X-ray satellites [23]. In addition to single particle detection, bolometric application is also underway for cosmic microwave background (CMB) [24]. A similar bolometric application to analytical science is promising, although there are no noticeable reports.

The SS thermal type achieves the fastest response time in a sub-nanosecond range. However, nano-structure inherently limits the sensitive area. In addition, it is a counting device type, not possessing energy resolution. Nevertheless, there are many applications that requires a fast-counting detector. The SS type with a delay-line format can have a sub-μm spatial resolution for X-ray photons, outperforming a spatial resolution of ~ 2 μm in the latest X-ray CCD cameras [13]. SSPD 2D-imagers are also pursued with a delay line scheme [25], a thermally coupled two-layer stack format up to 400,000 pixels [12, 26], row-column multiplex readout of 1024 SS-pixels [27], and 1024 SS pixels with an orthogonal time–amplitude multiplexing[28]. The promising SSPD 1D or 2D imagers can outperform CCD or CMOS cameras in both X-ray range and infrared-ultraviolet photon range.

In addition to the performance figures in Table 5.1, photon-number resolving detection is important in the quantum information and quantum computing areas, apart from analytical science. The measurement of the number of photons arriving at the same time in a telecommunication wavelength range is possible with the STJ and TES types by measuring absorbed energy; however, they are extremely slower than the SS type. We need to compare the superconductor detectors with semiconductor detectors for photon-number measurement. Photon-number resolving detection is also possible with semiconductor CMOS cameras. For example, a CMOS camera with 9,400,000 pixels has a sensitive area of 18.814×10.598 mm^2 [29]. The photon number sensitivity up to ~ 200 at – 35 °C is achieved by an ultra-low noise readout of 0.27 electrons and a low dark current of 0.006 electrons/pixel/s at quantum efficiencies above 30% between 300–900 nm. The 9.4-M pixel imaging camera outperforms small superconductor detectors at a cryogenic temperature. Nevertheless, SSPDs can outperform the semiconductor detectors in an infrared-ultraviolet range, where semiconductor detectors cannot cover.

5.3 System Development for Analytical Science

We learned in the preceding Chapters that detector physics can adequately explain the operation of several types of superconductor detectors, although there are still many unsolved/open questions on fundamental physics. Importantly, the experimental results ensure that the superconductor detectors outperform conventional semiconductor detectors with respect to some performance figures listed in Table 5.1. It is expected that the superconductor detectors revolutionize analytical science by overcoming the fundamental limits of conventional instruments. However, applications of the superconductor detectors to analytical science require further efforts on system development in addition to the detector development. The system development includes optics, cryogenics, and signal processing electronics. Practical use of the superconductor detectors is totally different from proof-of-concept research, requiring engineering endeavor, as described below.

Firstly, superconductor detectors are fabricated by using photolithographic techniques. It is one of the important research fields in the LTD community. We leave it to the proceedings of LTD conferences and references cited therein [30]. It is beyond the volume of this book.

Secondly, operation of the superconductor detectors requires cryogenic temperatures obtained by cryostats: ~ 3 K with a Gifford–McMahon (GM) or pulse tube type, ~ 0.3 K with a ^{3}He sorption type, ~ 0.05 K with an adiabatic demagnetization refrigerator (ADR) type, ~ 0.01 K with a ^{3}He-^{4}He dilution type. These refrigerators are available on the market; researchers and engineers install superconductor detectors, wiring leads from room temperature to a cold stage, and readout electronics. Recently, the supply of liquid cryogens such as nitrogen and helium became unnecessary; the users or analytical scientists without the knowledge of cryogenics appreciate the high performance of the superconductor detectors with automated cryostats without troubled by cryogen handling.

Liquid-cryogen-free cryostats accelerate practical use of superconductor detectors in analytical science beyond proof-of-concept experiments. However, cryogenic systems occasionally cost as much as analytical instruments themselves. In addition, the size of cryostats is often comparable with those of analytical instruments. These are bottlenecks for wide dissemination of the superconductor detectors in analytical science. Therefore, strong demands from analytical scientists are indispensable for installing superconductor detectors to analytical instruments. Chapter 6 describes some examples of such demands.

The heat capacity of superconductor detectors is extremely smaller than base plates of conventional cryostats; we do not require a high cooling power for bulky metal parts. A possible cooling route is to use palm-sized coolers in future [31]. If a cooling power meets a heat load of wiring leads, a miniaturized cooler can be utilized like a Peltier cooler for semiconductor detectors. A base-temperature range of micromachined coolers is currently 20–150 K in general. It is impossible to reach a temperature below 1 K at this moment; nevertheless, we are hopeful about the advances of the palm-sized coolers. For a sub-kelvin range, cooling is promissing with normal-metal/insulator/superconductor (NIS) junctions, which operate on transferring thermally excited electrons in the normal metal into the superconductor, so that the heat in the normal metal is removed due to bias voltage [32, 33]. For practical sub-kelvin coolers, a maskless direct laser writing lithography was developed to fabricate superconductor junctions on surfaces with a complex shape instead of perfectly flat wafers [34]. A combination of the palm-size micromachined cooler and the tunnel junction cooler has the potential of sub-kelvin refrigeration at a size of Peltier coolers in future.

With respect to temperature stability, superconductor quantum detectors stably operate when the temperature is kept below ~ $0.1T_c$. No peak shift or performance degradation occurs even when a temperature drift or fluctuation occurs. Contrarily, thermal detectors need a high temperature stability within ~ μK. Such stability can be achieved; nevertheless, a continuous calibration technique using a ^{55}Fe X-ray source is often crucial to compensate peak drift.

Thirdly, signal processing electronics is required for extracting necessary data from detector output. Real time signal processing without off-line processing is often mandatory in analytical science such that the condition of probe beams or optics can be optimized. An example of system development for the STJ and SS types appears in Ref. [5]. The TES and MMC types employ SQUID readout and multiplexing on a time, frequency, or code domain. The MKI type often uses frequency multiplexing. Those multiplexing techniques are effective to readout large-scale array detectors. The proceedings of LTD conferences and references therein should be referred to for the multiplex readout schemes [30].

Other important factors are long-term stability, reliability without malfunction, easy calibration, and easy operability. We need to consider all figures and factors described in this Chapter to select a suitable superconductor detector type to build a complete spectroscopic system. Chapter 6 describes some successful examples. As a practice for selecting a proper superconductor detector type, a possible application to atom probe tomography (APT) is discussed in Chap. 6, Sect. 6.4.5.

References

1. C.W. Fink et al., Appl. Phys. Lett. **118**, 022601 (2021)
2. Y.H. Kim, S.-J. Lee, B. Yang, Supercond. Sci. Technol. **35**, 063001 (2022)
3. K. Birkinshaw, Int. J. Mass Spectrom. **215**, vii (2002)
4. D.W. Koppenaal, C.J. Barinaga, M.B. Denton, R.P. Sperline, G.M. Hieftje, G.D. Schilling, F.J. Andrade, J.H. Barner IV., Anal. Chem. **77**, 418A (2005)
5. M. Ohkubo, J. Appl. Phys. **134**, 081101 (2023)
6. M. Kadin, M.W. Johnson, Appl. Phys. Lett. **69**, 3938 (1996)
7. IEC 61788-22-1:2017, Superconductivity—Part 22-1: Superconducting electronic devices—Generic specification for sensors and detectors: https://webstore.iec.ch/publication/26674
8. D.M. Schlosser et al., Nucl. Instrum. Methods Phys. Res. A **624**, 270 (2010)
9. L. Strüder, A. Niculae, P. Holl, H. Soltau, Microscopy Today **28**, 46 (2020)
10. H. Kraus, F. Feilitzsch, J. Jochum, R.L. Mössbauer, Th. Peterreins, F. Pröbst, Phys. Lett. B **231**, 195 (1989)
11. A.D. Semenov, P. Haas, B. Günther, H-W. Hübers, K. Il'in, M. Siegel, A. Kirste, J. Beyer, D. Drung, T. Schurig, A. Smirnov, Supercond. Sci. Technol. **20**, 919 (2007)
12. B.G. Oripov, D.S. Rampini, J. Allmaras, M.D. Shaw, S.W. Nam, B. Korzh, A.N. McCaughan, Nature **622**, 730 (2023)
13. C. Watanabe, M. Ukibe, N. Zen, G. Fujii, K. Makise, M. Ohkubo, Te-Hui Lee, and Di-Jing Huang. IEEE Trans. Appl. Super. **29**, 2400104 (2019)
14. I. Charaev et al., Nat. Nano. **18**, 343 (2023)
15. R.L. Merino, P. Seifert, J.D. Retamal, R.K. Mech, T. Taniguchi, K. Watanabe, K. Kadowaki, R.H. Hadfield, D.K. Efetov, 2D Mater. **10**, 021001 (2023)
16. M. Kurakado, K. Taniguchi, J. Instrum. **11**, C12071 (2016)
17. M. Kurakado, X-ray Spectrom. **29**, 137 (2000)
18. T.R. Stevenson, F.A. Pellerano, C.M. Stahle, K. Aidala, R.J. Schoelkopf, Appl. Phys. Lett. **80**, 3012 (2002)
19. D.E. Prober, J.D. Teufel, L. Frunzio, C.M. Wilson, R.J. Schoelkopf, Nucl. Instrum. Methods Phys. Res. A **559**, 676 (2006)

20. Shigetomo, Shiki Masahiro, Ukibe Ryutaro, Maeda Masataka, Ohkubo Yuki, Sato Shigeo, Tomita (2008) Energy resolution improvement of superconducting tunnel junction particle detectors with infrared-blocking filters Nuclear Instruments and Methods in Physics Research Section A: Accelerators Spectrometers Detectors and Associated Equipment **595**(2) 391–394. https://doi.org/10.1016/j.nima.2008.07.078
21. Masataka, Ohkubo Shigetomo, Shiki Masahiro, Ukibe Shigeo, Tomita Shigeo, Hayakawa (2011) Direct mass analysis of neutral molecules by superconductivity International Journal of Mass Spectrometry 299(2–3) 94–101. https://doi.org/10.1016/j.ijms.2010.09.027
22. A.B. Walter at al., PASP 132, 125005 (2020)
23. L. Gottardi, S. Smith, S., Transition-edge sensors for cryogenic X-ray imaging spectrometers, *Handbook of X-ray and Gamma-ray Astrophysics* ed. byC. Bambi, A. Santangelo (Springer, Singapore, 2023). https://doi.org/10.1007/978-981-16-4544-0_22-1
24. T. Matsumura et al., J. Low Temp. Phys. **176**, 733 (2014)
25. Q.-Y. Zhao, D. Zhu, N. Calandri, A.E. Dane, A.N. McCaughan, F. Bellei, H.-Z. Wang, D.F. Santavicca, K.K. Berggren, Nat. Photo. **11**, 247 (2017)
26. A.N. McCaughan, Y. Zhai, B. Korzh, J.P. Allmaras, B.G. Oripov, M.D. Shaw, S.W. Nam, Appl. Phys. Lett. **121**, 102602 (2022)
27. E.E. Wollman, V.B. Verma, A.E. Lita, W.H. Farr, M.D. Shaw, R.P. Mirin, S.W. Nam, Opt. Express **27**, 35279 (2019)
28. L.-D. Kong, H. Wang, Q.-Y. Zhao, J.-W. Guo, Y.-H. Huang, H. Hao, S. Chen, X.-C. Tu, L.-B. Zhang, X.-Q. Jia, L. Kang, J. Chen, P.-H. Wu, Nat. Photo. **17**, 65 (2023)
29. Hamamatsu ORCA-Quest CMOS Camera. https://www.hamamatsu.com/us/en/product/cameras/qcmos-cameras/C15550-22UP.html
30. Y-H. Kim, W. Cui, H-S. Jo, L. You (eds.), Special Issue: Low Temperature Detectors LTD20. J. Low Temp. Phys. **214**(3–4), **215**(3–4), **216**(1–2) (2024), and the previous proceedings series
31. H.S. Cao, H.J.M. ter Brake, Phys. Rev. Appl. **14**, 044044 (2020)
32. M. Nahum, T.M. Eiles, J.M. Martinis, Appl. Phys. Lett. **65**, 3123 (1994)
33. X. Zhang, P.J. Lowell, B.L. Wilson, G.C. O'Neil, J.N. Ullom, Phys. Rev. Appl. **4**, 024006 (2015)
34. S. Heiskanen, I. Maasilta, Appl. Phys. Lett. **117**, 232601 (2020)

Chapter 6
Analytical Applications

Abstract Chapter 6 describes examples of practical analytical applications of the superconductor detectors. The superconductor detectors allow surpassing fundamental limits of conventional analytical instruments. The analytical applications include X-ray spectroscopy and mass spectrometry. The novel analytical results contribute to a wide range of fields, such as dentistry, molecular biology, energy-saving society, planetary science, and prebiotic organic molecules in space. Potential applications to other fields such as atom probe tomography are also described.

Keywords Analytical science · X-ray spectroscopy · Mass spectrometry · SEM · TEM · Synchrotron radiation

6.1 General

In this Chapter, we describe novel results of analytical science realized by the superconductor detectors, mainly in X-ray spectroscopy and mass spectrometry. Note that the different communities use different terms: spectroscopy versus spectrometry. The objects to be measured are photons, ions, atoms, and molecules. By replacing a conventional detector, such as a semiconductor detector, with a superconductor detector, we can acquire novel data that the conventional detector cannot. This research direction rather belongs to the first group "(1) those whose primary interest is the detectors and the measurement" described in Chap. 5 [1]. Afterward, the superconductor detectors enable obtaining new insights into science and technology. This is the second group "(2) those whose primary interest is the science enabled by the detectors" [1]. The previous Chapters focus on the group (1), while this Chapter covers the movement from the group (1) to the group (2).

Applications of the superconductor detectors to analytical science were reviewed for MS in 1999 and for synchrotron radiation in 2006 [2, 3]. Material analysis with all the superconductor detector types was briefly reviewed in 2008 [4]. The applications of TESDs to X-ray or γ-ray spectroscopy also brought numerous excellent results [5–7]. We treat analytes in solid, liquid, and gas phases. Detailed discussions of

M. Ohkubo, *Superconductor Detectors*, https://doi.org/10.1007/978-3-031-84193-4_6

the intriguing results on fundamental physics are left to the literature, except listing the most recent topics: neutrino mass determination with STJDs using ^{7}Be electron capture [8, 9] and quantum electrodynamics test in highly-charged ions and muonic atoms with TESDs [10–12]. In addition, the LTD proceedings collect recent progress in the applications to X-ray astronomy, CMB, atomic physics, nuclear physics, and rare event search [13].

6.2 Introduction to Analytical Instruments

6.2.1 Combination of Probe and Spectrometer

In analytical science, analytical instruments generate measurement data to be analyzed. Analytical instrumentation consists of three components: a probe beam to evoke responses of samples, a spectrometer to separate the responses such as characteristic X-ray photons, photoelectrons, Auger electrons, ions, neutral particles as a function of energy, mass, or other physical parameters, and a detector to convert the responses such as particles or waves to electric signals. As described in Chap. 2, we focus on single particles. The combination with probe beam techniques is crucial in analytical instruments; the research and development on probes organize another community in analytical science. A particle count or wave intensity as a function of a parameter is called spectrum. Although the importance of detectors is often underestimated, as mentioned in Chap. 5, Sect. 5.1, we emphasize that detectors certainly determine the quality of spectra and affect the overall performance of analytical instruments. In contrast to the rare event search [14], it is generally preferable that the detectors for analytical instruments operate at an extremely high count-rate to realize high throughput analysis or quick imaging. Meanwhile, there are cases where an extremely high energy resolution is the highest priority at a low count rate. In this section, we examine a case of X-ray spectroscopy to explain the probe-spectrometer combination.

Figure 6.1 shows two X-ray elemental mapping instruments: electron probe micro analyzer (EPMA) and scanning electron microscope (SEM). EPMA and SEM are similar, possessing an electron gun with focusing optics and an X-ray spectrometer. The electron beam focused on a sample surface at a nano-scale ionizes atoms inside the sample. The ionized atoms emit characteristic X-rays as shown in Fig. 6.2, which will be explained later. An energetic electron creates a vacancy or a hole on one of the inner electron shells of an atom. The vacancy on a shell is filled with an electron by transition from an outer shell, emitting a characteristic X-ray photon at a certain probability. The energy of the X-ray photon corresponds to a difference in the binding energies of two shells. Major characteristic X-ray lines are called K, L, and M, which represent electron transitions from outer shells to the K, L, and M shells, respectively. One can find the detail of the X-ray emission lines and notation in Appendix B. There are useful websites such as NIST X-Ray Transition Energies Database and CXRO

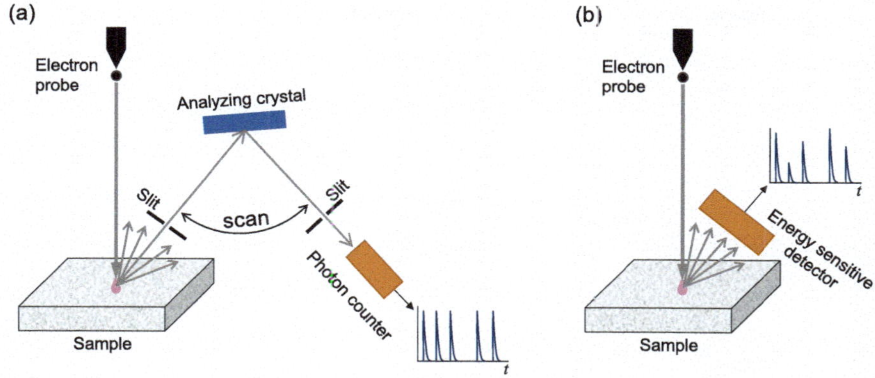

Fig. 6.1 Typical analytical instrument for X-ray spectroscopy: (**a**) electron probe micro analyzer (EPMA) and (**b**) scanning electron microscope (SEM). Analytical instrumentation consisting of a probe to excite a sample, a spectrometer to analyze electromagnetic waves or particles emitted from the sample. EPMA in (**a**) utilizes wavelength dispersive spectroscopy (WDS); SEM in (**b**) utilizes energy dispersive spectroscopy (EDS). The X-ray detector in (**a**) generates a constant pulse height, similar to GM counters and SSPDs, whereas one in (**b**) generates pulses, whose heights represent X-ray photon energies. Note that modern WDS spectrometers utilize a proportional counter or a scintillation counter, which has a certain level of energy sensitivity, to eliminate higher-order diffraction and background counts. Advanced WDS spectrometers even utilize an SDD, which has a superior energy resolution and a better photon counting performance. Superconductor EDS detectors surpass the advanced WDS spectrometers

X-ray data booklet [15, 16]. An X-ray spectrometer analyzes the characteristic X-rays. We can obtain elemental mapping or distribution images of atoms on sample surfaces by scanning the electron beam while analyzing characteristic X-rays.

In Fig. 6.1a, EPMA is equipped with an electron probe that excites characteristic X-ray lines, an analyzing crystal or grating, and an X-ray counter. This is called wave dispersive spectroscopy (WDS). Figure 6.1a has one set of the analyzing crystal and the X-ray counter for the sake of simplicity. For elemental mapping, one crystal-detector set usually detects one element, although a position sensitive X-ray counter allows covering a narrow energy range. A wide energy range can be covered by scanning the angle of the analyzing crystal or grating. A small solid angle is necessary for a high energy resolution, resulting in a low detection efficiency. EPMA is normally equipped with multiple WDS spectrometers for different elements. In Fig. 6.1b, the WDS spectrometer is replaced by an energy-sensitive detector, which is called energy dispersive spectroscopy (EDS). EDS requires no mechanical or electric scanning; it simultaneously covers a wide energy range or multiple elements at a large solid angle. One EDS detector can cover most elements from Be to U. However, conventional semiconductor EDS detectors normally have a ~ 10 times worse energy resolution than WDS spectrometers; some peak overlaps occur on K lines of light elements and L lines of heavy elements. Therefore, a better energy resolution beyond semiconductor detectors such as SDDs is a request from analytical science.

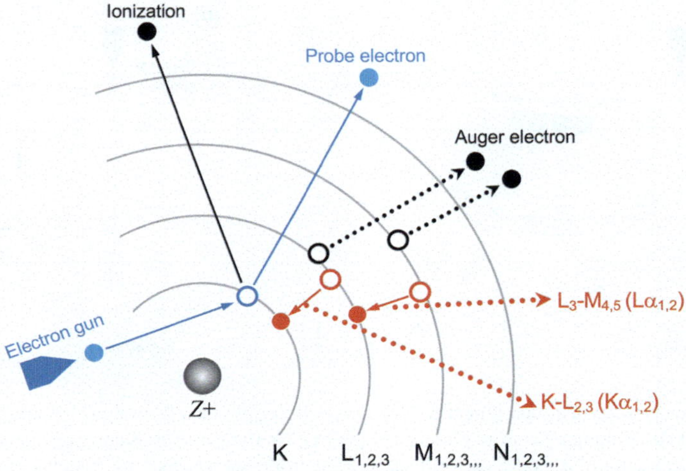

Fig. 6.2 Electronic excitation and characteristic X-ray emission. This diagram shows the generation of a vacancy or a hole (blue open circle) in the K shell and the succeeding transition of an atom. The blue closed circles denote the incident probe electron and the scattered probe electron. The black closed circle labeled "Ionization" denotes the electron emitted from the K shell to the vacuum level. The two black closed circles labeled "Auger electron" denote the Auger electrons; the two black open circles are the corresponding vacancies. A pair of the red closed and open circles represents the electron transition from the L shell to the K shell, emitting the K-$L_{2,3}$ X-ray. Another pair of the red closed and open circles represents the electron transition from the M shell to the L shell, emitting the L_3-$M_{4,5}$ X-ray. The gray sphere represents the nucleus with the atomic number Z. Two major relaxation processes occur: radiative X-ray emission (red doted arrows) and nonradiative Auger electron emission (black dotted arrows). The notation for the characteristic X-ray lines is based on the IUPAC recommendation with the Siegbahn notation in parentheses for reference. We assign the term "K-line analysis" to the hole in the K shell and the term "L-line analysis" to the hole in the L shell, respectively. Reproduced with permission from Ohkubo [17]. Copyright 2023 AIP Publishing LLC

As listed in Table 5.1, the superconductor EDS detectors have a high energy resolution comparable to WDS spectrometers for material analysis.

The electron beam creates a collision cascade cloud with a diameter of 0.3–5 μm at an acceleration voltage of 10 kV, depending on materials to be analyzed. Characteristic X-rays are emitted from the cloud region, which imposes the requirement of an electron beam energy that should be at least double the absorption edge energy of the target atom to obtain sufficient X-ray intensity. This limits the spatial resolution of elemental mapping images to ~ 1μm. The electron beam can be replaced with ion beam or synchrotron radiation (SR) beam.

Figure 6.2 shows the radiative and nonradiative processes after electron beam excitation. The electron impact on an atom inside a sample creates a vacancy or hole in one of the electron shells such as K, L_1, L_2, L_3, M_1, M_2, M_3, etc. The ionization cross-section σ values are reported in Ref. [18]. For elemental analysis, vacancies are often created in either K shell or L_3 shell. The K shell vacancy is frequently filled with an electron in one of the L shells, emitting a characteristic X-ray. The major characteristic lines are generated by transitions from the L_2 and L_3 shells to the K shell, which are labeled K-L_2 ($K\alpha_2$) and K-L_3 ($K\alpha_1$) lines, respectively, based on the IUPAC notation followed by the conventional Siegbahn notation in parentheses [19]. The IUPAC notation indicating the lower and upper shells for transition is more systematic compared with the Siegbahn notation. We follow the IUPAC notation. When the two lines are indistinctive in EDS, we use the notation K-$L_{2,3}$ ($K\alpha_{1,2}$) collectively. For L-shell ionization, the L_2 and L_3 shells have a considerably higherσ value than that of the L_1 shell. The major lines correspond to the transitions from M_4 and M_5 to L_3, which are labeled L_3-$M_{4,5}$ ($L\alpha_{1,2}$). The above-mentioned radiative X-ray emission processes are competitive with nonradiative Auger electron emission processes from L, M, or N. The radiative X-ray emission yield ω values are found in Refs. [20, 21]. The total efficiency of the X-ray emission is determined by multiplying σ and ω. We designate X-ray analysis with a hole in the K shell as the K-line analysis, and one with a hole in the L shell as the L-line analysis.

The two competitive processes are illustrated in Fig. 6.2. The K-shell vacancy induces either radiative X-ray emission of K-L_2 ($K\alpha_2$) or competitive nonradiative Auger electron emission form the L_3 shell denoted as KLL. Another transition is the K-L_3 ($K\alpha_1$) emission or the Auger electron from the M shell denoted as KLM. A transition from the M shell to the K shell is also possible, such as K-$M_{2,3}$ ($K\beta_{1,3}$) that is not depicted in Fig. 6.2. For the vacancy on the L_3 shell, possible transitions include L_3-M_4 ($L\alpha_2$), L_3-M_5 ($L\alpha_1$), or Auger electron emission from the M_5 shell or a higher shell. Other characteristic X-rays or Auger electrons due to transitions between any two shells are also possible. Cascade processes of X-ray and Auger emission continue as long as upper shells exist. They are called X-ray spectroscopy and Auger electron spectroscopy (AES), respectively. Additionally, photoelectrons emitted from electron shells are employed in X-ray photoelectron spectroscopy (XPS).

The difference between EPMA and SEM arises from X-ray spectrometers [22]. The WDS spectrometer for EPMA combines an analyzing crystal or grating with an X-ray photon counter as shown in Fig. 6.1a. GM counters were employed for WDS in the early years. The detector output pulse height is constant independent of X-ray photon energy. The GM counters operate up to ~ 10^5 count/s. Modern WDS spectrometers utilize a proportional counter or a scintillation counter. The photon-energy sensitivity of such counters enables elimination of higher-order diffraction and background counts, but the energy resolution is insufficient for acquiring X-ray spectra for material analysis. Precise acquisition of X-ray spectra is performed by scanning the analyzing crystal mechanically while the detector counts the number of X-ray photons. WDS employs the Bragg diffraction, based on the relations of

$$E_0 = \frac{hc}{\lambda} \text{ and} \tag{6.1}$$

$$2d \sin\theta = n\lambda, \tag{6.2}$$

where E_0 is the X-ray photon energy, h is the Planck constant, c is the light speed, λ is the X-ray wavelength, d is the crystal lattice spacing, θ is the angle of X-ray incidence, and n is an integer. X-ray detectors in WDS have no ability to precisely measure photon energy. The WDS spectrometer in Fig. 6.1a must have a collimator or slit to fix the diffraction angle, leading to a small solid angle. WDS spectrometers for EPMA have a typical energy resolution of ~ 10 eV for soft X-rays to obtain both a reasonable solid angle and a practical measurement time. At 10 eV, EPMA can separate most characteristic X-rays including K, L, M lines in a range of 100 eV and 10 keV. Note that a huge 10-m long spectrometer can achieve a $E_0/\Delta E$ value of ~ 20,000 compared with ~ 1000 in EPMA, although it requires a long measurement time even at synchrotron radiation beam lines. EPMA requires a high electron-beam current to generate a sufficient amount of characteristic X-rays to compensate a small solid angle, which can cause sample damage. Another disadvantage of EPMA is that the upper limit of the number of WDS spectrometers is normally five because of a geometrical limit. For elemental mapping, one WDS spectrometer detects one element, which restricts the number of simultaneously measurable elements to five.

SEM has one EDS detector as shown in Fig. 6.1b; it is normally a semiconductor detector. X-ray spectra of an entire energy range are acquired by the EDS detector without mechanical or electrical scanning. The detector output pulse height corresponds to the energy of each X-ray photon as described in the preceding chapters. In other word, EDS detectors possess energy resolution. An EDS detector can be located close to a sample surface such that it detects a large portion of X-ryas emitted at 4π steradian. EDS can have a significantly larger solid angle than WDS, although the efficiency of WDS is constantly improved [23]. X-ray analysis with SEM tolerates a low electron-beam current compared with EPMA; it is effective to avoid radiation damage especially in biological samples and polymers. Furthermore, the EDS detector enables simultaneous analysis of multiple elements more than five. The latest semiconductor EDS detectors such as SDDs have an energy resolution of 42 eV @ 277 eV and 124 eV @ 5.9 keV [24], which are often insufficient for separating K, L, and M lines from multiple elements in a sample. Light elements embedded in a matrix typically have characteristic line natural widths of 10–20 eV, broadened from those of isolated atoms by chemical bonds forming electronic band structure in addition to the natural broadenings of the electron shell levels. If an EDS detector on SEM has an energy resolution of ~ 10 eV, it can perform elemental mapping without peak overlap.

6.2.2 Early Applications of Superconductor Detectors

Superconductor detectors realize the combination of an EPMA-WDS-level high energy resolution beyond natural linewidths of elements and an SEM–EDS-level high detection efficiency. This is one of our motivations to employ superconductivity for advanced analytical instruments. In 1998, a single TES pixel with an Al/Ag bilayer thermometer and a 250-μm square Ag absorber, developed at NIST, recorded an energy resolution of 3–4 eV in an X-ray range below 2 keV [25]. Figure 6.3 shows spectral comparison between EDS with the TESD and WDS with a thallium acid phthalate (TAP) analyzing crystal. Their energy resolutions are similar; the TESD-EDS spectra are almost the same as those with WDS. For the Al/Ag-TESD, a poly-capillary X-ray lens increased detection efficiency. At present, the 240-pixel Mo/Cu-TESD spectrometer optimized for an energy range below ~ 2 keV has a ΔE value of 1.1–1.5 eV at 500 eV after integrating 240 pixels with a total sensitive area of 2.1 mm^2 and a count rate capability up to ~ 2 kcps [6, 7]. Above ~ 50 kcps, the 240-pixel TESD exhibits a rapid ΔE deterioration and a considerable count loss [7]. The sensitive area of 2.1-mm^2 is approaching 10 mm^2 of a high-resolution SDD. The STJ type also recorded an energy resolution comparable with WDS for EPMA: a 141-μm square Nb/Al STJD recorded 5.9 eV at 277 eV and 8.8 eV at 554 eV in 1998 [26]. Above ~ 50 kcps, one STJ pixel exhibited a rapid ΔE degradation [26]. The 100-pixel STJD has an integrated ΔE value of 6 eV at 400 eV as shown in Fig. 3.16 [17]. This energy resolution is worse than that of the TESD, but the STJ array detector spectrometer can receive X-ray photons up to ~ 200 kcps, as shown in Fig. 3.33. It is considerably lower than 5 Mcps expected from 50 kcps × 100 pixels, probably limited by insufficient bias supply at a high count rate. The STJD-EDS with a count rate over 1 Mcps, which is equivalent to that of SDDs, is a future challenge.

6.3 X-Ray Spectroscopy

6.3.1 Scanning Electron Microscopy (SEM)

6.3.1.1 Electron-Induced X-Ray Emission

SEM is one of the versatile imaging apparatuses by scanning a focused electron beam [27]. SEM-EDS analysis frequently utilzes K-lines from atoms inside samples because K-$L_{2,3}$ ($K\alpha_{1,2}$) lines are distinct compared with other many higher order lines. The ratio of radiative X-ray emission to nonradiative Auger electron emission rapidly increases as the atomic number Z increases except a small reduced region in $Z < 18$ [20, 21]. In other word, X-ray elemental analysis is advantageous for high Z elements. On the other hand, AES and XPS have higher efficiency for low Z elements or light elements. This is the common belief; however, we need to reconsider ionization efficiency of L shells. The electron-induced ionization cross-section depends on both electron energy and Z number [18]. For K vacancy generation, the curves of cross-section σ versus electron-energy E_0 have a sharp peak below 10 keV in $Z <$ ~ 14. The σ–E_0 curves in $Z >$ ~ 14 exhibit flatter E-dependence, still exhibiting a distinct maximum value below 50 keV in Z < ~ 32. For L vacancy generation, the σ–E_0 curves of relatively heavy elements resemble those of the K vacancy generation in light elements. Therefore, we can maximize the ionization efficiency by adjusting

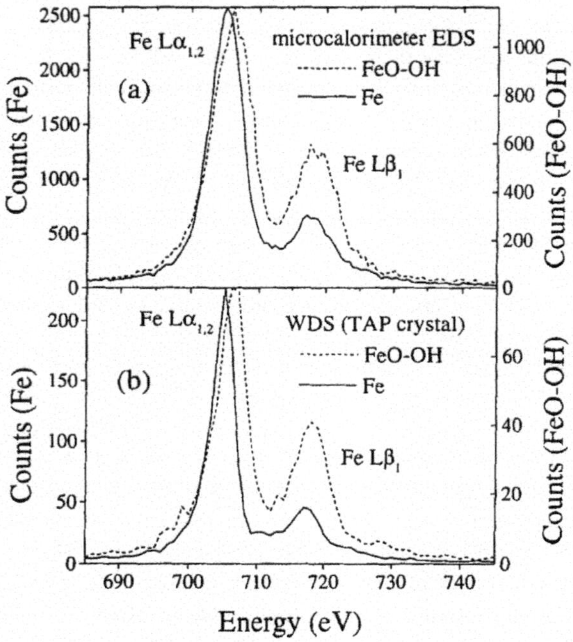

Fig. 6.3 Comparison between EDS with TESD and WDS with TAP analyzing crystal in 1998 [25]. The single TES pixel with an Al/Ag bilayer thermometer and a 250-μm square Ag absorber recorded an energy resolution of 3–4 eV in an energy range below 2 keV and a maximum count rate of 500 cps with a τ_{fall} value of 215 μs. The Al/Ag-TESD was cooled at 70 mK by an adiabatic demagnetization refrigerator (ADR). The measured samples were iron oxyhydroxide (FeO-OH) and pure iron (Fe). The Siegbahn notation corresponds to the following IUPAC notation: Fe $L\alpha_{1,2}$—Fe L_3-$M_{4,5}$ and Fe $L\beta_1$—Fe L_2-M_4. Reproduced with permission from Wollman et al. [25]. Copyright 1998 AIP Publishing LLC

electron beam energy. At the maximum ionization cross-section, the total efficiency ($\sigma \times \omega$) values for the L_3 shells are higher than those for the K shells in the entire Z range, as shown in Fig. 6.4. This means that the L line analysis can have a higher sensitivity for trace elements than the K line analysis. Considering collision cascade, an electron energy approximately double the energy corresponding to the maximum σ yields a sufficient efficiency for detecting L lines.

As an example, the parameters for Zn, which has an L-edge energy range of 1.0–1.2 keV, are listed in Table 6.1. Zn is one of the important elements in dentin, as described in Sect. 6.3.1.3. The effective cross-section for the Zn L-line analysis with 5-keV electrons is 5.8 times higher than that for the Zn K-line analysis with 20-keV electrons, even though the L_3 vacancy leads to the Auger electron emission at a probability of 98.8%. It should be noted that a small change of the electron energy in 3–5 keV causes a large change of the Zn L_3 X-ray emission yield, whereas it is negligibly small for the Zn K in 20–30 keV. Although precise electron-energy control is necessary, the L-line analysis in an electron energy range of 3–10 kV is convincingly

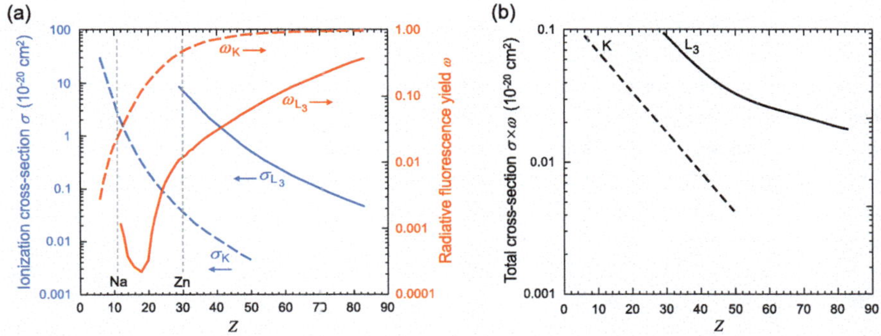

Fig. 6.4 Electron-induced ionization and X-ray emission [17]: (**a**) Z-number dependence of ionization cross section σ and relative fluorescence yield ω and (**b**) total cross section $\sigma \times \omega$ for characteristic X-ray emission in K-line and L_3-line analyses. Reproduced with permission from Ohkubo [17]. Copyright 2023 AIP Publishing LLC

more efficient than the K-line analysis in 20–40 kV. Besides considering the total efficiency $\sigma \times \omega$ at a specific electron energy, an actual analytical efficiency is influenced by multiple scattering during the energy loss process of probe electrons inside samples. We must adjust the electron beam energy to maximize the intensity of a specific L line, as the optimal energy varies depending on the Z number. It is worthwhile to pursue the L-line analysis with SEM in a characteristic X-ray energy-range below ~ 2 keV, where the STJ and TES types have ΔE values narrower than natural linewidths of heavy-element L lines in solid samples as well as light-element K lines.

Table 6.1 Comparison of K-line analysis and L-line analysis for Zn

Emission line (IUPAC/Siegbahn/energy)	Electron energy (keV)	Ionization cross-section σ (cm^2)	Radiative fluorescence yield ω	Total cross-section $\sigma \times \omega$ (cm^2)
Zn K-$L_{2,3}$/K$\alpha_{1,2}$/ 8.6 keV	30	3.7×10^{-22}	0.47	1.7×10^{-22}
Zn K-$L_{2,3}$/K$\alpha_{1,2}$/ 8.6 keV	20	3.3×10^{-22}	0.47	1.6×10^{-22}
Zn L_3-$M_{4,5}$/L$\alpha_{1,2}$/ 1.0 keV	5	7.7×10^{-20}	0.012	9.2×10^{-22}
Zn L_3-$M_{4,5}$/L$\alpha_{1,2}$/ 1.0 keV	3	8.6×10^{-20}	0.012	10.3×10^{-22}

The electron-impact ionization cross-sectionσ, radiative fluorescence yield ω, and total cross-section $\sigma \times \omega$ were tabulated at electron acceleration voltages of 3, 5, 20, and 30 kV. Zn is one of the important trace elements in dentin. There are no σ data for Zn L_3 in Ref. [18], so that the values of Cu L_3 are listed instead of Zn L_3. The radiative fluorescent yields are referred to in Ref. [20]. The effective cross-section of the Zn L-line analysis is more than 5 times larger than that of the K-line analysis

In addition to the total cross-section, the L-line analysis on SEM outperforms the K-line analysis with respect to spatial resolution for elemental mapping. The L-line analysis is performed at a lower electron energy than the K-line analysis. The cloud sizes of electron collision cascades in dentin samples are normally of orders of 0.1 μm at 5 kV for the L-line analysis and ~ 2 μm at 20 kV for the K-line analysis [28]. Therefore, when we correctly adjust electron beam energy, the L-line analysis is advantageous to both detection sensitivity of elements and spatial resolution for elemental mapping.

With SEM, both TESD and STJD can be utilized for elemental identification using L and M lines in a range below ~ 2 keV. SDD's energy resolution is insufficient to resolve K-lines of light elements and L- or M-lines of heavy elements, all of which are congested below ~ 2 keV. High energy resolution of the superconductor detectors solves this problem. With the L-line analysis, the spatial resolution is a few tens to hundreds of nm, which is far better than a few μm with the K-line analysis at ~ 10 kV [29, 30]. An energetic electron creates a collision cascade cloud with diameters of ~ 80 nm at 2 keV, ~ 300 nm at 5 kV, ~ 3 μm at 10 kV in Si [27]. X-rays are emitted from the cloud, whose size limits spatial resolution of SEM–EDS elemental mapping. Low voltage SEMs realize a sub-μm spatial resolution with the L-line analysis of relatively heavy elements as well as K-line analysis of light elements in $Z < \sim 14$. The STJ and TES type play a crucial role in elemental and chemical analysis with the L- or even M-line analysis at sub-μm spatial resolution, covering almost all the elements in a soft X-ray range below ~ 2 keV.

6.3.1.2 SEM Equipped with STJD

Figure 6.5 shows a low-voltage SEM equipped with an STJ array detector. This low-voltage SEM (JEOL JSM-7200F) has spatial resolutions of 1.6 nm at 1 kV and 1.0 nm at 20 kV for secondary electron imaging, an acceleration voltage range of 10 V–30 kV, and a current range of 1 pA–300 nA. The silver cylindrical chamber beside the SEM lens-barrel is the ^{3}He sorption cryostat. The cold snout connected to the 0.3-K cold stage holds the STJ array detector with 100-pixels, each of which has a sensitive area of 100×100 μm^2 [31]. A polycarpellary X-ray lens compensates the small 1-mm^2 sensitive area. Each STJ pixel is connected a charge-sensitive preamplifier and a field programmable gate array based digital signal processing (FPGA-DSP) circuit on the left-hand side of the ^{3}He cryostat. The cryostat has automatic functions of cooling to 0.3 K from room temperature and of recharging liquid ^{3}He; a one-shot holding time is 32 h [32]. While scanning the electron beam, the STJD and a 25-mm^2 SDD simultaneously measure X-rays emitted from a sample. A peak-height ratio of STJD/SDD for characteristic X-rays from light elements is approximately 0.17; however, 10 times sharper peaks can compensate this low value. The ratio will be increased to ~ 1.0 by increasing the number of the STJ pixels and adjusting optics. The STJD covers a range of 50 eV–2 keV at an energy resolution of 6–10 eV, while the SDD covers a range of 50 eV–100 keV. The SDD has an energy resolution of 122 eV at 5.9 keV. Both STJD and SDD simultaneously acquire X-ray spectra to construct elemental mapping images of the same field of view.

Figure 6.6 compares the STJ and SDD X-ray spectra. There are many cases that the SDD cannot separate full-energy absorption peaks of characteristic X-rays below ~ 2 keV. Biological samples include light elements as major compositions and metallic trace elements as functional agents. Figure 6.6 shows X-ray spectra of a hen eggshell and a porcine dentin. The X-ray spectra were simultaneously acquired for 10 min with both SDD and STJD, while scanning the field of views in (a) and (b) with an electron beam. The hen eggshell and dentin mainly consist of calcite ($CaCO_3$) and hydroxyapatite ($Ca_5(PO_4)_3(OH)$), respectively, reinforced by collagen fibers. The dentin is a tooth part located between enamel and dental pulp. Both biological samples have similar compositions of Ca, P, Mg, Na, O, N, and C. Mineral elemental compositions are slightly difference; they play important roles in these composite

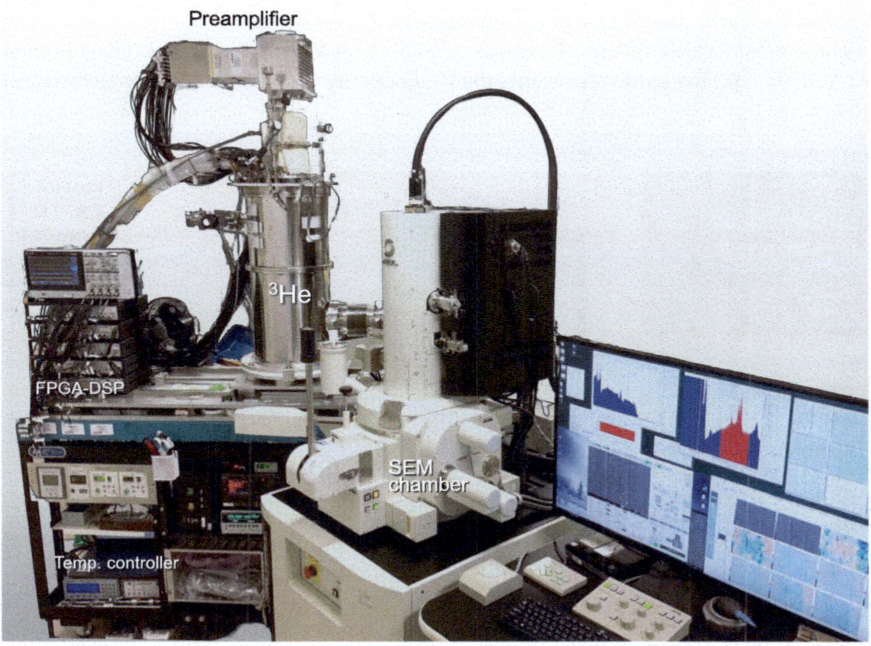

Fig. 6.5 Low-voltage scanning electron microscope (SEM) equipped with 1-mm^2 array STJD and 25-mm^2 SDD (STJD-SEM) [31]. The acceleration voltage for the electron beam ranges from 10 V to 30 kV. The best spatial resolutions are 1.6 nm for secondary electron and 3 nm for X-ray analysis in a very low energy range. The silver cylindrical metallic chamber on the left-hand side of the white SEM lens barrel is the liquid-cryogen free ^{3}He cryostat with a pulse tube cooler. The pulse tube cooler should be isolated from the SEM stage to avoid vibration propagation. The snout connected to the cold stage of the cryostat holds the STJD at the snout head. The large display at the far right is used to control SEM, SDD, and STJD, and to analyze elemental mapping images acquired with both STJD and SDD. Reproduced with permission from Ohkubo [17]. Copyright 2023 AIP Publishing LLC

structural biomaterials. The 25-mm^2 SDD spectrum of the eggshell is identical to that of the porcine dentin, which is not plotted. The broad peaks of C, N, O mask minor peaks of mineral elements, not enabling the distinction between the two samples. On the other hand, the STJD spectra reveal a small but distinguishable difference. In this soft X-ray range between 100 and 750 eV, several L lines from the trace elements appear in addition to the intense K lines of C, N, and O. Elemental mapping images are shown in Fig. 6.8 later. The STJD spectra for the eggshell and the porcine dentin show fine features due to Ca, Cr, and F, whereas the SDD spectrum exhibits only C, N, and O broad peaks with a subtle sign of F. The peaks of the trace elements in the STJD spectra are mainly assigned to L-lines. Although some peaks indicated by the question marks are unknown, the difference in the L lines enables the distinction between the eggshell and the porcine dentin. Especially, the L line peaks of the Cr L_2-$M_{2,3}$ line and Cr L-M,N complex line for the dentin are stronger than those for the eggshell. The K line analysis failed to detect the Cr K line at an acceleration voltage above 10 kV, but the presence of Cr was confirmed by MS. The spectral difference in Fig. 6.6 clearly demonstrates the advantage of STJD over SDD.

The FWHM value of the SDD peak at 525 eV, which is the O K-L (Kα) line, is 64 eV. This FWHM value represents the SDD energy resolution because the natural

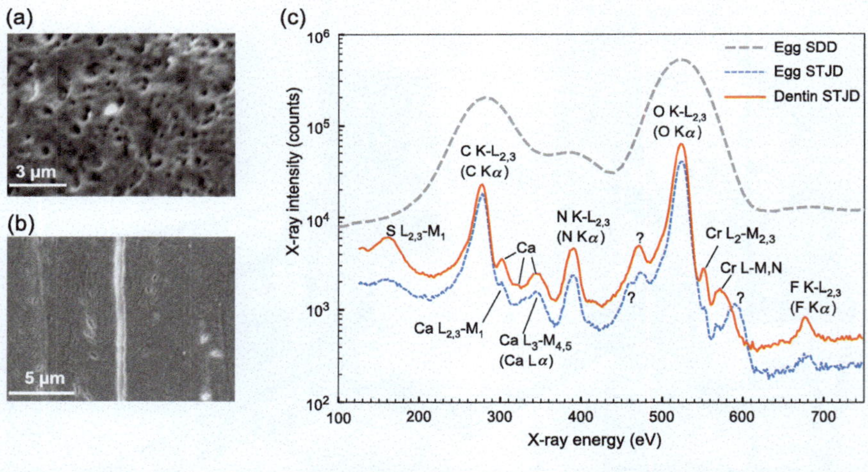

Fig. 6.6 Advantage of superconductor detectors over semiconductor detectors: (**a**) secondary electron image of outer surface of eggshell, (**b**) secondary electron image of cross-section of porcine dentin, and **c** comparison between STJD and SDD in X-ray spectra. The semiconductor detector is a 25-mm^2 SDD. The superconductor detector is a 1-mm^2 array STJD with a polycapillary X-ray lens. Pores appear on the eggshell surface in (**a**). A dentinal tubule runs vertically in (**b**). The electron beam had an energy of 5 keV and a current of 3.5 nA. The STJD and SDD spectra were simultaneously acquired for 10 min while scanning the fields of view with the electron beam repetitively. Assignment to the characteristic X-ray lines is based on the IUPAC notation recommendation (see Appendix B) with the conventional Siegbahn notation in parentheses. The eggshell and dentin have similar compositions of C, N, and O with slightly different compositions of trace heavier elements. Corresponding 2D elemental mapping images are shown in Fig. 6.8. Reproduced with permission from Ohkubo [17]. Copyright 2023 AIP Publishing LLC

linewidth is negligibly narrow. On the other hand, the STJD peak has an FWHM value of 16 eV; the natural linewidth of the O K-L line of the O atoms is evaluated at 15 eV, subtracting an STJD energy resolution of 6 eV according to the propagation law. An isolated oxygen atom should have a natural linewidth of less than 0.2 eV. The line broadening of 15 eV is typical for characteristic X-ray lines from light elements inside compounds such as $CaCO_3$ and $Ca_5(PO_4)_3(OH)$: line broadening stems from electronic band structure or outer electron shell overlapping in solids. The energy resolution of STJD is better than the natural linewidths; therefore, further improvement is unimportant for elemental mapping analysis in this energy range. However, a higher energy resolution is preferable for measuring line shapes that represent chemical states or electronic structures of the L, M, and higher electron shells. TESDs and MMCDs with an energy resolution approaching 1 eV are preferable for chemical state analysis with XES, although a measurement time is longer than that of STJDs.

6.3.1.3 SEM Elemental Mapping

SEM–EDS is used to identify elemental composition at local regions on a sample surface by acquiring X-ray spectra. Subsequently, the elemental mapping image for each element provides rich analytical information, comparing with secondary-electron morphological SEM images. Porcine dentin is a biological structural material that requires a superconductor detector and a sub-μm spatial resolution [33]. Dentin contains minerals (70%), organic substances (18%), and water (12%). The water causes a problem even after drying dentin samples. When an electron beam irradiates dentin samples, they deform due to further evaporation of remaining water. The deformation causes spatial resolution degradation. Therefore, a low electron beam current is crucial to mitigating the radiation-induced deformation.

Dentin has a mineralized tissue with fine hydroxyapatite crystals, which surround collagen fibers and non-collagenous proteins. The major elemental components of the dentin are C, N, O, P, and Ca. The minor trace elements are F, Na, Mg, Al, Si, S, Cl, Cr, Cu, and Zn. These elements were identified by the STJD-SEM shown in Fig. 6.5. In addition to radiation damage, another problem in analyzing biological samples is Na that is everywhere in tissues and other parts of animal bodies. The Na K-line masks Zn L-line on X-ray spectra when an SDD is utilized to analyze dentin samples. The trace elements play important roles in oral environment against dental caries. Fluorine is commonly known as an inhibitor for caries. In addition to F, the caries-prevention effect of Zn is reported [34–37]. It seems that Zn reduces collagen degradation and assists a repair process of demineralized dentin. In a natural environment, the trace elements are ingested from water, foods, and environmental pollution. Naturally, excessive ingestion of the trace elements causes diseases. Elemental distribution of the trace elements as well as their functions are not fully understood. The presence of Zn atoms in dentin was confirmed by whole area analysis; however, there was no easy approach to investigate which parts of dentin include Zn atoms because dentin has sub-μm structure.

Before elemental mapping, the first step is to acquire X-ray spectra to define region-of-interest (ROI) windows for elements. The STJD and SDD acquired X-ray spectra in Fig. 6.7 for 10 min, while scanning an area of $13 \times 17\ \mu m^2$ of the porcine dentin cross-section with electrons at either 5 or 20 kV. We expect sub-μm resolution with the L-line analysis at 5 kV to visualize the interior of the tubule walls, whose thickness is ~ 1 μm or less. The *y*-axis of Fig. 6.7 has units of photon count per second divided by both electron beam current and an energy bin of 1 eV. When we compare X-ray intensity in cps/nA/eV, the measurement at 20 kV naturally has a higher X-ray intensity than that at 5 kV except a range below the O K line. Although the total cross-section $\sigma \times \omega$ of the Zn L_3-$M_{4,5}$ line is maximum at 3 keV, an acceleration voltage of 5 kV was selected considering the electron cascade cloud. The higher statistical fluctuation at 20 kV than that at 5 kV stems from the lower beam current to avoid serious damage and deformation.

In Fig. 6.7, all the spectra are overlapped with the continuum events of Bremsstrahlung, whose maximum X-ray energy is equal to probe electron energy. The SDD red spectrum at 20 kV in (a) shows the weak Zn K-$L_{2,3}$ (8.63 keV) line. Therefore, the dentin sample certainly includes Zn atoms. However, the weakness of the Zn K-$L_{2,3}$ peak at 20 kV prohibits elemental mapping and moreover spatial resolution is insufficient for observing the inside of the tubule walls. Here, we expect that the Zn L_3-$M_{4,5}$ line is advantageous for imaging Zn. The red peak near 1.05 keV at 20 kV in (a) corresponds to either Na K-$L_{2,3}$ (1.041 keV) or Zn L_3-$M_{4,5}$ (1.012 keV); however, the SDD cannot distinguish these two lines. Similarly, the SDD spectrum at 5 kV (blue solid line) in (a) also has the peak near 1.05 keV in addition to the K lines of the light elements such as Si, Al, Mg, F, O, N, and C. No Zn K-vacancy is created at 5 keV, so that the Zn K-line is completely missing in the blue curves. The inset in (a) shows that the maximum energy of Bremsstrahlung is equal to the probe electron energy of 5 keV. In contrast to the SDD spectra, the STJD spectrum at 6 nA and 5 kV in Fig. 6.7c clearly displays two separated peaks; the gray line is the SDD spectrum for comparison. A similar peak separation should be possible with the STJD even at 20 kV in (b); however, the low electron-beam current of 60 pA at 20 kV to avoid serious radiation damage results in a low X-ray intensity and an unclear Zn L_3-$M_{4,5}$ line. In addition, the measurement spot at 20 kV is different from that at 5 kV, and has a less content of Zn atoms. Figure 6.7 apparently demonstrates that the STJD outperforms the commercial SDD in the X-ray spectrum measurement for 10 min. The ROI windows for Zn and Na can be selectively defined on the blue STJD spectrum in Fig. 6.7c; it is impossible on the gray SDD spectrum.

Let us consider Bremsstrahlung and an S/N ratio. The Bremsstrahlung X-rays are generated when probe electrons are deflected by an electric field near atomic nuclei, forming an X-ray background continuum up to the energy of the probe electrons. The S/N ratio, which corresponds to the characteristic X-ray intensity divided by the Bremsstrahlung intensity, is important for the detection limit of each element. We assume that the peak intensity of the Zn L-line is half of the Zn + Na peak. The S/N ratios in the SDD spectra are 0.7 for the Zn K-line peak at 20 kV, 0.5 for the Zn L-line peak at 20 kV, and 0.4 for the Zn L-line peak at 5 kV. These ratios for the Zn L-line at 20 and 5 kV are comparable and lower than that for the Zn K-line.

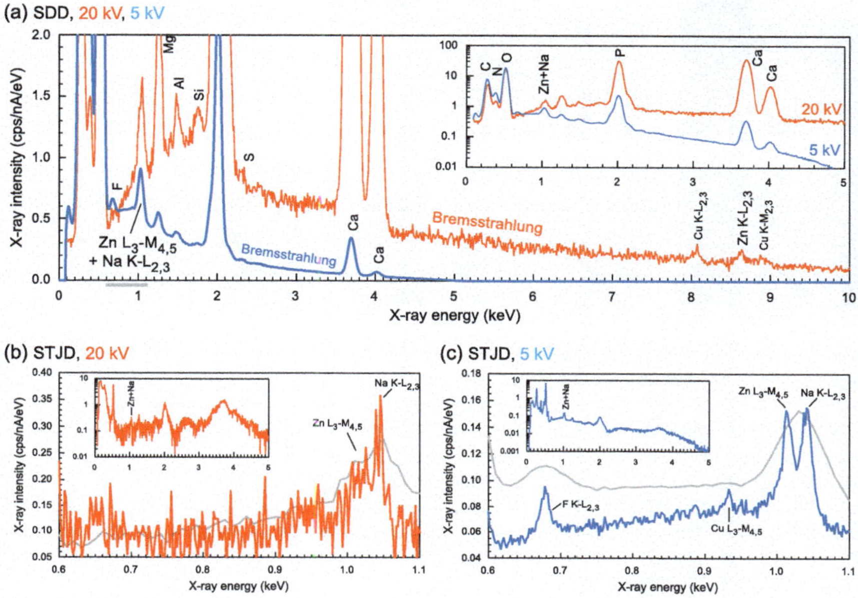

Fig. 6.7 Comparison between SEM X-ray spectra at 20 kV (red solid lines) and 5 kV (blue solid lines) for porcine dentin: (**a**) SDD spectra at 20 and 5 kV in a range between 0 and 10 keV with semilogarithmic plot in inset; and (**b**), (**c**) STJD spectra in a range between 0.6 and 1.1 keV. The spectra exhibit the full-energy peaks of F K-line (676.8 eV), Cu L-line (929.7 eV), Zn L-line (1.012 keV), Na K-line (1.041 keV), and other lines. The gray bar below the x-axis near 1 keV in (**a**) corresponds to the energy range in (**b**), (**c**). The X-ray intensities (count/s) are normalized by the electron current (nA) and the energy width of multichannel analyzer bins (eV): they are in units of cps/nA/eV. The electron beam currents were 60 pA at 20 kV and 6 nA at 5 kV. While scanning an area of $13 \times 17\ \mu m^2$ with the electron beams for 10 min, the X-ray spectra were simultaneously acquired with the 100-pixel STJD with an energy resolution of 6 eV at 400 eV and the 25-mm^2 SDD with an energy resolution of 129 eV at 5.9 keV. The measurement spot at 5 kV was different from that at 20 kV; therefore, the peak intensity patterns are different. The insets plot the spectra at a semilogarithmic scale. For comparison, the SDD spectra divided by 6 are also depicted by the gray solid lines in (**b**), (**c**). The background continuum stems from Bremsstrahlung up to the electron beam energies. Reproduced with permission from Ohkubo [17]. Copyright 2023 AIP Publishing LLC

Therefore, Zn K-line measurement at 20 kV should have a better detection lower limit than Zn L-line measurement due to the low Bremsstrahlung intensity at 8.64 keV. Nevertheless, the maximum total cross-section for Zn L_3-$M_{4,5}$ at 5 kV is 5.8 times higher than that for Zn K-$L_{2,3}$ at 20 kV, as discussed in Sect. 6.3.1.1. The Zn L_3-$M_{4,5}$ peak is stronger and more distinct than the Zn K-$L_{2,3}$ peak even with the higher Bremsstrahlung intensity near 1 keV. Therefore, the Zn L_3-$M_{4,5}$ line is advantageous for elemental mapping. Moreover, sub-μm spatial resolution at 5 kV is mandatory for observing tubules. Consequently, for imaging Zn atoms, the measurement at 5 kV is advantageous over that at 20 kV.

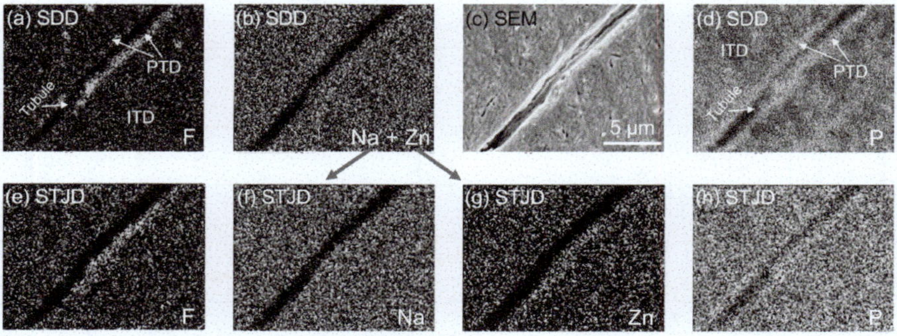

Fig. 6.8 SEM-EDS elemental mapping images of porcine dentin cross-section: (**a**), (**b**), (**d**) SDD images and (**c**) SEM image in the upper half panels, and (**e–h**)STJD images in the lower half panels. The field of view was scanned with an electron beam at 5 kV and 6 nA for 1 h. The elemental concentration is represented from black for the lowest count to white for the highest count without subtracting the Bremsstrahlung background. The STJD and SDD images were acquired simultaneously. One tubule is running diagonally in this field of view. The tubule is surrounded by the peritubular dentin (PTD) walls containing phospholipids. The tubule houses the odontoblastic process; the dried remain is observable inside the tubule in the SEM image in **c**. The parts other than the PTD region is called intertubular dentin (ITD). Reproduced with permission from Ohkubo [17]. Copyright 2023 AIP Publishing LLC

The next step is to measure elemental mapping images at 5 kV after defining the ROI windows for the Zn L_3-$M_{4,5}$ line and the Na K-$L_{2,3}$ line on Fig. 6.7c. Figure 6.8 shows the dentin cross-section images: SDD X-ray images and SEM image in the upper half panels, and STJD X-ray images in the lower half panels. The SEM image in (c) and the SDD image for P in (d) show that one tubule surrounded by the P-rich wall is oriented along the diagonal direction. The 5-keV electron beam repeatedly scanned the dentin cross-section area at 6 nA for 1 h. Before discussing the images, we must know the morphological structure of the mammalian dentin. Numberless tubules pass through the dentin from the odontoblast inside the dental pulp to the dentin-enamel interface or the dentin-cement interface. The tubule walls consist of the strongly mineralized-tissue sheath called peritubular dentin (PTD) containing phospholipid, which appears as the P-rich walls in Fig. 6.8d. The tubules house the odontoblastic process that connects the odontoblast cell and the end of the tubules at the above-mentioned two interfaces. The most space of dentin is filled with intertubular dentin (ITD), which has a collagen matrix containing small hydroxyapatite crystallites (sub-μm to nm). The odontoblast cells and the odontoblastic processes play an important role in forming and preserving the ITD structure and sensing physical and chemical stimulation on tooth surfaces. Beyond these structural data, we need to further investigate the structure and function of the tubules and PTD [38].

The ROI windows for the F K-line and the P K-line are unquestionable because these peaks are solitary in Fig. 6.7. On the other hand, the tails of the Zn L-line and the Na K-line overlap each other even on the STJD blue spectrum in Fig. 6.7c. Therefore, the ROI window for the Zn L-line covers the lower-energy tail excluding the peak top; the ROI window for the Na K-line covers the higher-energy tail. Off-line analysis allows us to reconstruct X-ray spectrum and peak separation by numerical fitting at each location; however, it is a time consuming task, which will be performed in future.

In the SDD and STJD images for phosphorus in Fig. 6.8d, h, the P enrichment inside the PTD wall is consistent with the known composition of PTD. The SDD image for Na + Zn in (b) suggests a slight enrichment of Na and/or Zn in the lower PTD wall, although it is ambiguous. Meanwhile, the STJD Zn and Na images in (f) and (g) reveal that the Na atoms are distributed homogeneously throughout both PTD and ITD, whereas the Zn atoms concentrate exclusively inside the lower PTD wall. It appears that the F and Zn enrichment occur at the same place inside the lower PTD wall. The upper PTD wall contains almost no F and Zn. The region containing both Zn and F has a length of 13 μm along the tubule orientation. The formation of ZnF_2 inside the PTD wall is most plausible. Observations of other locations in the dentin reveal that there are two types of elemental enrichment inside the PTD walls: the enrichment of F only and the enrichment of both F and Zn.

The STJD images contribute to investigating caries prevention or disease due to F and Zn. The world health organization (WHO) designates F and Zn as the root caries inhibitors. The caries-prevention effect of trace fluorides has been frequently reported [39]. However, it is still unidentified whether F is an essential trace element or not because there are some reports claiming no meaningful effect. Furthermore, excessive F ingestion from water and foods causes a disease called fluorosis, accumulating F in bones and teeth. Especially, infants swallow flavored fluoride toothpaste, which causes dental fluorosis on enamel surfaces. Therefore, WHO recommends not serving infants flavored fluoride toothpaste. The trace element Zn is contained in most metalloproteinases that catalyze proteolysis from proteins to polypeptides or amino acids. The caries prevention and restoration effects of Zn were reported [34–37]; however, similarly, excessive Zn ingestion also causes a disease of copper deficiency. The mechanisms of caries prevention and diseases due to F and Zn are still unelucidated. Therefore, it is necessary to investigate the F and Zn distribution in dentin and the flow of F and Zn inside dentin.

Based on our observation with STJDs, the enrichment patterns of F and Zn vary depending on locations, which should be related to the merits and demerits of F and Zn. The F atoms are enriched inside PTD at most locations, whereas the accumulation of Zn occurs at the limited locations inside the PTD walls. We infer that both F and Zn are supplied to the dentin through the odontoblastic processes and diffuse into the PTD and ITD regions. The Zn and F atoms accumulate together inside PTD at some localized locations, forming the stable ZnF_2 inside PTD. The ZnF_2 formation can be a sign of excessive ingestion, fixing both elements inside PTD. Further systematic studies are required. STJD elemental mapping with the L-line analysis at a sub-micrometer spatial resolution paves the way to understand the roles of F and Zn in teeth by avoiding the serious peak overlaps. The L-line analysis is better than the K-line analysis with respect to both the spatial resolution and the detection efficiency of elements especially in $Z > \sim 30$.

A TESD on SEM has an energy resolution of 2.0 eV at 1.49 keV (Al K-$L_{2,3}$) [40], which is better than 4.1 eV at 400 eV of the STJD. The TESD facilitates both chemical state analysis and elemental identification. It is possible to distinguish metallic U and U^{6+} in an oxide using the U M_5-N_7 (U $M\alpha_1$) line; and ammonium iron citrate ($C_6H_8O_7{\cdot}xFe^{3+}{\cdot}yNH_3$) used as a food additive and iron nitrate ($Fe(NO_3)_39H_2O$) used as a dye mordant using the Fe L_3-$M_{4,5}$ ($L\alpha_{1,2}$) line and Fe L-M,N (L_β) line [40]. The development of SEM with a TES array detector up to 128 pixels is in progress to realize non-destructive chemical-state determination for a nuclear safeguard [41, 42]. Precise peak shape analysis of characteristic X-ray lines is called XES, which is described in Sect. 6.3.3.4 in detail. STJDs can also perform XES [17]; however, further improvement of STJD energy resolution is necessary.

6.3.2 *Transmission Electron Microscopy (TEM)*

Another important microscopic instrument using electrons is transmission electron microscopy (TEM). Samples for TEM should have a thickness of ~ 100 nm or thinner in general. Electrons accelerated at 80–200 kV in most cases pass through a thin sample, so that the electron collision cascade cloud is negligibly small in contrast to ~ 1 μm in the K-line analysis with SEM. This leads to a high spatial resolution of even an atomic scale for X-ray imaging, which is unachievable with SEM. In addition, no background of Bremsstrahlung appears on X-ray spectra. X-ray imaging with scanning transmission electron microscopy (STEM) is possible at an atomic scale with a 100-mm^2 large SDD [27]; EDS detection of single atoms is reported for heavy atoms embedded in a carbon-based matrix [43, 44]. However, the theoretical energy-resolution limit of semiconductor detectors is insufficient to resolve the K-, L-, M-lines, similar to SEM–EDS with an SDD. The TES type was deployed on TEMs [45–48]. A 64-pixel TESD mounted on STEM has an energy resolution of 7 eV at 5.9 keV; a 224-pixel TESD enhances a detection efficiency assisted with a polycapillary lens for astromaterials of a sample-return mission [48].

One of the typical chemical-analysis methods on TEM is electron energy loss spectroscopy (EELS) at an atomic scale. Energetic electrons lose their energies through inelastic scattering in a substance. Measurement of the energy loss produces a similar absorption-edge structure to those obtained by X-ray absorption spectroscopy (XAS) in the next Sect. 6.3.3. Electron intensity as a function of electron energy represents the inner shell excitation, which corresponds to XAS, and the valence-band to conduction-band excitation, which corresponds to e–h pair generation. EELS is used for investigating chemical state analysis, plasmon excitation, and electronic structure. TESD's high energy resolution can complement EELS by performing chemical state analysis with EDS-XES.

6.3.3 X-Ray Absorption Spectroscopy (XAS) and X-Ray Emission Spectroscopy (XES)

6.3.3.1 Fundamentals and instruments for X-Ray Absorption and Emission

X-ray spectroscopy provides nanoscale elemental mapping images as described in the previous section. Furthermore, it enables the measurement of atomic configuration and even electronic structure of solids. This technique is known as X-ray absorption spectroscopy (XAS), which is divided into X-ray absorption near edge structure (XANES) and extended X-ray absorption fine structure (EXAFS). Figure 6.9 illustrates X-ray absorption and succeeding X-ray emission processes. X-ray photons perform a similar excitation as electrons in Fig. 6.2; however, the absorption process of X-ray photons is totally different from that of energetic electrons. The electrons induce the collision cascade, which results in energy loss of the primary electrons. Therefore, the probe electrons have diverse kinetic energies between initial E_0 and near zero, depending on depth from a sample surface. In very thin samples for TEM, quasi-constant electron energy is maintained. On the other hand, when a sample absorbs an X-ray photon, all of the photon energy is transferred to an absorbing atom or emitted electrons such as photoelectrons and Auger electrons; therefore, the photon disappears when it is absorbed. X-ray intensity or photon density at E_0 exponentially decreases due to photoelectric effect as a function of depth, following $I = I_0 e^{-\mu x}$, where μ is the linear attenuation coefficient and x is the depth from a surface or thickness of a substance. The photon energy is constant at any x values, independent of depth: we consider a photon energy range and a Z number range, where disturbances such as Compton scattering and Raman scattering are negligible. This feature enables the measurement of X-ray absorbance as a function of photon energy. If the sample is Si, a thickness of ~ 5 μm is sufficient to measure the K-edge absorption profile. We can measure even thick bulk samples when we use fluorescence yield as described later in this section. Photon energy scanning is performed by a monochromator using analyzing crystals or gratings, which is similar to the X-ray spectrometer in Fig. 6.1. Probe X-rays are frequently generated by a bending magnet or an undulator at synchrotron radiation facilities. The energy stability $\Delta E_0/E_0$ of the probe X-rays is ~ 10^{-4} or even better.

Photon energy above a threshold level is required to excite an electron in an inner shell to the conduction band. It is called the absorption edge. The photon energy at the absorption edge for inner shells such as K is in an X-ray range; it is more than 1000 times higher than gap energy in semiconductors and Fermi energy of metals. The absorption edge energy or the binding energy of the Si K 1s orbital is 1.839 keV, for example. X-ray absorbance as a function of probe photon energy exhibits fine structure, which is called XAFS. XAFS is divided into two regimes: XANES within ~ 50 eV from an absorption edge, and EXAFS covering a wide range up to 1000 eV above an absorption edge [49]. In XANES, a shell electron is excited to an unoccupied state in the conduction band when an X-ray energy is

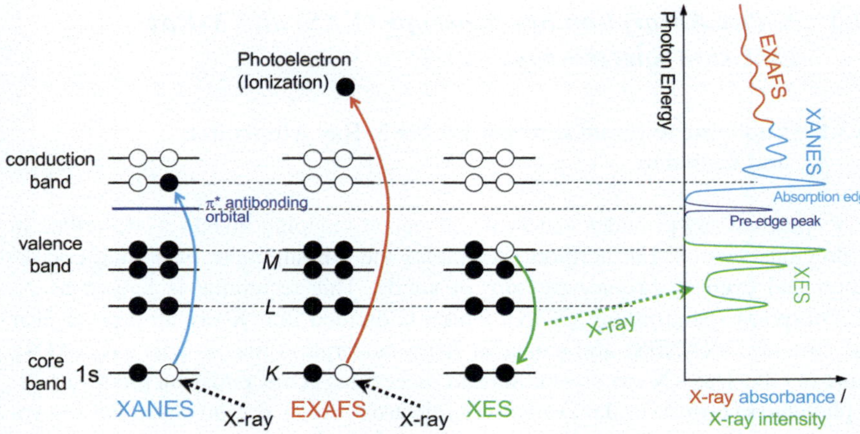

Fig. 6.9 Excitation and relaxation processes in XANES, EXAFS, and XES. XANES and EXAFS are collectively called XAFS or XAS. The excitation processes resemble those in electron excitation in Fig. 6.2. The apparent difference from electron excitation is that the X-ray photon energy is constant at any depths from a sample surface when we neglect disturbances such as Compton scattering and Raman scattering. An atom occasionally absorbs an X-ray photon, exciting an inner-shell electron to a conduction band in XANES, and to the vacuum level in EXAFS, which is ionization. The absorption edge energy and the XANES structure are useful to analyze charge states and chemical states of atoms. EXAFS provides information on symmetry of local atomic arrangement or interatomic distances, based on the interference between the photoelectron wave and the scattered electron waves. A pre-edge peak below an absorption edge occasionally appears due to transition from an inner shell to one of the unoccupied anti-bonding orbitals such as π^* in compounds or molecules. The pre-edge peak is utilized for investigating chemical bonds. After exciting an electron to one of the unfilled orbitals or to the vacuum, X-ray emission occurs due to electron transition from a higher orbital, which is the same as those in Fig. 6.2. Analysis of the characteristic X-ray line shape is a complementary method to deduce the electronic structure of outer shells (valence band) that are influenced by charge states or chemical bonds; this is called XES. Reproduced with permission from Ohkubo [17].

just above the absorption edge. The absorbance represents DOS in the unoccupied conduction band as shown in Fig. 6.9. Precise measurement of the edge energy provides information on ionic valence. Fine structure above the edge energy reflects chemical states of atoms of interest in a matrix or chemical bonds between atoms. On the other hand, in EXAFS, an excited electron has a sufficient energy to leave the atom; the electron becomes a photoelectron. It is also referred to as ionization, which occurs when the probe photon energy is increased by ~ 100 eV beyond the edge energy. The emitted photoelectron is scattered by neighboring atoms inside condensed matter. The interference between the primary photoelectron wave and the scattered electron waves affects the X-ray absorbance, which results in oscillation. Oscillating EXAFS spectra represent the symmetry of local atomic configuration or interatomic distances.

After absorbing an X-ray photon, the vacancy in an inner shell is filled with an electron from an outer shell, emitting a characteristic X-ray photon, or an Auger electron, as shown in Fig. 6.2. We utilize characteristic X-rays for elemental mapping in Sect. 6.3.1.3. Historically, a characteristic X-ray peak in X-ray spectra was called "line": see Appendix B for details. In addition, precise line-shape analysis is called XES. Figure 6.9 illustrates the relationship among XANES, EXAFS, and XES. In contrast to XANES representing unoccupied conduction-band structure, XES is a method to deduce the electronic structure or DOS of occupied outer shells that are influenced by surrounding chemical environment. XANES and XES are complementary to study the electronic structure of valence bands and conduction bands. EXAFS provides local atomic configuration and interatomic distances.

XAS is a method to provide information on the electronic and atomic-scale structure averaged over an area of ~ mm in general. An X-ray nanobeam enables XAFS imaging, although the beam scanning is a time-consuming process in addition to monochromator scanning. TESDs enable data acquisition of an entire energy range of interest, not utilizing a monochromator, which paves the way to lab-based compact XAS instrumentation [6]. For both XAS and XES, superconductor detectors contribute to building in-house time-resolved instruments apart from synchrotron radiation facilities [23]. The energy resolution of 1.25–1.58 eV in superconductor thermal detectors is approaching a resolving power $E_0/\Delta E_0$ of 10^4, which is comparable with that of ordinary monochromators. As mentioned above, XANES resembles TEM-EELS. The merit of XAS is that it does not require samples with a thickness of ~ 100 nm: even bulk samples can be measured.

X-ray absorbance can be measured by two methods: the intensity of X-rays transmitting a thin sample, and the intensity of characteristic X-rays or electrons emitted from a sample, while increasing the probe X-ray energy across an absorption edge. The transmission method requires a thin-film form, not as thin as TEM samples. One problem of the transmission method is that trace elements in matrices cannot be measured. Therefore, other methods equivalent to X-ray transmission are often used for analyzing bulk samples and trace elements. As shown in Figs. 6.2 and 6.9, the responses of a sample to probe X-rays are photoelectrons, Auger electrons, and characteristic X-rays: another route to shake-off electrons is unsuitable to material analysis. For XAFS measurement of bulk samples, there are two methods: electron yield (EY) XAFS using photoelectrons and Auger electrons, and fluorescence yield (FY) XAFS using characteristic X-rays. EY-XAFS is extremely surface sensitive in a few nm within escape depth of photoelectrons and Auger electrons; therefore, it is sensitive to surface states or surface cleaning. EY-XAFS cannot measure and reveal bulk properties. In contrast, characteristic X-rays can escape from a relatively deep region: a few 100 nm in the K line of the nitrogen dopant in ZnO, for example. Therefore, FY-XAFS is sensitive to bulk properties. Both EY and FY are divided into two regimes: "total" and "partial." The total yield method accepts the entire energy range of electrons or X-rays. The partial method accepts a selected energy range. The partial electron yield (PEY) method with an electron energy analyzer can select a depth range, where electrons are emitted from a sample surface within a few nm. The partial fluorescence yield (PFY) with an X-ray spectrometer can select elements of interest. Therefore, PFY-XAFS is suitable for analyzing trace elements diluted in a matrix. The trace elements are extremely important for both functional materials such as semiconductors and structural materials such as high-performance steels. Addition of impurities at 10–100 ppm drastically improves material performance. In PFY-XAFS, the energy resolution of X-ray detectors is crucial especially for trace light elements because the K lines of light elements are close to the L or M lines of heavy matrix elements. Semiconductor detectors, even the latest SDDs, are frequently insufficient to distinguish multiple characteristic X-ray lines in a soft X-ray range below ~ 2 keV.

Superconductor detectors are advantageous in a soft X-ray range below ~ 2 keV as shown in Figs. 6.3 and 6.6. LLNL developed the first PFY-XAFS instrument with an array STJD for advanced light source (ALS) in 2002 [50]. They realized V L-edge XAFS measurement on porphyrin and asphaltene with 300-ppm V as model samples to test the instrument. The application of superconductor detectors to synchrotron science was reviewed, comparing different superconductor EDS detectors and WDS in 2006 [3]. Figure 6.10 shows the performance figures of the X-ray detectors as of 2006 and the covering areas on the space of element concentration (sample concentration) versus energy resolution (line separation). The performance figures are currently improved as listed in Table 5.1. Overall, the energy resolution of superconductor detectors is 10–50 times better than those of semiconductor EDS detectors and the solid angle is approaching that of high-resolution SDDs. The superconductor detectors cover a region of trace element concentrations of less than ~ 1% and ΔE < ~ 100 eV, where the performance of both grating WDS and semiconductor EDS is inadequate.

(a)

Detector	Resolution ΔE_{FWHM} (eV)	Count rate (counts s^{-1})	Efficiency $\eta\Omega/4\pi$	P/B ratio
Ge (typical)	130	3×10^5	0.1	50:1
Ge (best)	60	3×10^4	0.03	200:1
STJ (typical)	20	10^5	10^{-4}	200:1
STJ (best)	10	10^6	10^{-3}	1000:1
Grating (typical)	0.5	10^5	10^{-6}	200:1
Grating (best)	0.2	10^6	10^{-5}	1000:1
Grating with optic	0.2	10^6	3×10^{-4}	200:1

(b)

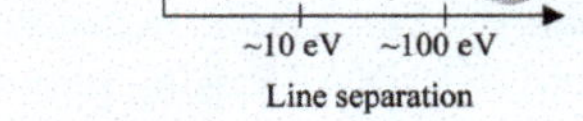

Fig. 6.10 Comparison among grating WDS, semiconductor EDS detectors, and superconductor (cryogenic) detectors for 500-eV X-rays as of 2006 [3]: (**a**) performance figures and (**b**) covering areas. The detectors are devoted to synchrotron science of partial fluorescence yield (PFY)-XAFS in a soft X-ray range. "Efficiency" in **a** denotes the solid angle coverage $\Omega/4\pi$ multiplied by quantum efficiency η. "P/B ratio" stands for peak-to-background ratio. "Line separation" in (**b**) is equivalent to energy resolution. The performance figures are currently improved as listed in Table 5.1. TESD and MMCD, not listed, have a better energy resolution of 1–3 eV. A 220-pixel TESD has an energy resolution of 1.5 eV, a practical count rate up to ~ 4 kcps, and a solid angle Ω of 2×10^{-3} sr [7]. STJDs have an energy resolution of 2.5 - 6 eV below ~ 1 keV and a practical count rate up to ~ 200 kcps; an STJ array with a polycapillary lens has 15×10^{-3} sr [31]. Reproduced with permission from Friedrich [3]. Copyright 2006 International Union of Crystallography

One of the detector development directions is increasing the number of pixels. For example, in 2006, a single Nb/Al junction pixel, developed at AIST, was used to measure an XAS spectrum of oxygen atoms in a high-*k* gate material, Hf-Al oxide, for MOS devices [51]. The oxygen was one of the major elements in the MOS devices. In 2009, the number of STJ pixels reached 100 [52].

Figure 6.11 shows an XAFS instrument equipped with an array Nb/Al-STJD with 100 STJ-pixels [52, 53]. Monochromatized synchrotron radiation X-ray beams normally have a $\Delta E_0/E_0$ value of 10^{-4} and a diameter of 1–2 mm. The probe beam irradiates a sample in the sample chamber in (a). At the head of the cold snout in (b), the 100-pixel STJD is cooled to ~ 0.3 K by a liquid-cryogen-free ^{3}He cryostat. A holding time at 0.35 K is 32 h [32], which is sufficient to acquire several XAFS spectra. Automatic re-cooling assists users. After amplifying output pulses from the STJ pixels, the stack of FPGA-DSP in (c) acquire pulse height spectra. A computer, not shown in the picture, controls pixel calibration, synchrotron radiation beam, data acquisition, and data analysis.

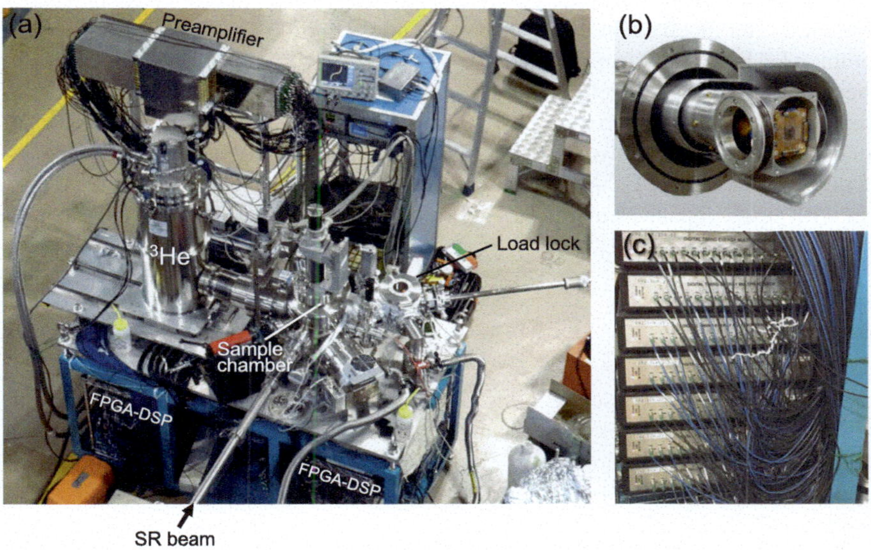

Fig. 6.11 STJD-XAFS: X-ray absorption fine structure instrument with superconductor tunnel junction detector [52, 53]: (**a**) fully-automated closed-cycle ^{3}He sorption refrigerator with pulse tube cooler, sample chamber, and load-lock chamber for exchanging samples without breaking vacuum at KEK-PF synchrotron radiation facility, (**b**) internal structure of cold snout for 100-pixel STJD, **c** field-programmable-gate-array based digital signal processors (FPGA-DSP) for deriving timing and energy data from output pulses. The pulse processing electronics is the same as that for SEM and MS. Reproduced with permission from Ohkubo [17]. Copyright 2023 AIP Publishing LLC

6.3.3.2 XAFS with Superconductor Detectors

General

Typical semiconductor X-ray detectors with an energy resolution of ~ 120 eV at 5.9 keV cover an entire energy range of the K lines from Li to U (~ 50 eV– ~ 100 keV). Meanwhile, superconductor X-ray detectors such as STJD and TESD with an energy resolution of ~ 1.5– 6 eV often cover a range below ~ 2 keV, focusing on the K-edges of light elements and the L-edges of heavy elements [7, 52]. Measurement of trace elements is a research area of the superconductor detectors [3]. One example of the early years was a biological sample; L-line XANES measurement of ~ 100 ppm Mn in photosystem II (PS-II) protein [54]. The L-edge analysis, which is related to chemical bonds with surrounding atoms, provides more information than K-edge analysis together with resonant inelastic X-ray scattering (RIXS) [55, 56]. PS-II has a cluster of four Mn atoms (Mn_4CaO_5), which play an important role in photosynthesis together with PS-I, cytochrome complex, and ATP synthetases. PS-II decomposes water into hydrogen and oxygen by using four photons with an energy of ~ 1 eV. The mechanism of entire photosynthesis processes has not been known completely. Biological samples require a low probe-beam dose to avoid damage. Therefore, high detection efficiency is crucial in addition to high energy resolution. High detection efficiency is realized by arraying pixels, increasing sensitive area. Such array detectors are also effective to measure diluted samples. In the subsequent sections, we describe PFY-XAFS with STJDs for wide-gap compound semiconductors with trace dopant elements [57–59] and lab-based time-resolved EDS-XAFS with TESDs [6, 55, 60, 61].

Nitrogen Dopant in ZnO

ZnO is a wide-bandgap semiconductor, which is promising for optoelectronics. Natural defects such as oxygen vacancies and interstitial Zn atoms lead to an n-type semiconductor; therefore, p-type doping is a challenging task. One of the p-type dopants is nitrogen. The doping of nitrogen was performed in-situ during the epitaxial growth of ZnO at a thickness of ~ 1 μm by molecular beam epitaxy (MBE) with a plasma radical nitrogen source at an N concentration of ~ 1% (~ 3×10^{20} cm^{-3}). The yellowish as-grown film became transparent after rapid thermal annealing at 800 °C for 3 min. The structural transformation was studied by PFY-XAFS. The N K-edge at 397 eV is lower than the O K-edge at 533 eV. Therefore, the XANES measurement for the N K-edge was uninfluenced by characteristic X-rays from the matrix elements: O K-$L_{2,3}$ (525 eV) and Zn L_3-$M_{4,5}$ (1.012 keV). Strong Zn L-line and O K-line are far apart from the N K-line (392 eV). However, the N K-line is often buried under a noise level when we use a Si or Ge detector that is commonly used at synchrotron radiation beamlines. Exceptionally, a windowless Ge detector at ELETTRA is able to detect N K-line. An LLNL-STJD at ALS has a better S/N ratio as well as an energy resolution better than that of the Ge detectors [57].

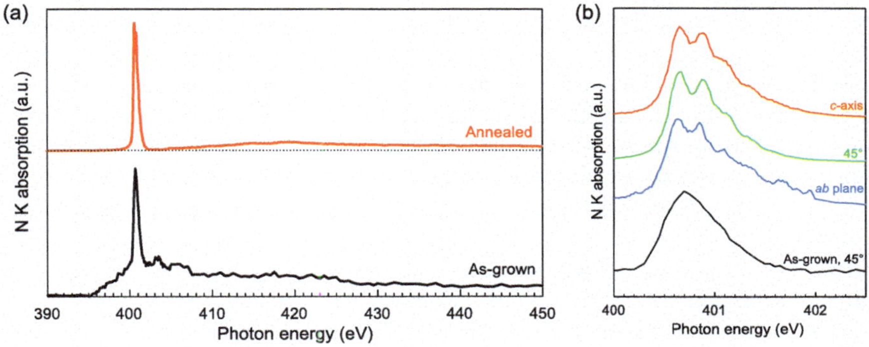

Fig. 6.12 XANES spectra of nitrogen-doped ZnO (ZnO:N) film [57]: (**a**) wide scan and (**b**) fine scan near 400.7 eV in as-grown state (black) and after rapid thermal annealing at 800 °C for 3 min (red, green, and blue). The LLNL STJD easily measured the N K-line peak above a noise level at the ALS beamline. The as-grown yellowish film with a nitrogen concentration of 1% exhibits the fine structure from the pre-edge region to ~ 430 eV in (**a**), indicating the O-site substitution with the N atoms. After the annealing, the vibrational fine structure in (**b**) at the different polarization directions indicates the N_2 bubble formation. Reproduced with permission from Fons et al. [57]. Copyright 2006 American Physical Society

Figure 6.12a shows the PFY-XANES spectra of the N-doped ZnO (ZnO:N) sample, acquired by the LLNL-STJD [57]. In the as-grown state, the XANES black spectrum in (a) has the sharp peak at 400.7 eV, accompanied by the wide fine structure from a pre-edge region below 400 to ~ 430 eV. An ab initio multiple-scattering calculation with the FEFF code [62] reveals that the N atoms substitute for the O site in the as-grown state. The ab initio calculation turns XAFS into a sophisticated atomic-scale structural analysis tool. After the rapid annealing, the fine structure almost disappears except the sharp peak at 400.7 eV on the red spectrum in (a). The high-resolution vibrational structure in Fig. 6.12 b ensures that N_2 molecules are formed inside the ZnO crystal. Almost no polarization-direction dependence is recognizable in (b), which implies that the N_2 molecules are not oriented to a specific crystallographic orientation of the ZnO crystal. This elucidates that the N substitution for the O site as the acceptor is a metastable phase. N atoms are not fully soluble in ZnO; therefore, they are excluded from the matrix crystal in an equilibrium state after annealing. The STJ-XAFS confirms how the p-type doing into ZnO is problematic.

Nitrogen Dopant in SiC and ROI Window for XAFS

SiC is another important wide-bandgap semiconductor for energy-saving power electronics. The SiC devices have reached the stage of practical uses, such as for electric train inverters or electric airplanes. However, the binary system SiC still has difficulty in impurity doping compared with Si. The dopants of N and P are promising for n-type, while Al and B for p-type. In Si, the recovery from damage due to dopant ion-implantation easily occurs because it contains only the single matrix element. A combination of dopant ion-implantation into Si at room temperature and post thermal annealing activates the dopant atoms at 100% while healing the implantation damage. However, such combination process is inadequate for SiC. In SiC, ion implantation should be performed at an elevated temperature; post annealing is effective in activating the dopant atoms, but not at 100% Lattice damage recovery in SiC is more complex than that in Si because SiC has approximately 250 polytypes. Two widely-utilized polytypes for SiC diodes and transistors are the cubic 3C structure with a plane sequence of ABC and the hexagonal 4H structure with ABCB.

PFY-XAFS measurement of the N atoms in SiC faces a serious problem in contrast to the case of the N-doped ZnO: the strong C K-line at 277 eV interferes with detection of the extremely faint N K-line at 392 eV. Due to this difficulty, no XAFS measurement of N-doped SiC (SiC:N) was conducted before an experiment with the STJD spectrometer in Fig. 6.11. Nevertheless, there were some reports on amorphous mixtures of Si–C-N–O with a nitrogen concentration of a few 10% [63, 64]. A PFY-XAFS experiment with the STJD was conducted on N ions implanted into a 4H-SiC substrate at a uniform dose of 4×10^{19} cm^{-3} (800 ppm) in a depth range of 0–200 nm at 500 °C. The N dopant activation and the recovery of the implantation damage were performed by post annealing at either temperature of 1400 or 1800 °C for 5 min. Figure 6.13 shows the X-ray fluorescence spectrum of the SiC:N at a probe beam energy of 453 eV, acquired with the AIST 100-pixel STJD at the beam line BL11 of KEK PF. This spectrum is used to define a ROI window for the N K-line [58]. The STJD spectrometer distinguishes the faint N K-line from the 1000 times stronger C K-line at a superior energy resolution, which becomes apparent by comparing with the broad SDD peaks labeled "Egg SDD" of C, N, and O in Fig. 6.6c. The O K-line in Fig. 6.13 excited by the second-order beam at 906 eV is attributed to an oxidized surface layer. The peak between N K-line and O K-line originates from elastically scattered first-order probe beam photons. While increasing the probe energy across the N K-edge, the elastically scattered probe beam crosses the N K-edge. The scattering intensity is theoretically zero along the direction perpendicular to the beam direction; however, experimentally, it cannot be eliminated completely in this low photon energy range. Therefore, the ROI window depicted by the black bar covers an entire XANES scan-range from 390 to 460 eV. The first XANES measurement for SiC:N is shown in Fig. 6.17 later. The wide ROI window is adequate for the SiC:N sample, but an ROI window considerably wider than a characteristic X-ray line causes artifact features on XANES spectra due to some impurity peaks. In addition, when we need to scan a range wider than 390–460 eV, a sudden reduction of the scattered beam intensity at the upper edge of the ROI window precludes proper XAFS spectrum acquisition. Elimination of elastically scattered probe beams is possible with TESDs at an energy resolution of ~ 2 eV [6].

Figure 6.14 shows the appearance of a 240-pixel hybrid TESD, developed at NIST, which has two types of TES pixels: 350-μm square pixels for a high energy range and 124-μm square pixels for a low energy range. At the best energy resolution of 1.1 eV in a soft X-ray range, the TESD can exclude the effect of the elastically scattered

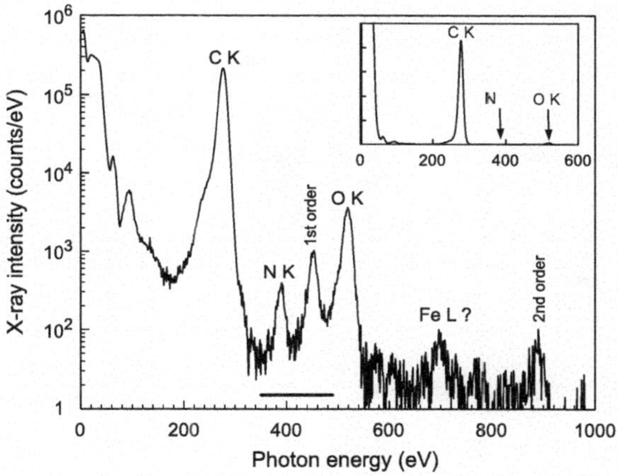

Fig. 6.13 X-ray pulse height spectrum with synchrotron radiation at 453 eV for N-doped SiC (SiC:N), acquired with array STJD at KEK PF [58]. The matrix crystal is the compound semiconductor 4H-SiC; N ions implanted at 500 °C and at multiple energies of 10–120 keV uniformly distribute in a depth range of 0–200 nm. The N concentration is 4×10^{19} cm^{-3} (800 ppm). The inset shows the linear plot, in which no N K-line is recognizable. The elastically scattered probe beam appears at 453 eV for the first order and 906 eV for the second order. The black bar is the region-of-interest (ROI) window for measuring PFY-XANES spectra of the N K-edge. Semiconductor detectors with a ~ 10 times worse energy resolution cannot distinguish the faint N K-line masked by a tail of the strong C K-line. Reproduced from Ohkubo et al. [58]. Licensed under CC BY-NC-SA 3.0: http://creativecommons.org/licenses/by-nc-sa/3.0/

probe beam because the K line energies of light elements are 5–10 eV lower than the edge energies [6]. On the other hand, heavy elements have a large energy difference between a K-line and a K-edge, such as 6.403 keV of the Fe K-L_3 line versus 7.11 keV of the Fe K edge, so that no interference between the emission line and the scattered probe beam occurs even when we use an SDD. Figure 6.15 shows the X-ray pulse height spectrum of a gold sample with an organic surface contamination. At a probe beam energy of 297.8 eV, which is higher than the C K-edge energy of 284.2 eV, the C K-line appears at 277 eV due to the organic contaminant. In this measurement, the energy resolution is 2.46 eV at 297.8 eV, which is independent of photon energy. At this high energy resolution, the C K-line is distinctly apart from the probe beam. Figure 6.16 shows the scatter plot of the detected events on the space of X-ray-energy versus probe-beam-energy. The sample is the NIST Standard Reference Material 1216-I, containing C in an unknown chemical state at 0.7% by mass inside the matrix of porous micro-particulate silica beads with a diameter of ~ 20 μm. The scatter plot in (b) is an enlargement of the region near the C K-edge in (a). The horizontal event group is due to the C K-line. The slanting event group is due to the elastically scattered probe beam that scans a range of 265–325 eV. By extracting the events enclosed by the violet line, it is possible to eliminate the effect of the scattered probe beam completely. We can recognize the elimination effect in (c) by

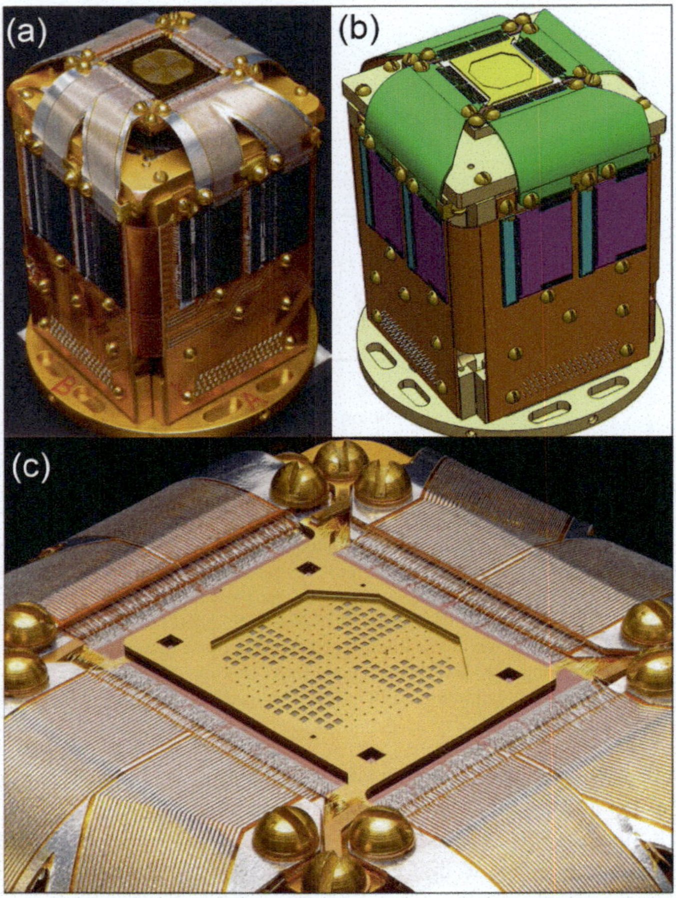

Fig. 6.14 TESD spectrometer with 240-pixel for analytical science [6]: (**a**) TESD package mounted at snout head, (**b**) computer-aided design model, and **c** aperture mounded over TES chip. The TES pixels have a sensitive area of either $350 \times 350\ \mu m^2$ or $124 \times 140\ \mu m^2$. The aperture has two kinds of square holes fitted to both types of TES pixels to eliminate artifact signals from the area other than the X-ray absorbers. Eight time-division SQUID multiplexer chips are mounded on the package side walls. The snout is cooled to 65 mK. The 124-μm square pixels record an energy resolution of 1.1 eV at 500 eV. Reproduced with permission from Doriese et al. [6]. Copyright 2017 AIP Publishing LLC

comparing the total yield corresponding to a wide ROI window and the pure C K-yield. By dividing the C K-yield with the probe beam intensity, we can obtain the PFY-XANES spectrum for the C K-edge in (d).

Returning to the SiC:N sample, Fig. 6.17a plots the sum of the N K-yield and the scattered probe beam yield within the ROI window, normalized by the probe beam intensity, as a function of probe beam energy. The STJD cannot exclude the

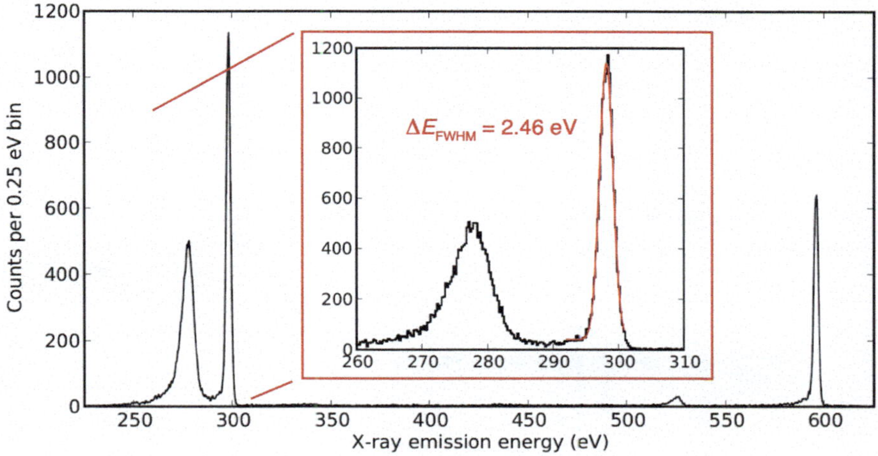

Fig. 6.15 Pulse height spectrum of gold sample with carbon surface contamination. The probe beam energy is 297.8 eV, 13.6 eV higher than the C K -edge energy of 284.2 eV. In addition to the C K-line at 277 eV, the elastically scattered probe beam appears at 297.8 eV. The peak near 600 eV is due to the second order probe beam. The energy resolution is 2.46 eV at 297.8 eV. Reproduced from Doriese et al. [6]. Copyright 2017 AIP Publishing LLC

scattered probe beam; nevertheless, the XANES spectra are adequate for comparing with theoretical calculations. The XANES spectra represent the as-implanted state in green and the states after 5 min annealing at 1400 °C in blue and 1800 °C in red. Figure 6.17b shows the ab initio calculation results with the FEFF code for the N atoms substituting for either the Si site (red dashed curve) or the C site (blue solid curve). We calculate two cases, 3C-SiC and 4H-SiC polytypes, because the 3C phase can be readily formed during annealing for damage recovery. A staking fault between 4H and 3C is the probable transformation. XANES cannot distinguish a difference between the 4H and 3C phases. In addition, the 4H polytype has two non-equivalent sites for Si and C: the small difference is depicted by the curves in light colors in the lower panel in Fig. 6.17b, although it is almost unrecognizable. The calculation spectra have two distinct peaks at 399 and 405 eV. The 399-eV peak in the red dashed curve originates from the 1s-to-π^* antibonding-orbital transition, which is attributed to the C-N bonds for the Si-site substitution. The 405-eV peak in the blue solid curve originates from the antibonding-orbital transition on the Si-N bond for the C-site substitution. No 399-eV peak appears on the experimental spectra in (a). Consequently, the N atoms occupy the C site, which is commonly expected. The Si site substitution suggested occasionally is ruled out in this SiC sample. Figure 6.17c depicts the Si and C sites in the 4H-SiC crystal for the N dopant.

The SiC lattice recovery from the implantation damage state occurrs while the post-annealing proceeds. An ion channeling experiment using MeV He ions reveales that a lattice damage rate of ~ 7% in the as-implanted state decreases to an undetectable rate after annealing at either 1400 or 1800 °C [58]. In contrast to this drastic

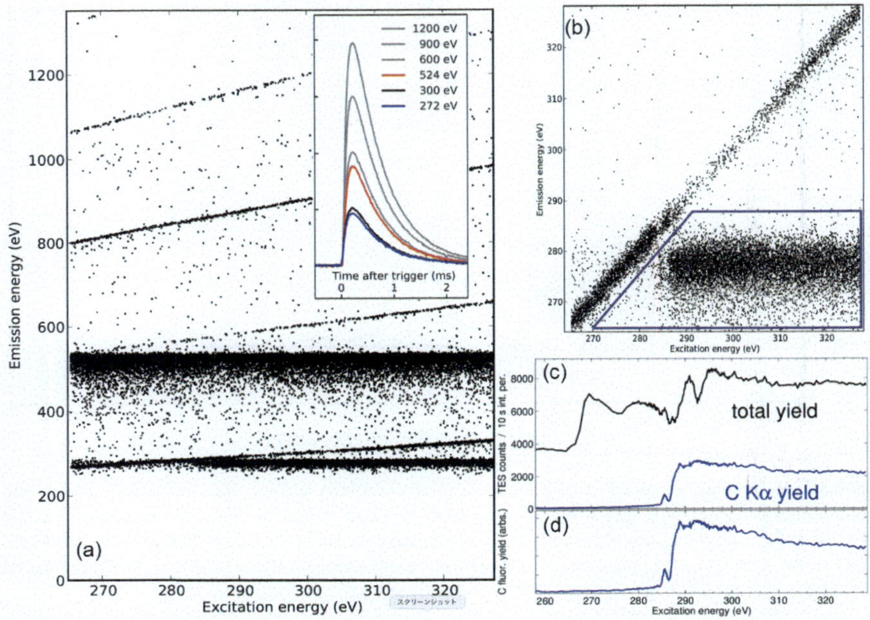

Fig. 6.16 Elimination of elastically-scattered probe beam with array TESD: (**a**) scatter plot of detected events on X-ray-photon-energy versus probe-beam-energy space, (**b**) enlargement near C K-edge, **c** total event yield and C K-line yield, (**d**) XANES spectrum of the C K yield normalized by probe beam intensity. The sample is the NIST Standard Reference Material 1216-I, containing C in an unknown chemical state at 0.7% by mass inside a matrix of porous micro-particulate silica beads. In (**a**), horizontal event groups are assigned to the O K-line at 525 eV and the C K-line at 277 eV, which are independent of the probe beam energy. The slanting event groups are due to the elastically scattered probe X-rays with primary and higher orders. In (**b**), an enlarged scatter plot for the C K-edge shows the slanting probe beam events and the horizontal C K-line events, which are enclosed by the violet line. In (**c**), the C K-line events are extracted from the total yield, avoiding the interference by the probe beam. In (**d**), the C K-line yield is normalized by probe beam intensity, constructing the XANES spectrum of the C K-edge and completely avoiding the interference from the scattered probe beam. Reproduced with permission from Doriese et al. [6]. Copyright 2017 AIP Publishing LLC

recovery, no meaningful change occurs on the XANES spectra. This signifies that the local atomic arrangement around the N atoms in the as-implanted state is the same as those after annealing at the high temperatures. This indicates the importance of the hot ion-implantation; the N atoms locally substitute for the C sites in the as-implanted state at 500 °C while the ion implantation produces the damage. The local substitution with lattice damage is essential for the N-dopant activation. The post-annealing exclusively removes the macroscopic damage of the host SiC crystal; the N atoms substitution for the C sites is essentially unaffected by the annealing. Post-annealing after ion implantation at room temperature cannot realize N-dopant activation in SiC. The XANES with the array STJD reveals the reason why the two-step process of hot

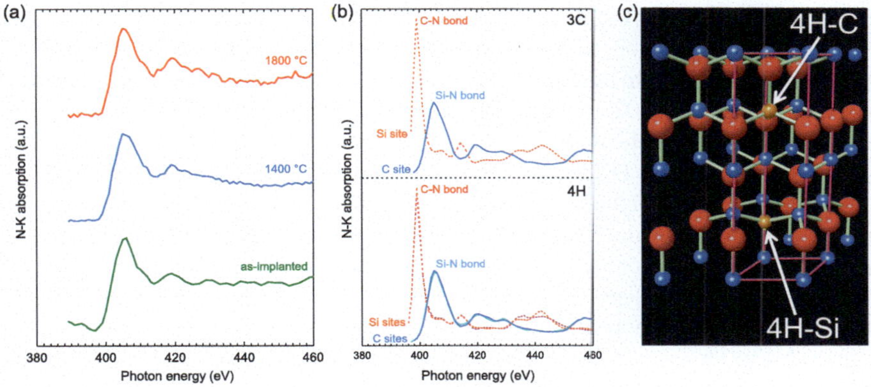

Fig. 6.17 XANES spectra of N-doped 4H-SiC at 800 ppm [58]: (**a**) experimental spectra in as-implanted state (green) and states after annealing at 1400 °C (blue) and 1800 °C (red) for 5 min, (**b**) ab initio calculation with FEFF code for N atoms substituting for either Si site (dashed red line) or C site (blue solid line), and (**c**) crystal structure of 4H-SiC with possible two substitution sites on Si (4H-Si) and C (4H-C). The N ions were impanated at 500 °C. Two non-equivalent sites for either Si or C in 4H-SiC lead to the slightly different calculation results in (**b**), although a difference between the heavily and lightly colored curves is negligibly small. The calculation on 3C-SiC was also performed because 4H-to-3C transformation easily occur during annealing. This is the first successful XANES measurement of SiC:N. Ohkubo et al. [58]. Licensed under CC BY-NC-SA 3.0: http://creativecommons.org/licenses/by-nc-sa/3.0/

ion implantation and post-annealing is indispensable. Unlike ZnO:N, no nitrogen gas bubbles are formed in the SiC:N samples even after annealing. This is explainable from an electronegativity point of view, considering the discussion on Si[65] and SiGe alloys [66]. The details of the difference in N doping between ZnO:N and SiC:N can be found in Ref. [17].

How low concentration can we measure a trace element in a matrix with the STJD-XAFS instrument shown in Fig. 6.11? SiC is also known as an advanced structural material with excellent mechanical strength, high oxidation resistance, and high radiation hardness. The importance of the combination of electronic and structural properties is revealed in porous SiC for gas sensing and catalyst supporting. One method to fabricate mesoporous SiC is anodization assisted with high-energy ion irradiation lattice defects. The sample for low concentration measurement is a 3C-SiC film epitaxially grown on a Si substrate with chemical vapor deposition (CVD). The film has a thickness of ~ 2.5 μm and a nitrogen dopant concentration of 1×10^{18} cm^{-3} at 20 ppm, which is 40 times lower than the previous 4H-SiC sample. Si ions irradiated the SiC film at 5.1 MeV up to a dose of 1.57×10^{15} cm^{-2} at 400 °C to create defects. A porous SiC region with a pore size of several μm was prepared by anodizing the irradiated sample [59]. Figure 6.18 shows that the STJD-XAFS instrument has the capability of measuring the N dopant at a concentration of as low as 20 ppm. We compare the XANES spectra at two locations with and without the ion irradiation. The N K-edge spectrum of the 3C-SiC with the ion irradiation resembles those of the 4H-SiC in Fig. 6.17. On the other hand, in the as-grown state without the ion irradiation, we can recognize the two peaks at 402 and 405 eV. These two peaks are assigned to the two paths of the 1s-to-π^* antibonding-orbital transition on the C-N bond for 402 eV and one on the Si–N bond for 405 eV, respectively, as explained for the 4H-SiC in Fig. 6.17. The two peaks are also observed in the amorphous Si–C-N CVD films [63, 64]; they represent the substitution of the N atoms for the Si site and the C site, respectively. The red curve in Fig. 6.18 indicates that the CVD 3C-SiC film has the N substitution for both Si and C sites equally in the as-grown state in contrast to the previous 4H-SiC:N. This difference from the 4H-SiC case probably stems from the 3C structure. The C-N bond peak for the Si-site substitution almost disappears after the ion irradiation; the Si–N bond peak for the C-site substitution is unchanged. The experimental results demonstrate that both Si- and C- site substitutions are possible in 3C-SiC; the N atoms on the C site are more stable than the N atoms on the Si site. The successful acquisition of the N-K-edge XANES spectra of the N dopant atoms at 20 ppm demonstrates the usefulness of the STJD-XAFS instrument for analyzing trace light elements.

Time-Resolved XAFS

Time-resolved XAFSis one of the pump-probe techniques to analyze transient states such as oxidation reduction reactions at a time scale of a few hundreds of ps or even sub-ps. Most XAFS measurements need a long acquisition time of several hours to scan monochromator energy, which is similar to WDS. In addition, the pump–probe XAFS technique requires repetitive excitation and delayed XAFS measurements. This measurement process is time-consuming even when we employ synchrotron-radiation. The XAFS acquisition time can be shortened by using a white-source and EDS-detector combination that covers an entire energy range of interest without scanning monochromatic probe energy. The EDS-detector should have a sufficient energy resolution such as a few eV for EXAFS measurement. XANES measurement is harder than EXAFS, which requires a sub-eV resolution. In early years, a single-pixel TESD was utilized to perform X-ray absorption measurement by using a 9-V battery powered white X-ray source. A small jump of X-ray absorption was recognized at the Ti K-edge and Co K-edge in 2006 [60]. Figure 6.19 shows the experimental setup, where a thin sample is placed between a white-like X-ray source and an EDS detector that measures EXAFS spectra without scanning probe-beam energy.

If there is a bright X-ray source and an EDS detector with an energy resolution of ~ 1 eV, we can realize lab-based EXAFS measurement without using a monochromator at huge facilities such as synchrotron radiation and free electron laser. The lab-based table-top EXAFS measurement is performed by using a laser plasma X-ray source and a 22-pixel TESD with an integrated energy resolution of 3.14 eV at 5.9 keV [61]. Figure 6.20 displays an X-ray absorption spectrum of ferrocene ($Fe(C_5H_5)_2$) with a Cu filter. The part (a) shows the pulse height X-ray spectrum in a range of 5.5–9.2 keV, which contains the peaks due to a ^{55}Fe calibration source, fluorescent X-rays

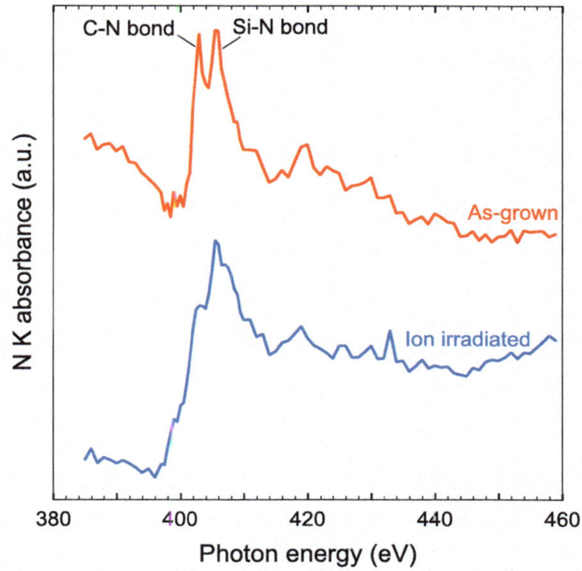

Fig. 6.18 XANES spectra of 3C-SiC film at an N dopant level of 20 ppm [59]: as-grown state in red and post-irradiation state in blue. The red curve is shifted upward for clarity. The 3C-SiC was epitaxially grown on Si with chemical vapor deposition (CVD) at an N concentration of 1 $\times$ 10^{18} cm^{-3} (20 ppm). The 5.1-MeV Si ions at 1.57 $\times$ 10^{15} cm^{-2} irradiated the selected area at 400 °C. Although the N concentration is 40 times lower than that in Fig. 6.17, the STJD successfully acquires the PFY-XANES spectra of the N K-edge, avoiding interference with the strong C K-line. The peak at 402 eV and 405 eV are assigned to the 1s-to-π^* antibonding-orbital transitions of the C-N bonds and the Si–N bonds, respectively. The N substitution for both Si and C sites contrasts with the substitution exclusively for the C site in the hot implantation into the 4H-SiC in Fig. 6.17. The MeV ion irradiation causes a preferential loss of the C-N bonds, which indicates that the N substitution for the Si site is unstable. Reproduced with permission from Maeda et al. [59]. Copyright 2020 American Chemical Society

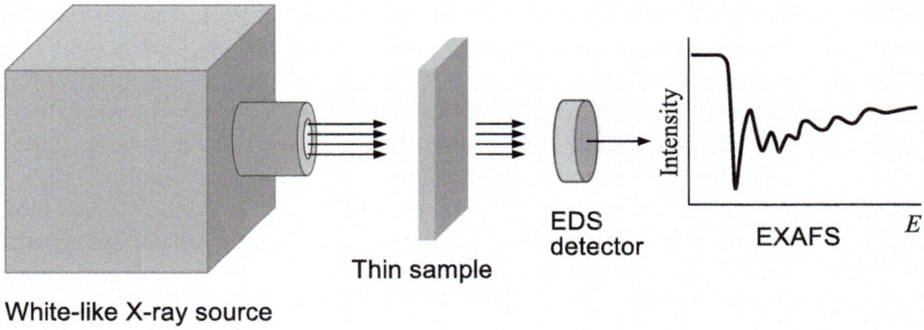

Fig. 6.19 EDS-EXAFS: XAS measurement with white X-ray source and EDS detector [60]

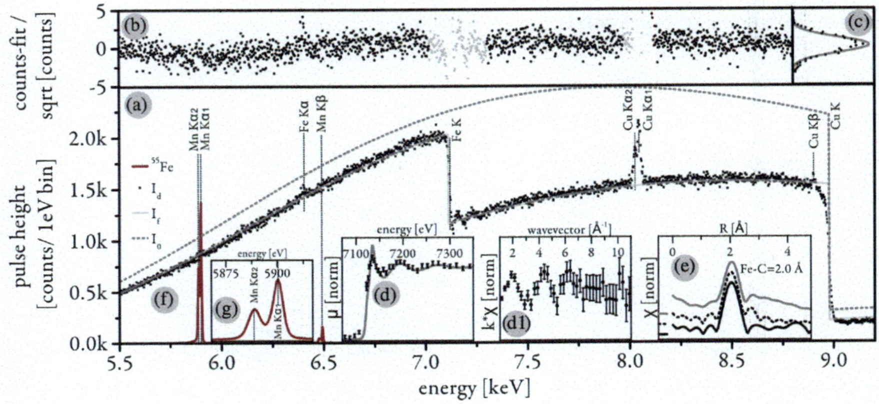

Fig. 6.20 EDS-EXAFS: X-ray absorption spectrum with laser plasma white X-ray source and TESD detector [61]. Note that "pulse height" is usually used for the *x*-axis representing photon energy of X-ray spectra; the *y*-axis is X-ray intensity in counts/channel or counts/bin. The "pulse height" should read X-ray intensity. The sample is ferrocene ($Fe(C_5H_5)_2$) with a Cu filter. The part (**a**) shows the TES-EDS X-ray spectrum, in which *x*- and *y*-axes are the detected X-ray photon energy and X-ray intensity in count per 1-eV bin. The part (**a**) is a conventional pulse height spectrum, but it includes EXAFS oscillation. The EXAFS spectrum of Fe is enlarged in (**d**). In addition to the Fe and Cu K-edges, the pulse height spectrum includes the peaks due to a ^{55}Fe calibration source and the fluorescent X-rays from Mn, Fe, and Cu. Reproduced with permission from Uhlig et al. [61]. Copyright 2013 American Physical Society

from the Fe atoms in the sample and Cu in the filter. It is possible to observe clear EXAFS oscillation of Fe and Cu simultaneously at an energy resolution of 3.14 eV. The white X-ray source and TESD combination enables EXAFS measurement of multiple elements simultaneously. This experiment paves a way to lab-based time-resolved XAS measurement.

The first time-resolved EXAFS measurement was performed for the photoreduction reaction of ferrioxalate $[Fe^{III}(C_2O_4)_3]^{3-}$ complexes [67]. The ferrioxalate existing in natural water or aerosol produces $CO_2^{\cdot-}$ radicals by photoreduction. The photolysis of the complexes results in the consumption of oxygen in natural water. It is used as chemical actinometer. The photoreduction process has not been elucidated completely. The Fe K-edge EXAFS spectrum of the ferrioxalate was acquired with the 240-pixel TESD with an energy resolution of 6.5 eV at 6.93 keV (Co K-L_3 line), at a delay time of 100 ps after 400-nm pump pulses. The sample is a 440-mM aqueous solution of ferrioxalate forming a water jet form with a diameter of 100 μm. The EDS-EXAFS experiment identifies an intermediate state of Fe^{II}. It is concluded that ligand-to-metal charge transfer is prompt, which is consistent with other experiment with a free electron X-ray laser. However, the acquisition time for the EXAFS spectrum in a range of 7–8 keV was 154 h, which was settled by a photon count rate of the TESD for EDS-EXAFS. A similar table-top time-resolved EXAFS was performed by WDS with a flat Si(111) analyzing crystal, a sub-ps X-ray source, and a liquid-nitrogen cooled X-ray CCD [68]. The WDS spectrometer, which covers an

EXAFS range of ~ 1000 eV at a time, has an energy resolution of 20 eV on a setup at a sample-to-crystal length of 120 mm and a crystal-to-CCD length of 100 mm. The energy resolution of 20 eV that is worse than 6.5 eV of the TESD is to obtain a practical detection efficiency. The energy resolution of WDS can be improved to a sub-eV range, but it results in a long acquisition time. They did not report an acquisition time; therefore, we cannot directly compare TESD-EDS-EXAFS and CCD-WDS-EXAFS in total efficiency. Nevertheless, it seems that WDS-EXAFS and TESD-EXAFS requires a similar measurement time when the WDS spectrometer has a similar energy resolution to that of the TESD. We should consider total throughput as well as energy resolution. The non-scanning CCD-WDS is a competitor of superconductor EDS detectors. Practical applications of superconductor detectors to analytical science are different from proof-of-concept experiments and require further development for practical use.

We need to consider energy resolution, count rate, covering energy range, and effective solid angle for analytical instruments. STJDs satisfy the count rate requirement at a moderate energy resolution of 2.5–6 eV below 500 eV, but they cannot cover an X-ray range of 2–10 keV. X-ray photon detection efficiency of STJDs is as low as ~ 5% at 6 keV. TESDs have a high energy resolution of 1–2 eV at 6 keV; however, the count rate of TESDs at such high energy resolution does not satisfy the requirement for high throughput analysis. Furthermore, the 1-2 eV resolution is occasionally considered as medium resolution. When sub-eV or meV resolution is mandatory, analytical scientists employ WDS. We need to consider Table 5.1 to compete or cooperate with existing detectors or conventional spectroscopic methods to realize advanced analytical instruments.

6.3.3.3 Particle Induced X-Ray Emission (PIXE)

Electron and X-ray probes are common to induce characteristic X-rays. Another probe beam is proton or helium ions, which can avoid Bremsstrahlung background compared with electrons. No such distinct peak due to an elastically scattered X-ray beam appears in ion probes, but we must take into account of backscattered probe ions, which is another analytical method called Rutherford backscattering. The background events limit the detection sensitivity of trace elements as shown in Fig. 6.7. X-ray-probe induced events interfere characteristic X-ray detection on pulse height spectra as shown in Figs. 6.13 and 6.15. The analytical method with proton or helium ions is called particle induced X-ray emission (PIXE). TESDs were deployed to a Van de Graaf accelerator facility at University of Jyvaskyla [69, 70]. A 160-pixel NIST-TESD with time-division multiplexing SQUID readout is installed into an analyzing chamber. They performed PIXE experiments in a range of 1–15 keV by using a 2.01-MeV proton beam and the 30–60 TES-pixels. The best pixel has an energy resolution of 3.06 eV; the tolerable count rate of the array TESD is ~ 1 kcps. They measured lead pigments: lead red of Pb_3O_4 and lead white of $2PbCO_2{\cdot}Pb(OH)_2$, in addition to an industrial-grade bulk Pb sheet as a reference material. Figure 6.21 shows PIXE spectra of thee Pb-related samples. The blue TESD spectra identify impurity elements that are unrecognizable on the red SDD spectra. The Pb red pigment contains Pb and Au. The Cu peak stems from the sample stage. On the other hand, the industrial-grade Pb sheet and the Pb white have many impurity elements such as Ba, Ca, Ti, Fe, and Ni at 100–9000 ppm. In addition to elemental identification, the TESD-PIXE also enables chemical shift analysis of Ti at different forms of metallic Ti, TiN, and TiO_2 by comparing Ti K-$L_{2,3}$ and Ti K-M_3 line shapes [70].

Besides TESDs, an array AIST-STJD is deployed for PIXE to a Van de Graaf accelerator facility at University of Tsukuba [71, 72]. The 500-pixel STJD was installed at the focused ion beam line [72]. The performance of one of the pixels was investigated for light element analysis at this moment. A 4-MeV proton-beam irradiated model samples such as SUS304, aluminum sheet, PTFE, and polyimide. They investigated STJD artifact events such as substrate events due to high energy X-rays, back-scattered protons, and devised an artifact-event rejection protocol with pulse rise-time discrimination to reduce background events in a low energy range [72].

6.3.3.4 X-Ray Emission Spectroscopy (XES)

Characteristic X-ray line shapes represent the electronic structure of valence-bands, as depicted in Fig. 6.9. Normally, XES is performed by a combination of synchrotron radiation and WDS that has sub-eV energy resolution in most cases. A 192-pixel TESD with an energy resolution of 4 eV at 4.93 keV was used to analyze valence-to-core XES for Ti-based compounds with simple ligand geometries and different oxidation states: Ti metal, $Ti^{II}O$, $Ti^{III}N$, $Ti^{IV}C$, and two forms of anatase and rutile $Ti^{IV}O_2$ [73]. The array TESD and a commercial X-ray tube in a geometry of X-ray fluorescence analysis enables the lab-based measurement of necessary energy ranges for Ti K-$L_{2,3}$ (4.504–4.510 keV) and Ti K-M or K-N (4.931–4.966 keV). Figure 6.22 shows lab-based XES spectra of the six Ti-based compounds. The TESD-EDS-XES can capture differences among the six samples in K-M to K-N lines representing the valence-to-core transitions in an energy range near 4950 eV. Although the 4.0-eV energy resolution is approximately 10 times worse than WDS (~ 0.5 eV), the valence-to-core XES features acquired by the TESD is consistent with the data by a combination of a synchrotron radiation beam and a WDS spectrometer. Currently, the TESD acquisition time is 100 times longer than that of WDS with synchrotron radiation. As is in the time-resolved TESD-EDS-EXAFS, improvement of count rate by a factor of 10–100 is desirable to realize practical analysis.

The 100-pixel Nb/Al-STJD with an average energy resolution of 6 eV can modestly measure the shapes of K-lines of light elements and L-lines of heavy elements at a count rate up to ~ 200 kcps [Ref. 17]; however, STJDs cannot cover above ~ 5 keV due to a low X-ray absorption efficiency of ~ 12% and a low energy resolution of ~ 15 eV following the $\sqrt{E_0}$ dependence. Ta/Al-STJDs have a better energy resolution and a higher detection efficiency, although they are still insufficient for an energy range over ~ 5 keV. A large array with a small junction size or

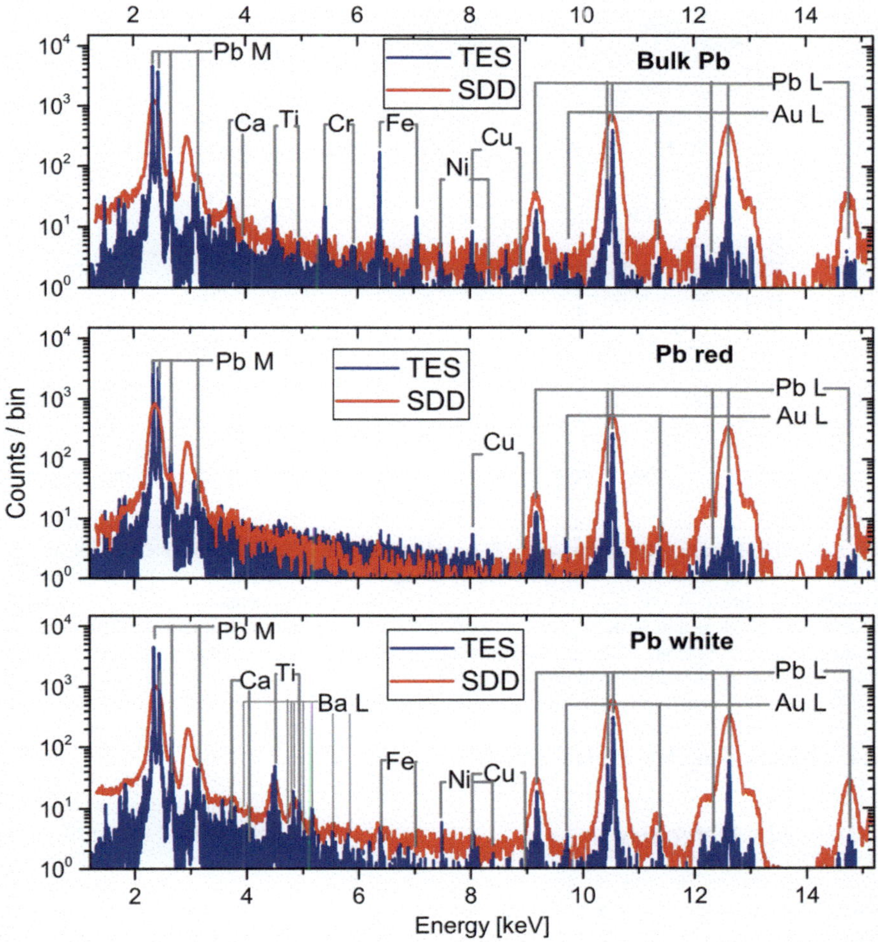

Fig. 6.21 PIXE spectra of three Pb-related samples [70]: industrial-grade Pb sheet, Pb red pigment Pb_3O_4, and Pb white pigment $2PbCO_2 \cdot Pb(OH)_2$. The red spectra were acquired for 10 min with the SDD with an energy resolution of 130 eV at 5.9 keV. The blue spectra were acquired for 50 min with the array NIST-TESD with an energy resolution of 3.06 eV for the best pixel. The Pb sheet and the Pb white pigment contain many impurity elements at 100–9000 ppm as well as Pb, whereas the Pb red contains only Pb and Au. The Cu peak is due to the sample stage. Reproduced with permission from Palosaari et al. [70]. Copyright 2016 American Physical Society

aperture to reduce spatial inhomogeneity can satisfy the requirement for table-top XAFS and XES in a soft X-ray range below a few keV, focusing on the K-lines of light elements and the L-lines of heavy elements.

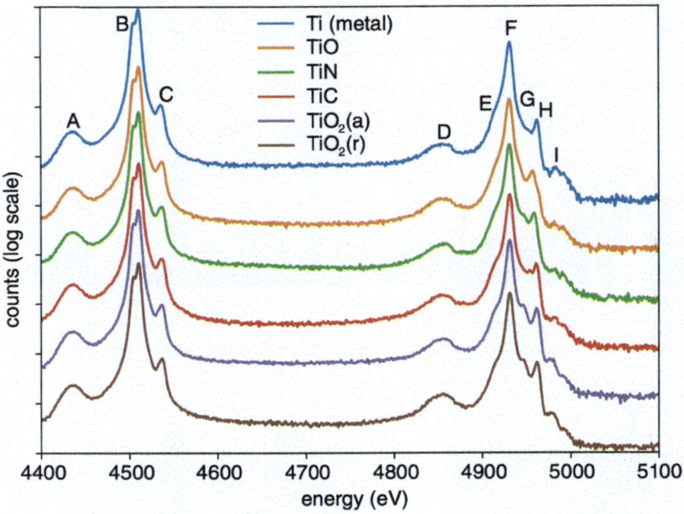

Fig. 6.22 TESD-EDS-XES spectra of six Ti-based compounds [73]: Ti metal, $Ti^{II}O$, $Ti^{III}N$, $Ti^{IV}C$, and two forms of anatase and rutile $Ti^{IV}O_2$. The capital letters denote the peaks assignments: A—detector artifact, B—Ti K-$L_{2,3}$, C—K-L satellites, D—detector artifact, E—$K_{\beta'}$, F—K-$M_{2,3}$, G—$K_{\beta''}$, H—K-$M_{4,5}$, and I—K-M satellite. Peak shifts and line shapes represent the differences in valence-to-core transitions of the six compounds. Reproduced with permission from Miaja-Avila et al. [73]. Copyright 2021 John Wiley & Sons, Ltd

6.4 Mass Spectrometry (MS)

6.4.1 MS Detectors

Mass spectrometry (MS) began with the study of "Ray of Positive Electricity and Their Application to Chemical Analysis" by Thomson in 1913 [74]. He observed that positive rays consisted of particles with varying masses, unlike cathode rays, and thus realized that the positive rays were useful to analyze chemical compounds such as carbonic acid, marsh gas, cyanogen, and helium. The chemical compounds in a gas discharge became positive ions. The ions were deflected by electric or magnetic force, and detected by photography, which was an imaging ion detector. No solid-state ion detectors existed in the 1900s. Twenty-seven years afterward, the term "Mass Spectrometer" was coined in a patent by Hoover, Jr. in 1940. Currently, mass spectrometers in mass spectrometry, including peripheral equipment, analytical methods and databases, are the pivot of analytical science not only for chemists but also for physicists. MS also deals with negative ions after Thomson's positive rays.

Mass spectrometers consist of three components: ionization, separation, and detection, similar to other analytical instruments. As an example of the separation types, Fig. 6.23 illustrates time-of-flight (TOF) mass separation. In addition to the three components, the relative importance of peripherals is substantial. Especially, analyzing software packages and databases are extremely important for omics such as proteomics and metabolomics that target peptides and proteins. It is noteworthy that some MS analysts rarely observe mass spectra directly. This causes less attention to MS detectors in the MS community [1, 75]. However, undoubtfully, new detectors innovate mass spectrometers in a similar way to other analytical instruments.

Figure 6.23 shows different ion source types: electron impact (EI) and fast atom bombardment (FAB) for cases in which dissociation is not a serious problem; and matrix-assisted laser desorption/ionization (MALDI) and electrospray ionization (ESI) for peptides and proteins without dissociation. EI and FAB induce fragmentation of molecules. In contrast, MALDI and ESI are categorized as soft ionization that induces no fragmentation even for large biomolecules such as proteins and protein complexes bonded by non-covalent force. MALDI and ESI enable analysis of intact biomolecules. Before the invention of MALDI and ESI, ionization of biomolecules caused dissociation, which is called in-source decay (ISD). Ions generated in an ion source are accelerated by a static voltage U in vacuum, acquiring a kinetic energy of

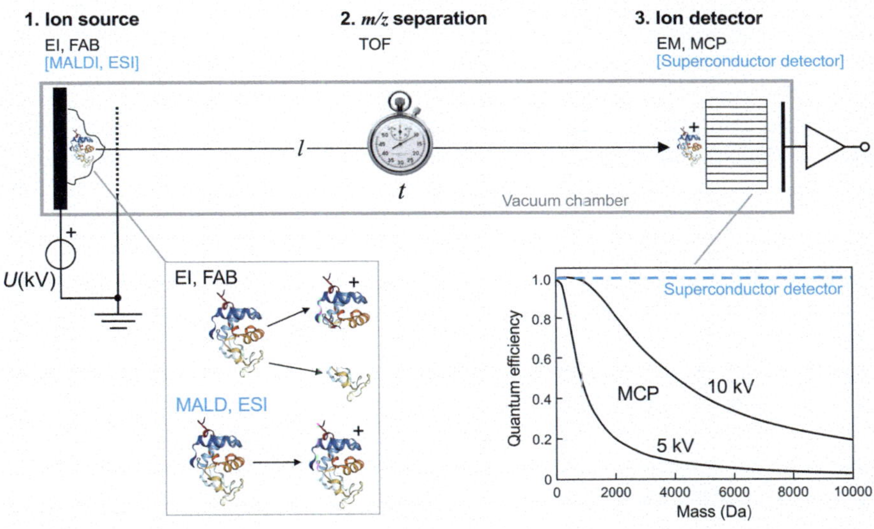

Fig. 6.23 Time-of-flight mass spectrometer (TOF–MS): 1. ion source, 2. mass separation or *m/z* separation, and 3. particle or ion detector. Analytes are ionized using ionization methods such as EI, FAB, MALDI, and ESI. The ionized molecules are accelerated at U (kV). Flying for a duration of t over a distance of l, the ions are detected by a detector. The time interval between an ionization or acceleration trigger and a detector output corresponds to the time-of-flight (TOF) that is t. Microchannel plate (MCP) detectors, a conventional type in MS, exhibit low detection efficiency for high mass proteins [76, 77]. The lower-right inset shows the data for polystyrene oligomers accelerated at either 5 or 10 kV [76]. Superconductor detectors have mass-independent detection efficiency [78]. More importantly, superconductor detectors solve two inherent MS problems: "*m/z* overlap" and "neutral loss." The lysozyme molecular structure in the lower left panel are based on the RCSB PDB archive at https://www.rcsb.org under CC BY-SA 3.0. Reproduced with permission from Ohkubo [17]. Copyright 2023 AIP Publishing LLC

zeU, where z is the number of ionic charge. The ions fly toward the detector, which is called the linear mode; they are detected by an MS detector. Metastable ions excited in ion sources dissociate during free flight, which is a process known as post-source decay (PSD).

MALDI normally utilizes a pulsed UV laser to be absorbed by matrix molecules mixed with analyte molecules. Each laser pulse starts a clock and detector signals stop it. One laser pulse generates numerous ions with a diversity of m values. The time-of-flight t of each ion is expressed by

$$t = \frac{l}{v} = l\sqrt{\frac{1}{2eU}\frac{m}{z}}, \tag{6.3}$$

where v is the flight velocity of the ion. By measuring t, we can calculate m/z. The ion count or ion intensity (ion flux) as a function of m/z forms a mass spectrum. Modern commercial TOF–MS instruments have more sophisticated configuration, such as the reflectron mode to improve a mass resolution and tamdem MS for molecular structural analysis by using fragmentation. Besides TOF, there are many ion separation techniques: sector type that is similar to the instruments by Thomson, Fourier transform ion cyclotron resonance (FT-ICR), ion trap, and ion mobility.

The difference between the ion count and the ion intensity resembles the difference between the calorimetric and bolometric detector types. The former is called "pulse mode" or "pulse counting mode," on which the number of individual ions within a short time-interval (time bin or m/z bin) is counted. On the pulse mode, the TOF value of each ion is converted to a digital value by using a time-to-digital converter (TDC). The TOF values are then converted to m/z values according to Eq. 6.3. A histogram of the ion count against the x-axis divided into m/z bins forms a mass spectrum, similar to the pulse height spectrum of X-rays. The other mode is called "current mode" or "analog mode." On the current mode, detector output current is continuously recorded against the elapsed time after ionization or acceleration. This corresponds to flux measurement in a similar way to the thermal detection of IR radiation. The pulse mode has a high sensitivity due to single ion detection but dynamic range is narrow; counting loss occurs at high ion flux. On the other hand, the current mode has a low sensitivity but dynamic range is wide because the detector does not count individual ions. In this book, we focus on the pulse mode, which is also called single particle MS or single molecule MS [79].

There are several types of MS detectors [75]. A common ion detection mechanism involves secondary electron emission induced by ion impact on a detector surface; typical detectors are electron multipliers (EM) and microchannel plates (MCP). The electron multiplication increases the number of electrons from initial one electron to ~ 10^6 electrons, which can activate semiconductor devices such as FET transistors. Ion detection efficiency or quantum efficiency of MS detectors affects total sensitivity of MS instruments. The quantum efficiency is predominantly proportional to the ion velocity v, although it depends on many factors such as kinetic energy, mass, molecular composition, and detector operation condition [80]. Therefore, at a certain acceleration voltage, the quantum efficiency is roughly inversely proportional to $\sqrt{m}$.

The lower right inset in Fig. 6.23 plots the MCP quantum efficiency as a function of mass at either acceleration voltage of 5 or 10 kV for polystyrene oligomers with a mass of 2500 Da [76]. Beyond 100 kDa, the MCP detection efficiency is nearly zero [76, 77]. Proteins such as 14.305-kDa lysozyme and 66.430-kDa bovine-serum-albumin (BSA) are still detectable at an extremely low sensitivity, whereas 1-MDa immunoglobulin M (IgM) is below the detection limit.

The initial motivation of the superconductor MS detector development was a mass-independent quantum efficiency, as depicted by the blue dashed line in the lower right inset of Fig. 6.23 [Ref. 81]. The flat quantum efficiency is expected from the sensitivity to phonons or the meV energy gaps of superconductors. The phonon energies are higher than $2\Delta_g$, as shown in Fig. 3.4. The ion surface-impact in a keV range generates a Ω_D phonon bubble; the succeeding relaxation and quasiparticle generation processes are the same as those in Fig. 3.3 b. Figure 6.24 illustrates ion or molecule impact on a detector surface [82]. If the ion is an MeV α-particle and the detector is Si, the α-particle penetrates deep into the Si crystal, generating collision cascades and e–h pair. The Si detector enables kinetic energy measurement of α-particles without counting loss; such detectors are called charged particle detectors in nuclear physics. On the other hand, if the ion is an IgG biomolecule with an energy of 5–10 keV and the detector is an MCP, the particle completely stops on the detector surface, and one secondary electron is occasionally emitted every ~ 100 collisions. The IgG is an immunoglobulin with a mass of 146 kDa. The secondary electron is multiplied by a factor of ~ 10^6 just like photomultipliers (PMTs). A peak of IgG can be formed on a mass spectrum; however, counting loss is large. When the detector is Nb in a superconducting state, the IgG molecule that stops on the Nb surface generates the Ω_D phonon bubble, exciting numerous quasiparticles. If we assume that the 1-keV IgG contains carbon atoms only, each carbon atom conveys a kinetic energy of 82 meV, which is sufficient to generate a Debye phonon at $\Omega_D = 24$ meV. Therefore, the quantum efficiency of superconductor MS detectors is 100% without counting loss. Moreover, Nb/Al-STJDs enable kinetic energy measurement besides ion counting. Even when a 1.6-μm thick SiO_2 layer covers the Nb surface, the STJD generates a distinct peak on a pulse height spectrum for 3-keV CH_3CO ions [83]. This demonstrates that STJDs have a quantum efficiency of 1.0 for large molecules, enabling kinetic energy measurement through the phonon propagation.

The first application of the STJ type to MS was proposed and demonstrated by Twerenbold in 1996 to solve the problem of the low quantum efficiency for high-mass proteins [81, 84]. The STJ MS detector with an area of 20 × 80 μm^2 consisted of a layer structure of Sn/SnO_x/Sn. The NIS type was also applied to MS by Hilton et al. in 1998 [85]. An NIS MS detector with a layer structure of Ag/AlO_x/Al attached to an 200-μm square Ag absorber distinguished different charge states at a better energy resolution than the STJ MS detector. Both STJ and NIS types have similar output pulse parameters of $\tau_{rise} = 1.2$ μs and $\tau_{fall} = 17$ μs. The works on superconductor MS detectors before 1999 were reviewed by Frank in Ref. [2]. Large-molecule detection beyond 1 MDa at 10 times faster response was realized by a 16-pixel Nb/Al-STJD by Wenzel et al. in 2005 [86]. The STJ and NIS types can detect high-mass molecules up to ~ 1 MDa at a high sensitivity in the proof-of-concept experiments.

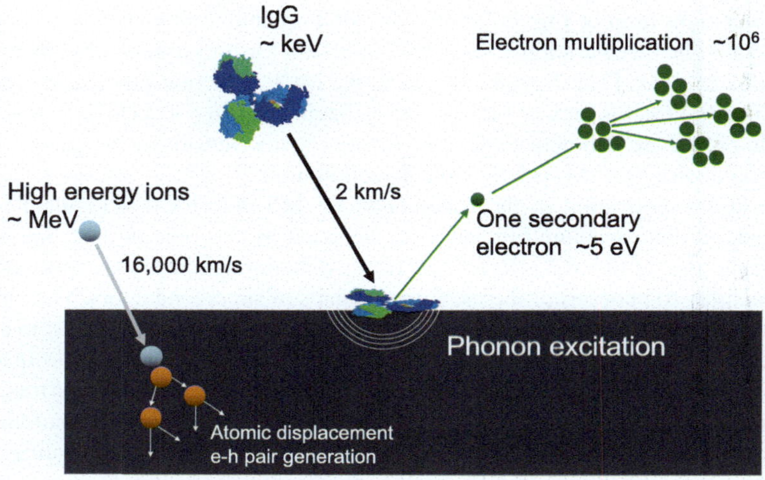

Fig. 6.24 Comparison between secondary-electron-emission detection and phonon-sensitive superconductor detection [82]. The illustration also shows a difference between keV particle impact that creates a secondary electron in MS detectors and MeV particle impact that creates a cascade of atomic collisions in charged particle detectors for nuclear physics. The secondary electron is multiplied by a factor of ~10^6 for a semiconductor amplifier. The atomic collision cascade generates e–h pairs when the detector material is a semiconductor. The charged particle detectors can measure kinetic energies of energetic particles, but have no sensitivity to molecules in a keV range. Reproduced with permission from Ohkubo [82]. Copyright 2008 Elsevier B.V

However, both STJ and NIS types are extremely smaller than a typical ion beam size of ~ cm. Fast-response-type MCPs have a diameter of ~ 2.5 cm. Furthermore, for detecting high-mass molecules, conversion dynode detectors or ion conversion detectors with a sensitive area of ~ cm enables the detection of macromolecules up to ~ 2 MDa or even higher: it operates at room temperature and is sufficient to acquire mass spectra with a large sensitive area. The conversion dynode detectors utilize secondary ion emission and post-acceleration instead of secondary electron emission and multiplication in MCPs [87, 88] Even when the quantum efficiency of the conversion dynode detectors is much less than 1.0, the large sensitive area compensates the low quantum efficiency. The mass-independent detection efficiency of superconductor MS detectors [78, 89] still have a merit for quantitative comparison of mass-peak heights in a wide mass range. However, currently, MS analysts do not need to compare peak heights precisely. Therefore, the mass-independent quantum efficiency of superconductor detectors unfortunately became less important in the MS community.

Superconductor detectors still have superiority over conventional MS detectors. Besides simply detecting macromolecules at the mass-independent sensitivity, we can solve two fundamental problems of mass spectrometry by using kinetic energy sensitivity. The IUPAC defines MS as an analytical method to characterize gas phase ions with mass number (m) and charge number (z) [90]. More preciously, it is a value of m divided by z, which is m/z. Equation (6.3) discloses two inherent MS problems: "m/z overlap" and "neutral loss."

The m/z overlap occurs because ions are separated according to m/z. The spectroscopic axis (x-axis) of mass spectra is not m but m/z. Strictly speaking, therefore, conventional mass spectrometry should not be called mass spectrometry but "m/z spectrometry." Note that the dimensionless number m/z means a ratio of the unit-less mass number m to the unit-less charge number z. It is often called mass-to-charge ratio, but the MS community recommends simply calling m/z (ém óʊvər zi:); it does not have the unit of Da/C. The ion separation along the m/z axis implies that multimer ions at different charge states overlap on mass spectra such that singly-charged monomer (m/z), doubly-charged dimer ($2\ m/2z$), triply-charged trimer ($3\ m/3z$), and so on have the same m/z value. There is always ambiguity on z values in MS because all the ionization methods potentially produce multiply charged ions: especially ESI produces highly charged ions. A typical example of the m/z overlap is observed in the lysozyme protein that tends to aggregate and readily form multimers. The m/z overlap can be overcome with superconductor detectors that determine z values m/z, realizing true MS.

The neutral loss occurs in tandem mass spectrometry, which is abbreviated as MS/MS. In MS/MS, primary ions or precursor ions are selected by the first MS. The precursor ions are fragmented by a dissociation method. The second MS analyze the fragments. If necessary, this sequence is repeated. MS/MS is utilized for identifying molecules and determining molecular structure from fragmentation patters in addition to simple m/z measurement. Membrane proteins are not crystallized; therefore, MS/MS is applicable instead of X-ray crystallography. The problem of MS/MS is that the fragments do not necessarily have charge. When the fragments are neutral ($z = 0$), Eq. (6.3) becomes infinity. Therefore, neutral fragments cannot be analyzed and lost inside the ion separation section, which is called "neutral loss." The superconductor MS detectors can also overcome the neutral loss, realizing neutral MS.

6.4.2 MS Instrumentation with Superconductor Detector

Figure 6.25 shows one of the TOF–MS instruments equipped with an array STJD [32]. The ESI source and acceleration electrodes are located at the top of the vertical flight tube. The ESI source is milder than the MALDI, which allows ionizing even non-covalent protein complexes without ISD under atmospheric pressure. Meanwhile, it produces highly charged ions, which makes mass peak assignment complicated. Charge-state discrimination with superconductor detectors simplifies the peak assignment. The ions are transferred into the vacuum chamber through small orifices. Start signals for TOF are the triggers for high voltage pulses at 7 kV. The reflectron at the bottom of the flight tube reflects ions, focusing ions with the same m/z but different kinetic energies temporally, which drastically improves mass resolution. This is called the reflectron mode. The MCP detector detects the reflected ions.

When the reflectron is not engaged, the ions pass through the reflectron and impinge the surface of a 100-pixel STJD cooled at 0.3 K underneath the TOF tube. This is the linear mode. Each pixel has an area of $200 \times 200\ \mu m^2$. The 100 STJ-pixels are arranged in a 0.4-mm wide and 1-cm long area, designed for the beam shape due to the orthogonal acceleration in Fig. 6.25b. Unlike X-ray photons, ions need open space between an ion source and an MS detector; IR radiation from room temperature causes a serious temperature-rise of the STJD. To block the IR radiation, the honeycomb IR blocking filter cooled at 3 K is placed in front of the STJD [91, 92]. The filter having numerous holes with a diameter of 3 μm provides the open space for ions, reflecting the IR radiation. The STJD readout is the same as that for STJD-SEM and STJD-XAFS. Mass spectra are acquired with either STJD or MCP. This mass spectrometer is utilized to analyze hemoglobin that is a non-covalent hetero tetramer [17, 93]. MALDI is also a soft ionization method; however, it fragments non-covalent complexes. Other samples analyzed by several types of MS instruments at AIST are polystyrene polymers [94], proteins in a mouse cerebrum [95], C60 fullerenes [96, 97], and metal clusters [98].

Considering crosstalk and bandwidth for timing measurement with the STJD, the TOF–MS is equipped with a combination of Cu-alloy micro-flex coaxial cables inside

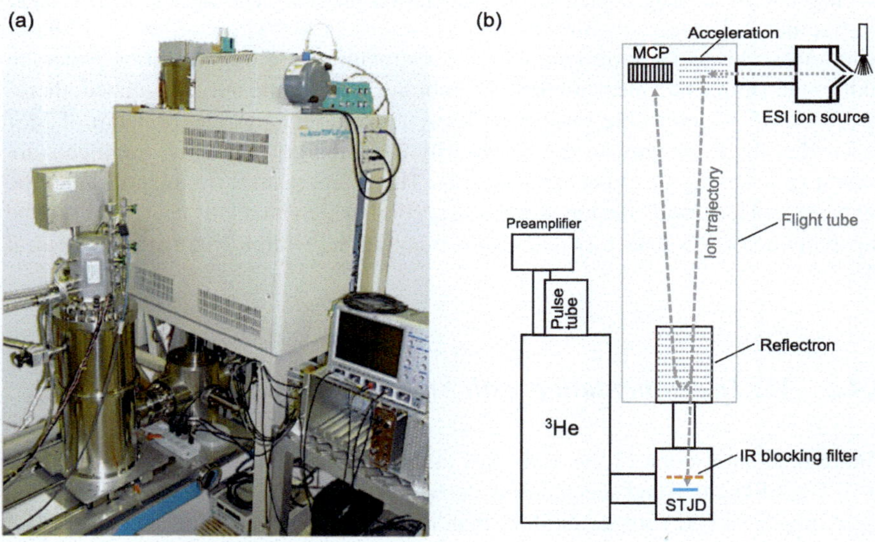

Fig. 6.25 ESI-TOF mass spectrometer equipped with STJD having 100 pixels of 200-μm square STJs [32, 93]: (**a**) ^{3}He cryostat standing next to MS enclosure and (**b**) component layout. Ions produced by electrospray ionization (ESI) under atmospheric pressure continuously flow into the vacuum chamber through small orifices and ion optics. Voltage pulses at 7 kV accelerate ions. Each TOF value is measured between a voltage pulse trigger and an STJ output pulse on the linear mode, or an MCP pulse on the reflectron mode. Ions passes though the IR blocking filter in front of the STJD [91, 92]. Reproduced with permission from Ohkubo [17]. Copyright 2023 AIP Publishing LLC

the cryofinger at 0.3 K, superconductor NbTi semi-rigid coaxial cables between 0.3 and 3 K, and CuNi semi-rigid coaxial cables between 3 K and room temperature. Measurement of thermal conductivity of all the coaxial cable types ensures the installation of these wiring leads, avoiding a serious heat load to the STJD [99–102]. The base temperature of 0.314 K holds for 32 h after installing the 100 coaxial cables. It was shorter than 130 h at 0.295 K without the cables and the cryofinger. When all the liquid ^{3}He evaporates, an automatic recharge process liquefies ^{3}He and maintains the base temperature. NbTi twisted-pair wiring is a suitable option for X-ray spectroscopy with X-ray windows.

6.4.3 *True MS Eliminating* m/z *Overlap*

6.4.3.1 *m/z* Overlap

A simple example of the *m/z* overlap appears in mass analysis of the air ionized by EI and accelerated at 3 kV. Figure 6.26 plots the ion count as a function of *m/z*, which is a conventional mass spectrum [103]. The atoms and molecules including isotopes in the air form the mass peaks. However, the peak at *m/z* 14 likely contains both $^{14}N^+$ and $^{14}N_2^{2+}$ ions. The *m/z* values of these two ions are exactly the same as 14.0025; therefore, it is impossible to separate these two ions even with any high-resolution MS instruments such as Fourier-transform mass spectrometer (FT-MS). Electron impact (EI) dissociates N_2 molecules; it produces $^{14}N^+$ as well as $^{14}N_2^+$ ions that should be the strongest peak at *m/z* 28 due to the largest ionization cross-section. In addition to singly-charged ions, EI produces doubly-charged ions. Therefore, we cannot rule out the formation of $^{14}N_2^{2+}$. One can speculate the presence of doubly-charge N_2 ions by observing $^{14}N^{15}N^{2+}$ (*m/z* 14.5), which is an isotope molecule, between the $^{14}N^+$ and $^{15}N^+$ peaks. However, the abundance of ^{15}N is only 0.364% in the air. No peaks appear at *m/z* 14.5 and 15.0 in Fig. 6.26. The situation of the oxygen molecule is more complicated than that of the nitrogen molecule. An OH^+ peak overlaps with $^{16}O^{18}O^{2+}$ (*m/z* 17) due to the presence of water in air; therefore, the presence of $^{16}O^{18}O^{2+}$ is ambiguous. Consequently, no mass spectrometer has confirmed the presence of the doubly-charged ions of $^{14}N_2^{2+}$ and $^{16}O_2^{2+}$ before an experiment using an STJD.

The detection of doubly-charged diatomic ions is one of the important topics in planetary science [104]. The doubly-charged diatomic ions and other positive ions are related to the planetary atmospheric evolution in the solar system. No exact experimental confirmation has been accomplished for the doubly-charged ions in the ionosphere. When the metastable N_2^{2+} and O_2^{2+} dissociate into singly-charged atomic ions, the N^+ and O^+ ions acquire a kinetic energy of several eV through a kinetic energy release (KER) process. It is believed that such atomic ions escape against the gravity of planets such as Venus, Earth, Mars and Titan. Therefore, the doubly-charged diatomic ions as well as other positive ions influence the evolution

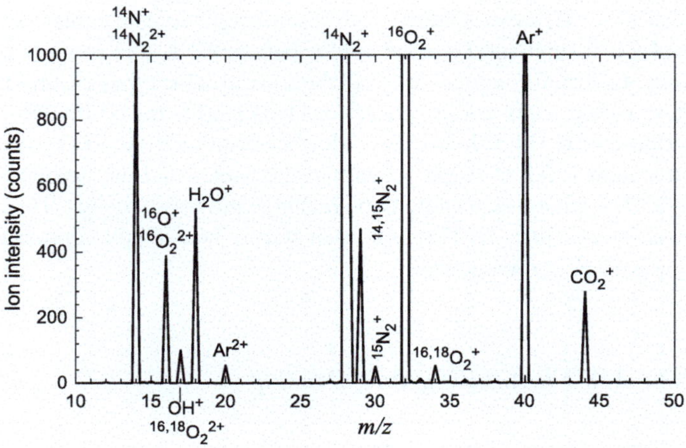

Fig. 6.26 Conventional mass spectrum of air ionized by EI at 70 V. The ions were accelerated at 3 kV. The *m/z* overlap can occur for the pairs of $^{14}N^{+}$–$^{14}N_2^{2+}$(*m/z* 14), $^{16}O^{+}$–$^{16}O_2^{2+}$(*m/z* 16), and OH^{+}–$^{16}O^{18}O^{2+}$(*m/z* 17). No MS instruments can separate these ions with the same *m/z* values because conventional mass spectrometry is *m/z* spectrometry to be precise. The intense peaks of the singly-charged atomic ions always mask the possible faint peaks of the doubly-charged diatomic ions. This results in no experimental report on the evidence of $^{14}N_2^{2+}$ and $^{16}O_2^{2+}$ ionization. Reproduced with permission from Shiki et al. [103]. Copyright 2008 John Wiley & Sons, Ltd

of planets. In addition to the simple diatomic molecules, the *m/z* overlap also occurs in monomers and multimers of peptides, proteins, and synthetic polymers.

The ambiguity of mass peaks stems from the fundamental limitation that z is unknown because the response of ions to electromagnetic force is determined by *m/z*. This inherent MS problem that is "*m/z* overlap" is soluble, if we can measure z independently. Ions accelerated at U, which is usually in a range of 3–30 kV, acquire a kinetic energy of zeU. Therefore, it is possible to determine z from kinetic energy measurement. However, conventional EM and MCP detectors are usually triggered by one secondary electron emitted from a sensitive surface. They have no sensitivity to z in a low charge-state range of 1+, 2+, 3+, and so on, although rough discrimination of highly-charged ions between 50+ and 100+ can be performed with MCP [105]. Here, STJDs enable kinetic energy measurement of ions in a keV range. STJDs are sensitive to phonons created by ion impact as shown in Fig. 6.24 and Fig. 3.3b.

6.4.3.2 Doubly-Charged Diatomic Ions

A kinetic-energy sensitive STJD was applied to the analysis of metastable doubly-charged diatomic ions [103]. Naturally, the peak intensity of the singly charge ions is higher than those of doubly-charged ions. The double-ionization potentials are substantially higher than single-ionization potentials. The intense peaks of singly-charged atomic ions obscure the possible presence of faint doubly-charged diatomic ions.

Figure 6.27 plots the ion impact events on a *m/z* versus kinetic-energy space [103]. When we count all the events for every *m/z* value, it forms the conventional mass spectrum in Fig. 6.26. Note that two events appear at *m/z* 14.5 corresponding to $^{14}N^{15}N^{2+}$. These two events are not dark counts. The STJD completely eliminates dark counts in this high-energy keV range. All the ion impact events generate distinctive pulses, the heights of which are significantly larger than a noise level and proportional to the kinetic energy, similar to the X-ray pulse in Fig. 3.14 b. The distinctive pulse generation by STJDs is in striking contrast to conventional MCP or other MS ion-counting detectors. These conventional MS detectors generate non-negligible dark counts because random electron emission from a sensitive surface readily occurs; we cannot distinguish between true events and noise events. On the other hand, noise events of STJDs have extremely lower pulse heights than those of the true events of keV ions. Therefore, the two dots at *m/z* 14.5 are surely attributed to $^{14}N^{15}N^{2+}$, which suggests the presence of the $^{14}N^{14}N^{2+}$ ions. However, further analysis such as estimation of ionization cross-section using mass peak intensity is impractical with two counts.

Figure 6.27b shows the histogram or pulse height spectrum of all the events at *m/z* 14 in the area surrounded by the dotted line in **a**. The kinetic energy values of the

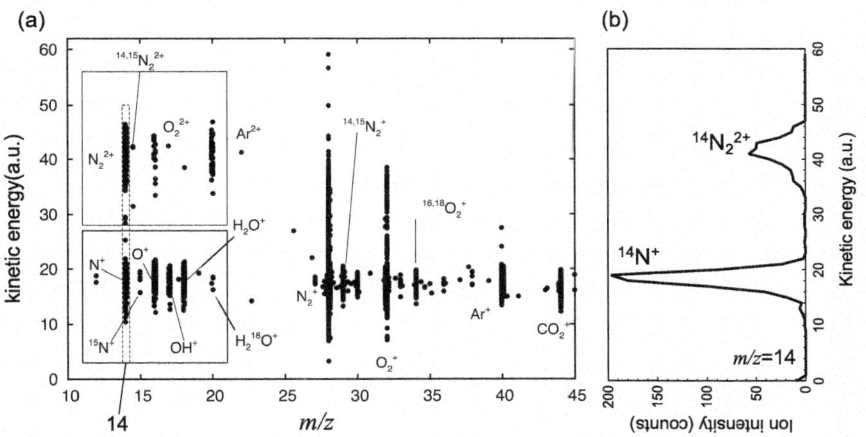

Fig. 6.27 Kinetic energy measurement with 200-μm square STJD for ions produced by ionizing the air with EI at 70 eV [103]: (**a**) *m/z* versus kinetic energy scatter plot, and (**b**) kinetic energy spectrum of the events surrounded by the dotted line at *m/z* 14 in (**a**). The ions were accelerated at 3 kV. The spectrum in (**b**) is equivalent to pulse height spectra in X-ray spectroscopy. The resolving powers $E_0/\Delta E$ are 6.2 at 3 keV and 7.8 at 6 keV. The $^{14}N_2{}^{2+}$ mass peak is masked by the intense $^{14}N^+$ mass peak. Such doubly-charged ions of homonuclear diatomic molecules play an important role in the atmosphere escape from planets [104]. Reproduced with permission from Shiki et al. [103]. Copyright 2008 John Wiley & Sons, Ltd

ions should be 3 keV for $^{14}N^+$ and 6 keV for $^{14}N_2{}^{2+}$; therefore, we can assign $^{14}N^+$ to the lower peak and $^{14}N_2{}^{2+}$ to the higher peak along the kinetic energy axis. This is the apparent evidence of the $^{14}N_2{}^{2+}$ presence. It is surprising that no experiment on the detection of this simple doubly-charged homonuclear diatomic ions has been reported before the experiment in Fig. 6.27. The reason is discussed below.

We can analyze charge states by using charge detection mass spectrometry (CDMS) [79, 106]. However, the performance of charge-number sensitive detectors used in CDMS is insufficient for separating 1+ and 2+. Another charge-state sensitive method is ion mobility spectrometry—mass spectrometry (IMS-MS) [107]. IMS-MS utilizes ion separation in a drift buffer gas such as helium. The scattering of ions with the buffer gas under an electric field causes a drift time difference depending on the molecule-size/charge-number ratio. If molecular sizes are comparable, it is possible to determine z values of the ions from drift times. In IMS-MS, stable ions can drift without dissociation. However, the gas-phase collision excites molecules, creating metastable ions and leading to dissociation in the course of multiple scattering in the drift gas. Concerning nitrogen molecules, the ionization-potential difference between $N_2{}^{2+}$ and $N_2{}^+$ is 27.30 eV that is higher than the first ionization potential of 24.59 eV for He. Therefore, even in the inert gas of He, the reaction of $N_2{}^{2+} + He \rightarrow N_2{}^+ + He^+$ occurs; $N_2{}^{2+}$ ions disappear inside the drift tube. Therefore, we cannot use IMS-MS to detect metastable $^{14}N_2{}^{2+}$ ions. Furthermore, metastable $N_2{}^{2+}$ ions spontaneously dissociate through a path of

$$N_2^{2+} \rightarrow N^+ + N^+. \tag{6.4}$$

The lifetime of $N_2{}^{2+}$ ranges from 3 s for the ground state to several μs for highly excited states. In the STJD experiment, a flight time of the 3-keV $N_2{}^{2+}$ ions is ~ 10 μs, which is considerably shorter than the lifetimes of $N_2{}^{2+}$ in most cases. Therefore, we can ignore the spontaneous dissociation. The clear peak separation in Fig. 6.27b enables the ion yield measurement of $^{14}N_2{}^{2+}$, avoiding interference from the strong $^{14}N^+$ peak in a similar way to the XAFS experiment for SiC:N. From the ion intensity or peak area, the electron impact ionization cross-section of $^{14}N_2{}^{2+}$ is evaluated for the first time: 1.32(± 0.073) × 10^{-18} cm^2 for the 70-eV electron impact together with the cross-sections of $^{16}O^{2+}$, $^{16}O_2{}^{2+}$, $^{40}Ar^+$, and $^{40}Ar^{2+}$ [103].

The peaks in Fig. 6.27b are considerably broader than those in X-rays spectra. The ΔE values are 485 eV for the 3-keV $^{14}N^+$ ions ($E_0/\Delta E = 6.2$) and 768 eV for the 6-keV $^{14}N_2{}^{2+}$ ions ($E_0/\Delta E = 7.8$). The ΔE values are ~ 30 times worse than that for X-rays. The ratio of 768 eV to 485 eV is 1.58, which is larger than a $\sqrt{E_0}$ prediction of 1.41. This discrepancy can be partly explained by the spatial inhomogeneity that is pronounced at higher particle energies for the 200-μm square STJD with the 50-nm thick Al layers, as shown in Fig. 3.29. In addition, the broad peaks are related to ion-surface collision processes [78, 96]. Nonzero-mass particles partially transfer their kinetic energy to a superconducting electrode in contrast to the full-energy absorption peaks of zero-mass X-ray photons. Molecule impact on the Nb surface induces fragmentation of incoming molecules, secondary electron emission, and secondary ion emission, all of which cause energy loss [96]. This energy loss leads to the statistical fluctuation. The efficiency of energy transfer to the superconductor absorber depends on ionic species, kinetic energy, and molecular weight. The transfer efficiencies measured with an NIS calorimeter are 59% for bovine serum albumin (BSA, 66.4 kDa) and 74% for sinapic acid (224.21 Da) at 20 keV [85]. The superconducting energy gap in STJDs causes additional loss into phonons less than $2\Delta_g$; therefore, the transfer efficiency is only 28% for human serum albumin (HSA, 66.4 kDa) at 25 keV [89]. Molecular-weight dependence of pulse height are divided into the following three regimes [95]. In a range below ~ 2000 Da, the energy transfer efficiency rapidly decreases with increasing *m* due to secondary electron emission [78, 96]. Between ~ 2000 Da and ~ 100 kDa, the efficiency gradually increases with increasing mass, which stems from the reduction of secondary electron emission. Above ~ 100 kDa, the efficiency is constant independent of *m* [108]; large molecules with a low velocity enters a soft-landing mode without emission of secondary electrons or ions. The residuals of soft-landing are observed on an STJ surface [82]. Molecular-weight dependence curves should be referred to in Ref. [78] up to ~ 35 kDa and in Ref. [95] up to ~ 200 kDa. STJDs hold a fairly good energy linearity for protein ions up to 30 keV [17, 78, 109] as shown in Fig. 3.18. Therefore, mass separation by kinetic energy measurement with STJDs is effectual when we do not analyze ions with very different masses; for example, ~ 2000-Da ions and ~ 200-kDa ions.

6.4.3.3 Atmosphere Escape from Planets

The dissociation path (6.4) is associated with KER based on a difference in ionization potentials. The N_2 molecule has ionization potentials of 14.534 eV for the first and 42.88 eV for the second. The difference 13.8 eV is equally distributed to the two N^+ fragments when dissociation occurs, so that each N^+ ion has a kinetic energy of 6.9 eV. In addition to KER, the coulomb repulsion of the two N^+ ions adds an extra kinetic energy, leading to a broad kinetic energy distribution with an FWHM value of 6 eV at a peak energy of 9 eV [110, 111]. When N^+ ions acquire a kinetic energy of above 8.6 eV, the N^+ ions can escape from the Earth against the gravity, which is called atmosphere escape [104]. A similar escape also occurs in $O_2{}^{2+}$. The atmosphere escape originating from the doubly-charged metastable ions affects the atmospheric evolution of Venus, Earth, Mars, and Titan, as illustrated in Fig. 6.28. The typical escape energy values for C^+, N^+, O^+ are tableted in Ref. [104]; the escape energy values for N^+ are 8.6, 7.5, 1.8, 0.32 eV for Earth, Venus, Mars, Titan. For the Earth and Venus, the escape threshold energy and the energy of N^+ are comparable, so that the occurrence of atmosphere escape is on the borderline. For the Mars and Titan, on the other hand, it considerably exceeds the escape energy, which indicates that the atmosphere escape occur when doubly-charged ions exist.

In the ionosphere, probable processes for double ionization are double photoionization of neutral molecules, double electron-impact ionization of neutrals, electron ionization of singly-charged molecules, high-energy cosmic particle impact, and Auger processes [104]. There is no report on the detection of $N_2{}^{2+}$ and $O_2{}^{2+}$ in the

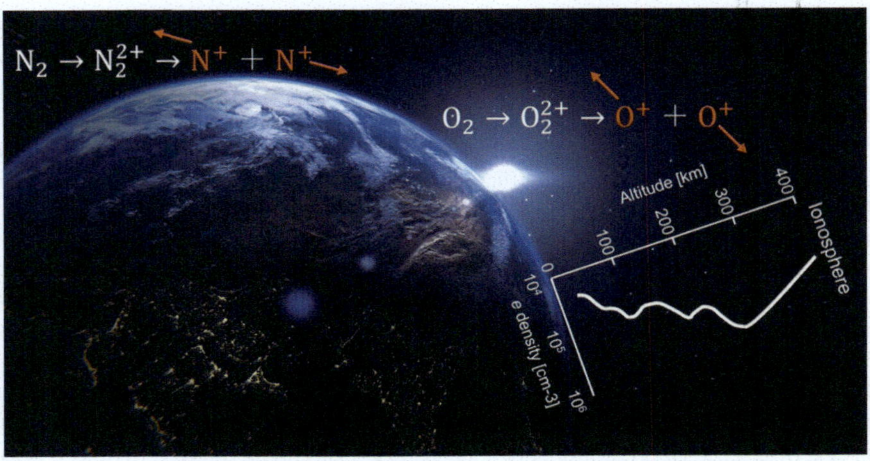

Fig. 6.28 Illustration of atmosphere escape from planet against gravity through ionization and dissociation of diatomic molecules in the ionosphere. The altitude of the inset for the ionosphere structure, altitude dependence of electron density, is not scaled to the size of the Earth. The electrons likely ionize atoms and molecules in the ionosphere. A high abundance of N_2 is popular on the planets in the solar system. Possible routes of atmosphere escape are shown for N_2 and O_2. Detection of doubly-charged diatomic ions is important for planetary science

ionosphere due to the *m/z* overlap. The detection of N_2^{2+} in the ionosphere requires a certain mass resolution in addition to a capability of kinetic energy measurement. If CO exists in the ionosphere, the mass resolving power should be better than ~ 3000 to separate CO^{2+}(*m/z* 13.9969) and N_2^{2+} (*m/z* 14.0025). It is necessary to execute the detection of doubly-charged ions in the ionosphere, the investigation of ionization-dissociation processes, and the measurement of ionization cross-section values on the ground. Superconductor ion detectors in the ionosphere pave the way for the elucidation of the atmosphere escape from the planets in addition to ground-based experiments described in the preceding sections.

In addition to the doubly-charged ions, electron–ion dissociative-recombination (DR) processes can also contribute to the planetary atmosphere evolution. DR produces energetic neutral atoms or molecules through KER. Naturally, most abundant ions are singly-charged, so that DR produces neutral fragments that likey have a kinetic energy above the threshold for escape. Conventional MS instruments cannot analyze neutral atoms or molecules as they are. This is called neutral loss. Superconductor MS detectors can also solve the problem related to the neutral loss, which will be described in detail in Sect. 6.4.4.

Considering instrumentation in the ionosphere or space, STJDs require a rather large ^{3}He cryostat, although launching a large cryostat is possible; for example, sounding rocket missions [112]. Another possibility for a compact instrument is to employ charge-stage discrimination with SSPDs, which operates in a higher temperature range between 4 and 30 K. Superconductor materials used for SSPDs are Nb, NbN, MgB_2, and BSCCO. An MgB_2-SSPD successfully detected biomolecule ions, lysozyme [113]. The MgB_2 -SSPD is attractive in detecting doubly-charged ions at ~ 20 K in space. In addition, a BSCCO-SSPD successfully detected single photons at temperature up to 20–25 K [114, 115]. The high temperature operation is advantageous to experiment in space, although the size of the BSCCO-SSPD is currently only $5 \times 5\ \mu m^2$. SSPDs cannot measure kinetic energy of individual ions; however, the lower threshold kinetic energy can be adjusted by changing bias current, as described in Sect. 6.4.3.5 [116]. By eliminating singly-charged ions, it is possible to selectively detect doubly-charged ions with SSPDs.

6.4.3.4 Charge State Discrimination and Fragmentation Analysis

Charge-state discrimination with STJDs and SSPDs is also demonstrated with biomolecules in addition to simple atoms and molecules. Figure 6.29a shows the charge-number measurement with the 100-pixel STJD, which simultaneously acquires m/z values from TOF and z values from kinetic energy for lysozyme multimer ions ($14.3 \times n$ kDa), where n is an integer [116]. Even though the lysozyme enzyme is 511 times heavier than $^{14}N_2$, the STJD can discriminate its charge states. Each junction has a layer structure of Nb(100)/Al(50)/AlO_x/Al(50)/Nb(100) in nm. Because energetic molecules stop on the Nb surface without significant penetration, a Nb thickness of 100 nm is sufficient in contrast to X-ray detection. The molecule impact transfers only 30–70% of the kinetic energy to a superconductor; the large fluctuation leads to broad peaks on pulse height spectra as described before. Therefore, we select the 200-μm square junctions with the 50-nm thick Al layers for a large sensitive area rather than pursuing high energy resolution. The lysozyme molecules are ionized by MALDI, which predominantly produces singly-charge ions with a small fraction of doubly- and triply-charged ions. The m/z values are calculated from TOF values using leading edge discrimination of output pulses, and the kinetic energy is determined from the pulse height. The FPGA-DSP modules in Fig. 6.11 c execute this pulse processing. Figure 6.29a plots the impact events of individual ions according to m/z values and detected kinetic energies. The kinetic energy axis is the new third spectroscopic dimension added to conventional mass spectra with ion intensity versus m/z. The evens between the red dashed lines A and B are due to the doubly-charged ions. The singly charge ions generate the events between B and C. The numbers in (a) beside the event groups denote the multimerization degrees of the singly-charged ions in red and the doubly-charged ions in orange.

MALDI generates a large amount of ions after each pulse laser shot, which results in simultaneous arrival of the intact ions and the PSD fragments at a detector surface. The simultaneous impact causes counting loss. Nevertheless, the STJD with small STJ pixels avoids the counting loss because the probability of multiple hits is negligibly small for each 200-μm square STJ pixel.

Figure 6.29b is a conventional mass spectrum for all the charge states and even neutral fragments, collecting all the events with any charge numbers between the red dashed lines A and C and the PSD events between C and D. Retrieving the events between A and B, the mass spectrum in **c** exclusively contains the doubly-charged

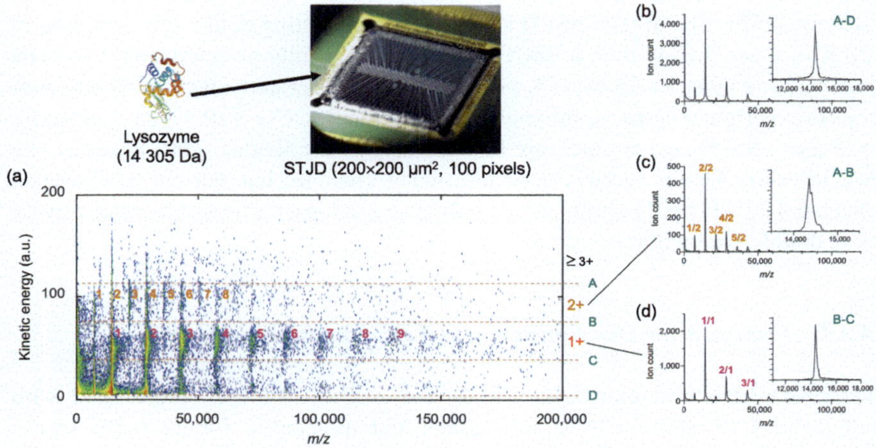

Fig. 6.29 Charge state discrimination with 100-pixel STJD [116]: (**a**) scatter plot of detected events on *m/z* versus kinetic-energy space, (**b**) *m/z* spectrum for all charge states and the PSD fragments between the red dashed lines A and D (conventional mass spectrum), (**c**) *m/z* spectrum exclusively for the doubly-charged ions between A and B, and (**d**) *m/z* spectrum exclusively for the singly-charged ions between B and C. The lysozyme enzyme protein (14,305 Da) was ionized by MALDI and accelerated at 17.5 kV. In (**a**), the numbers represent multimerization degrees: the orange numbers from 1 to 8 for the doubly-charged ions and the red numbers from 1 to 9 for the singly-charged ions. The event numbers are represented by blue, green, and red from the lowest count to the highest count. In (**c**) and (**d**), the mass peaks are labeled by *m* divided by 14,305 / *z*; for example, 1/2 means the monomer that is doubly-charged. The *m/z* overlap is observed for 1/1 and 2/2, and so on in (**b**). The STJD solves the problem of the *m/z* overlap or *m/z* spectrometry by realizing true MS that determines *m* without ambiguity of *z*. Reproduced with permission from Suzuki et al. [116]. Copyright 2010 John Wiley & Sons, Ltd

ions. Similarly, the mass spectrum in (d) contains the singly-charged ions only. The faction numbers denote the multimerization number/charge number. The *m/z*-overlap is recognized by comparing (b), (c), and (d). The STJD realizes true MS or *m* spectrometry by determining the charge states. Moreover, the events between C and D are produced by PSD during free flight after acceleration. The PSD events have the same TOF values as those of precursor ions, forming the vertical event lines at the *m/z* values of intact ions in (a). All the PSD fragments hold the same velocity as the intact ions. The PSD fragments have mass distribution; therefore, they have kinetic energy distribution along the *y*-axis in (a). The STJD can exclude the PSD events by kinetic energy discrimination, selecting the events above the dashed line C. Concomitantly, the PSD fragmentation patterns provide structural information of the precursor ions, which provides a type of MS/MS. Figure 6.29 demonstrates that the STJ type solves the *m/z* overlap problem and moreover enables MS/MS on the kinetic energy axis.

Recently, the PSD fragmentation analysis with an STJD was performed for metastable $Au_{10}(TBBT)_{10}$ catenane nanoclusters [117], where TBBT stands for the 4-tert-butylbenzenethiolate. The TBBT is also expressed by a combination of the letters “S” and “R” for residual. The gold nanoclusters surrounded by thiolates have unique optical, electronic, and magnetic properties due to quantum confinement effects. The $Au_{10}(TBBT)_{10}$ nanoclusters were ionized by MALDI. The STJD on the liner mode detected fragments produced by both ISD and PSD. Figure 6.30a shows a mass spectrum only for singly-charged ions, selected by the same procedure as that in Fig. 6.29d. Even though MALDI is a soft ionization method, no intact $Au_{10}(TBBT)_{10}^+$ with a mass of 3.819 Da is observed; the strongest peak is assigned to $Au_{10}(TBBT)_9^+$ with a mass of 3.456 kDa; one TBBT ligand is missing due to ISD. ISD also causes fragmentation in a m/z range below $Au_{10}(TBBT)_9^+$. The peaks above $Au_{10}(TBBT)_9^+$ are assigned to $[Au(SR)]_n$ and $[Au(SR)]_n(S)_m$ adduct ions produced by gas phase reaction in the ion source. In Ref. [117], the x-axis of the mass spectrum has a unit of "Th," which originates form the name of Thomson and represents "m/z." The unit of m is normally "Da," which originates from the name of Dalton. The mass of ^{12}C is 12 Da. The x-axis is usually expressed by m/z, but some journals adopt "Th."

Figure 6.30b shows the scatter plot of all the events after ISD, PSD ions, and PSD neutral fragments. The series of horizonal event groups along the red line contain singly-charge precursor ions. Below the led line, there are three groups of PSD fragments depicted by the blue lines. This feature implies that each precursor singly-charge ions has a unique PSD pattern. The fragmentation patters are explained by considering density functional theory (DFT) calculations and ab initio molecular dynamics (AIMD) simulations. The MALDI-TOF-MS with STJD (MALDI-STJD-TOF-MS) can acquire rich information on the stability or instability of numerous precursor ions from a single measurement [117]. STJDs simultaneously measure m/z and kinetic energy (z) regardless of ions or neutrals. This is important for the topic of neutral loss in Sect. 6.4.4.

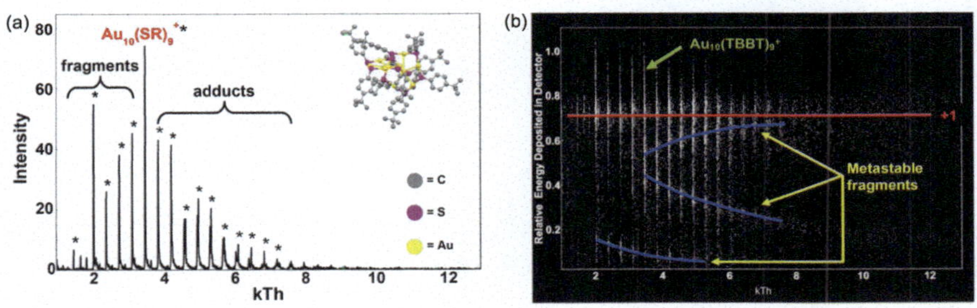

Fig. 6.30 Post source decay (PSD) analysis with STJD for $Au_{10}(TBBT)_{10}$ catenane nanocluster (TBBT = 4-tert-butylbenzenethiolate) [117]: (**a**) $Au_{10}(TBBT)_{10}$ structure at the upper right and MALDI-TOF-MS spectrum only for singly-charged ions selected by the STJD and (**b**) scatter plot of detected events on kinetic energy versus m/z space. TBBT is consists of S and R units. The x-axis has a unit of "Th," which originates form Thomson and represents "m/z." In (**a**), the peaks are assigned to the fragments due to in-source decay (ISD) and adducts produced in the ion source. In (**b**), the event groups along the red line are due to the singly-charged ions. The intact metastable ions dissociate into fragments during free flight, which is called PSD. The PSD patterns reveal the stability or instability of the precursor molecules. Reproduced with permission from Plath et al. [117]. Copyright 2022 ACS Publications

6.4.3.5 Application to Non-covalent Protein Complex: Hemoglobin

Non-covalent protein complexes perform indispensable functions in our bodies. Hemoglobin is one of the typical complexes, and has non-covalent bonds between four heme units and four globin units, forming a tetramer in Fig. 6.31a. Each heme has a Fe atom that conveys oxygen in the bodies of vertebrates. The hemoglobin tetramer of $\alpha^*_2\beta^*_2$ has a molecular weight of ~ 64,500. The "*" qualifiers indicate that one heme attaches to each subunit, adding 616 Da. The generating process of hemoglobin is still under discussion. Furthermore, hemoglobinopathy is known as a hereditary syndrome caused by inadequate molecular structure.

A typical inspection method for non-covalent protein complexes is electrophoresis. Alternatively, direct MS analysis provides a higher mass resolution, a shorter measurement time, and a less sample amount. As mentioned before, ESI is a

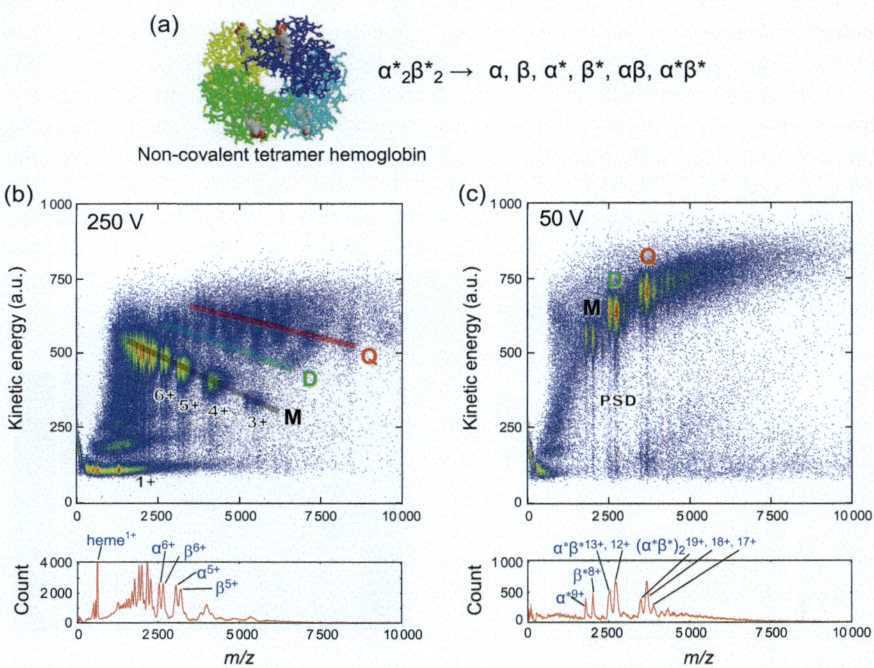

Fig. 6.31 Analysis of hemoglobin, non-covalent complex, with ESI-STJD-TOF-MS [93]: (**a**) non-covalent hetero tetramer structure $\alpha^*_2\beta^*_2$ of hemoglobin and fragmentation patterns due to non-covalent bond scission, (**b**) kinetic-energy versus *m/z* scatter plot and mass spectrum at an orifice voltage of 250 V, and (**c**) scatter plot and mass spectrum at 50 V. Figure 6.25 shows the appearance of the MS instrument with a 100-pixel STJD and ESI. The "*" qualifiers denote that a heme unit attaches to a globin unit. The number of events is represented by a rainbow gradient, with blue for the lowest count to red for the highest count. The letters M, D, and Q denote monomer, dimer, and tetramer, respectively. The fragmentation due to ISD occurs in (**b**), whereas the intact hemoglobin tetramers are observed in (**c**). The events along the vertical lines in (**c**) are due to PSD

fragmentation-free ionization method for non-covalent complexes; MALDI easily breaks all the non-covalent bonds. However, ESI generates highly charged ions over 20+ or more, which makes mass spectra complicated and causes *m/z* overlap. Analysis of ESI–MS spectra frequently faces a difficulty in identifying *m/z* peaks. In addition, even with ESI, ionization and transportation of intact hemoglobin complexes require a careful optimization of instrumental condition. Hemoglobin easily dissociates into the monomers of the α and β globin units and the αβ dimers. Figure 6.31 demonstrates that ESI-STJD-TOF-MS performs the optimization of the ESI source and ion transport condition for ISD-free analysis of non-covalent protein complexes.

The scatter plots in Fig. 6.31 show the data at two different orifice voltages of 250 V in (b) and 50 V in (c). The orifice is located between the spray nozzle for liquid samples at atmospheric pressure and the pinhole interface into the vacuum chamber in Fig. 6.25b. The lower panels depict the conventional mass spectra (*m/z* spectra) integrating all the events. It is apparent that the mass spectra at 250 and 50 V are totally different. The mass spectra are complicated by fragmentation into subunits due to ISD and diverse charge states due to ESI. The STJD scatter plots enable charge state discrimination, which allows peak assignment of multiply-charged ions. At an orifice voltage of 250 V in (b), the heme, which conveys oxygen in bloodstream, is apart from all the globin units by ISD and no intact hemoglobin [$\alpha^*{}_2\beta^*{}_2$(Q)] remains. The fragmentation occurs during the air-to-vacuum transport process. The kinetic energy measured by the STJD indicates that the detached heme ions are always singly charged. The solid lines in (b) along the M, D, and Q series indicate different charge states of monomer, dimer, and tetramer, respectively. At an orifice voltage of 50 V in (c), on the other hand, the scatter plot pattern is simple and completely different from that in (b). The intact tetramer $\alpha^*{}_2\beta^*{}_2$(Q) is recognized as 19+, 18+, and 17+ ions. In addition to the intact $\alpha^*{}_2\beta^*{}_2$(Q) tetramers, there are the $\alpha^*\beta^*$(D) dimers and the monomers(M) of α^* and β^*. No heme ions are observed; all the subunits hold a heme molecule without dissociation.

ESI-STJD-TOF-MS provides a method to analyze abnormal structure or defects of hemoglobin variants caused by mutations. By comparing scatter plots with and without ISD, it is possible to analyze molecular structure of non-covalent protein complexes. PSD analysis also contributes to the structural analysis. Recently, an STJD was also deployed on a linear quadrupole ion trap (LIT) MS instrument with ESI to investigate ESI operation and analyze generated ions [118].

6.4.3.6 Charge State Discrimination with SSPD

SSPDs operating at 4.2 K or higher also perform charge-state discrimination, which is more easily operated compared with STJDs at 0.3 K. Although it is not kinetic energy measurement of single ions, we can adjust a kinetic-energy threshold level for ion detection by varying bias current. The advantage of the SS type over the STJ type is shown in Fig. 6.32a. STJDs generate output signals with a pulse duration of a few μs, which is ~ 1000 times faster than those of TESDs and MMCDs. However, TOF–MS often requires a nano-second response time. Figure 6.32a shows an output pulse shape comparison between an STJD and an SSPD for 17.5-keV lysozyme ions. The 200-μm square STJD with "Al 50-nm" generates an output current pulse with $\tau_{\mathrm{rise}} = 400$ ns and $\tau_{\mathrm{fall}} = 1.1$ μs. The rise time is limited by a semiconductor current amplifier bandwidth; nevertheless, a constant-fraction discrimination module contributes to improving the temporal resolution to 90 ns. The intrinsic rise time of STJ pulses should be in a sub-ns range or shorter. The slow response results in a low mass resolution in TOF–MS. One technique to improve response time is to install a single flux quantum (SFQ) readout circuit [119]. Another ingenious detector design is a combination of STJD kinetic energy measurement and first trigger timing with a parallel Josephson junction [120]. The simultaneous time and energy measurement was confirmed by using X-rays that emulate ions; timing output pulses with a resolution of a few ps were generated by the Josephson junction [120].

The red two pulses in Fig. 6.32a were generated by an NbN-SSPD that is 200-μm square, 10-nm thick, 800-nm wide, and 40-mm long. The pulses exhibit $\tau_{\mathrm{rise}} = 1$ ns and $\tau_{\mathrm{fall}} = 27$ ns. These time values are considerably faster than those of the 200-μm square STJD. The sub-ns temporal response is adequate for TOF-MS. At this ultrafast response, the SSPD can discriminate the charge states of lysozyme ions: doubly-charged ions at I_b/I_c 0.65 in Fig. 6.32c and both singly- and doubly-charged

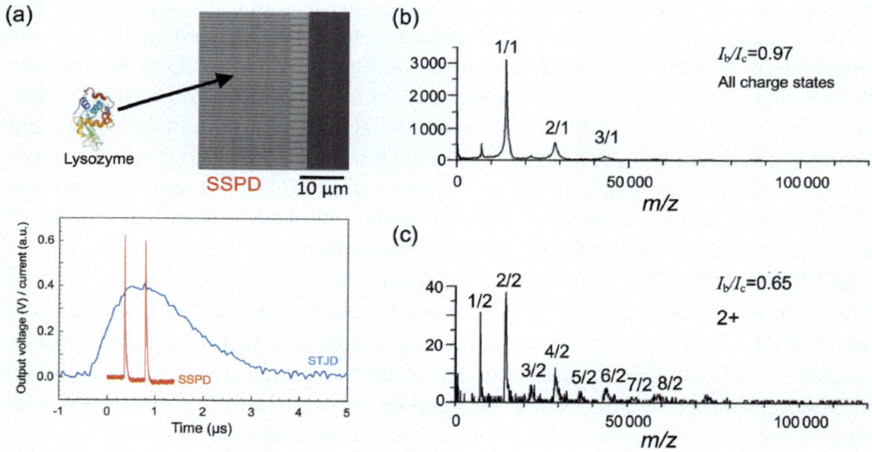

Fig. 6.32 Charge state discrimination with NbN-SSPD for multimers of enzyme protein lysozyme (14,305 Da) [116]: (**a**) photograph of a part of SSPD and comparison between SSPD output pulses in red and an STJD pulse in light blue, (**b**) SSPD mass spectrum for all charge states at $I_b/I_c = 0.97$, and (**c**) SSPD mass spectrum of mostly doubly-charged ions at $I_b/I_c = 0.65$. The lysozyme ions were generated by MALDI and accelerated at 17.5 kV. The red pulses of the 200-μm square SSPD with $t_f = 10$ nm, $w = 800$ nm and $l = 40$ mm exhibit fast response of $\tau_{\mathrm{rise}} = 1$ ns and $\tau_{\mathrm{fall}} = 27$ ns, whereas the blue pulse of the 200-μm square STJD exhibits $\tau_{\mathrm{rise}} = 400$ ns, which is limited by a readout circuit, and $\tau_{\mathrm{fall}} = 1.1$ μs. The 1-mm square parallel Nb-SSPD records faster response of $\tau_{\mathrm{rise}} = 0.52$ ns and $\tau_{\mathrm{fall}} = 0.92$ ns [121]. Reproduced with permission from Suzuki et al. [116]. Copyright 2010 John Wiley & Sons, Ltd

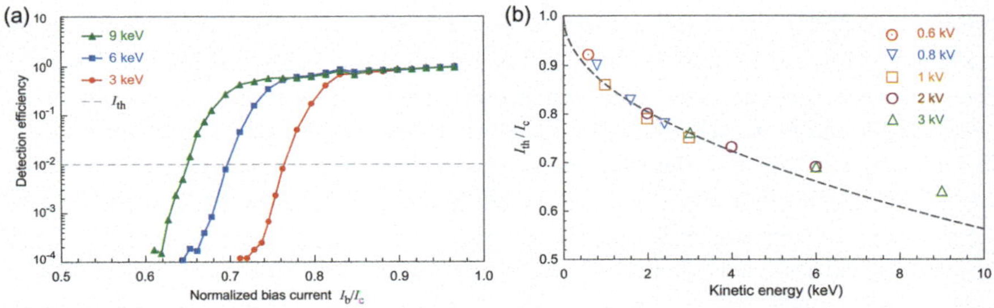

Fig. 6.33 Charge-state discrimination of NbN-SSPD [122]: (**a**) ion detection efficiency as a function of relative bias current I_b/I_c for Ar ions and (**b**) normalized threshold bias current as a function of Ar kinetic energy. We assumed in (**a**) that the crossing points between the dashed line I_{th} and the detection efficiency curves for 3, 6, and 9 keV represent the detection threshold bias currents. In (**b**), the Ar-kinetic-energy dependence of I_b/I_c agrees with the dashed line expected from the hot spot model in Figs. 4.16 and 4.19. Reproduced with permission from Suzuki et al. [122]. Copyright 2011 The Japan Society of Applied Physics

ions at I_b/I_c 0.97 in Fig. 6.32b. The SSPD charge-state discrimination for lysozyme ions is consistent with that by the STJD in Fig. 6.29.

The charge-state discrimination with SSPD is further validated by an experiment using Ar ions in an energy range of 500 eV–9 keV, avoiding ambiguity related to the structural complexity of large biomolecules [122]. Figure 6.33 indicates that we can control the lower detection threshold level by adjusting bias current. At $I_b/I_c = 0.74$, it is possible to almost completely eliminate the impact events of 3-keV Ar^+ ions and exclusively detect the 6-keV Ar^{2+} ions. This SSPD discrimination ensures the charge-state discrimination of the doubly-charged diatomic ions in the ionosphere.

The superconductor detector size of 200 × 200 μm^2 is extremely smaller than an ion beam diameter of ~ cm: the area ratio is 2500. The small size precludes practical applications to MS, even when superconductor detectors realize a quantum efficiency of 1.0 and charge state discrimination. Even when superconductor detectors have a 10 times better detection efficiency, the measurement time is still 250 times longer, which is unacceptable to practical analysis. The chapter "IV. advances for practical use" in Ref. [17] describes remarkable enlargement of the SSPD size, holding a nano-second response time. A 1-mm square Nb-SSPD with a parallel configuration recorded 520–920 ps for 17.5-keV angiotensin-I ions [121]. Even 5-mm square SSPD is possible, which covers most area of ion beam.

6.4.4 Neutral MS Eliminating Neutral Loss

6.4.4.1 Neutral Loss

Neutral loss is observed in MS/MS, which is one of the most important methods for investigating gas-phase molecular reaction, identifying molecules, and analyzing molecular structure. It is often called top-down proteomics to analyze proteins in life science. Precursor ions with a certain *m/z* value are selected by the first MS. The precursor ions encounter fragmentation by processes such as collisionally activated dissociation (CAD)/collision induced dissociation (CID), electron capture dissociation (ECD), and electron transfer dissociation (ETD). Molecular identification and structural analysis are performed by analyzing fragmentation patterns by second MS. Further addition of third MS or more is often implemented. In CAD, metastable molecules dissociate over their lifetimes by breaking weak bonds such as non-covalent bonds and hydrogen bonds, after precursor ions are activated through gas phase collisions. On the other hand, in ECD and ETD, electron attachment to multiply-charged ions leads to breaking strong chemical bonds such as peptide backbone; cleavages do not necessarily occur at weak bonds. ECD is a dissociation process induced by interaction between a positive ion and a low energy electron. ETD is similar to ECD; an electron is transferred from a negative ion or a neutral atom with a low electron affinity to a positive ion. Normally, dissociation in ECD and ETD of peptides occurs promptly without delay unlike CAD following the lifetimes of excited ions. Fragmentation patters, which are analyzed at the second MS stage, are characteristic of molecular species. Therefore, we can identify complex biomolecules and their molecular structure without X-ray crystallography. Structural analysis of proteins is often performed in synchrotron radiation facility by using X-ray diffraction, which requires biomolecule single crystals. This method is ineffective to membrane proteins because they do not crystallize. The MS/MS can analyze membrane proteins without using single crystals.

In conventional MS/MS with ECD or ETD, precursor molecules should have multiple charges such that one of the fragments carries a charge after electron attachment and thereby the second MS can analyze the fragment ions. ESI generates multiply charged ions: liquid droplets in a sprayed aerosol containing molecules are highly charged. In other ionization methods such as MALDI and EI, the abundance of singly-charged ions is the highest due to the lowest ionization potential; therefore, the amount of doubly- or triply-charged ions are small. When we consider a singly-charged ion encountering electron capture or transfer, as shown in Fig. 6.34, a metastable neutral molecule is produced and then dissociate into fragments, all of which are neutral. Neutral fragments do not respond to electromagnetic force, so that they are lost, which is called "neutral loss." Therefore, fragmentation analysis in MS/MS commonly requires multiply-charged precursor ions.

It is possible to ionize the neutral fragments and analyze them by another MS, which is called neutralization-reionization mass spectrometry. However, there is a possibility of further fragmentation in the course of the re-ionization process. Therefore, quantitative information of dissociation paths is lost. Another approach is neutral mass spectrometry using small resonators of nano-electromechanical systems (NEMS) [123]. However, NEMS can analyze gigantic particles only, such as 4.5-MDa tantalum metal clusters or 105-MDa bacteriophage T5 capsids [124]; analysis of proteins is impossible. NEMS-MS cannot cover a mass range below kDa, which is the typical range for analytical scientists. Note that 13-MDa bacteriophage HK97

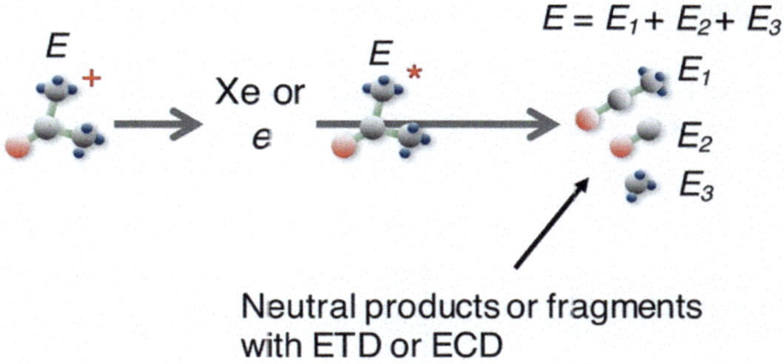

Fig. 6.34 Neutral loss in MS/MS with ETD and ECD. The first MS selects precursor molecules with a certain *m/z* value. This figure shows a difficulty in MS/MS for singly-charged precursor ions. In ETD, a precursor ion encounters electron transfer from a negative ion or a neutral atom with low electron affinity (Xe in this figure). In ECD, a precursor ion captures an electron from a low energy electron beam. The "*" qualifier denotes a metastable state. The Xe gas competitively induces either ETD or CAD. When the precursor ion has one charge only, all fragments are neutral. Such neutral fragments are lost inside MS apparatus because electromagnetic force has no effect on the trajectory of neutral molecules. Superconductor detectors measure kinetic energy regardless of ions or neutrals, enabling neutral MS based on the energy conservation expressed by $E = E_1 + E_2 + E_3$

capsid ions are detectable with an STJD-TOF-MS on the linear mode; superconductor detectors cover an extremely wide mass range from atoms to gigantic particles [125].

Figure 6.34 for ETD or ECD is connotative of competitive dissociation processes of CAD and ETD, when the targe gas is Xe. The electron affinity of Xe is lower than most elements, so that ETD occurs, which is charge-exchange neutralization. Additionally, it is well known that the collision with neutral Xe atoms also induces collisional activation that excites ions electronically. When we consider a singly-charged precursor ion with three units of M, N, and O, possible fragmentation patters with CAD are shown in Reaction (6.5). The letter "T" symbolizes a target atom for collisional reaction. The mark "*" denotes an electronically excited metastable state. Only one fragment can carry the charge. Conventional MS/MS analyzes ionic fragments only, which faces difficulty in distinguishing the two reaction paths of $M^+ + NO$ and $M^+ + N + O$. The most serious case is Reaction (6.6) for ETD, in which all the fragments are neutral. The same reaction occurs in ECD using an electron shower. This ECD process is similar to DR for prebiotic organic molecules in space, which is the topic in Sect. 6.4.4.3. In most cases, the abundance of singly-charged ions is the highest owing to the lowest ionization potential. Conventional MS cannot identify the paths in Reactions (6.5) and (6.6) completely.

The kinetic energy of a precursor ion is distributed to fragments, the energy of which is proportional to the fragment mass as shown in Fig. 6.34. Therefore, superconductor detectors enable neutral MS by measuring the kinetic energy of each neutral fragment.

$$\mathrm{MNO^+ + T \rightarrow MNO^{+*} \rightarrow \begin{cases} MNO^{+*} \\ MN^+ + O \\ M^+ + NO \\ M^+ + N + O \\ \ldots, \end{cases} \quad (CAD)} \tag{6.5}$$

$$\mathrm{MNO^+ + \mathit{e} \rightarrow MNO^* \rightarrow \begin{cases} MNO^* \\ MN + O \\ M + NO \\ M + N + O \\ \ldots. \end{cases} \quad (ETD\ or\ ECD)} \tag{6.6}$$

6.4.4.2 Kinetic-Energy Domain MS/MS with STJD

Figure 6.35 shows a schematic drawing of STJD-MS/MS[83]. The first sector-type MS-I selects precursor ions at a certain *m/z* value. The second MS-II is realized by the STJD kinetic-energy measurement, which is the solid-state MS in place of conventional vacuum-based MS. The STJD-MS/MS revelas rich information on neutral species, which have been conventionally analyzed by neutralization-reionization MS and charge inversion MS [126]. A precursor ion selected by the MS-I collides with a Xe target atom in the collision chamber. To retain single-collision condition, the Xe gas pressure is regulated such that the reduction of the precursor ion yield is less than 20%. The metastable ions or the neutralized metastable molecules dissociate into fragments at their lifetimes during free flight. The fragments carry the energy of the precursor ion, depending on the mass of each fragment. The STJD as the MS-II separates different fragments on the kinetic energy axis. We can detect both ionic and neutral fragments or only neutral fragments by disengaging or engaging the ion deflector. Comparison of the two spectra with and without the ion deflection reveals full reaction paths in Reactions (6.5) and (6.6).

The abundance of organic molecules was found in the giant molecular cloud, Sagittarius B2, near the center of the Milky Way. The interstellar organic molecules such as acetone, acetic acid, and formic acid can be formed through DR [127]. The abundance of organic molecules is still a mystery of interstellar physical chemistry [128]. Among the interstellar organic molecules, acetone was analyzed by STJD-MS/MS in Fig. 6.35, ionized at 70 eV and accelerated at 3 kV. The lab-based MS/MS experiment simulates radical formation in interstellar space. The radicals can be a source for amino acid formation. The MS-I selected the precursor ions, either acetone radical ions $(CH_3)_2CO^{+\cdot}$ (*m/z* 58) or acetyl ions CH_3CO^+ (*m/z* 43) that are the fragments created by ISD, where "·" denotes the radical state with an unpaired electron.

Figure 6.36a shows the kinetic-energy-domain MS-II spectra of all the detected events related to the precursor acetyle ions (CH_3CO^+). The largest peak

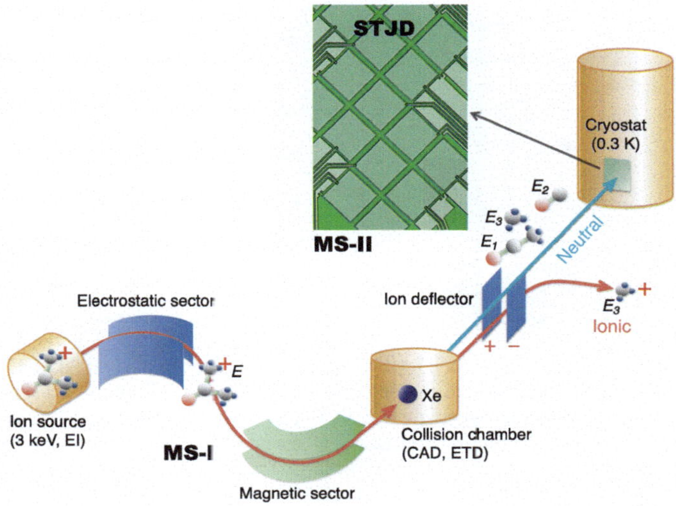

Fig. 6.35 Layout of STJD-MS/MS. The first sector-type MS-I (double focusing) selects precursor ions with a specific *m/z* value [83]. The precursor ions collide with a Xe target atom. Either CAD or ETD competitively occurs, producing ionic and neutral fragments. The Xe gas pressure is regulated for the single-collision condition. The precursor ions or precursor neutral molecules, both of which are metastable, dissociate into fragments at their lifetimes. Note that ECD or ETD of peptides is prompt. The STJD MS-II (solid state MS) measures the masses of the fragments on the kinetic energy axis. The ion deflector is used to exclusively detect the neutral fragments. Reproduced with permission from Ohkubo et al. [83]. Copyright 2011 Elsevier B.V.

at 3 keV is due to the precursor CH_3CO^+; most precursor ions pass through the collision chamber without colliding with a Xe atom. In addition to the intact CH_3CO^+ ions, a small amount of the metastable neutral $CH_3CO^{*\cdot}$ molecules are recognized by engaging the ion deflector. The CH_3CO^+ ions are neutralized by electron transfer and arrive the STJD surface before a lifetime. The peak of the $CH_3CO^{\cdot}$ molecules is small at the scale of Fig. 6.36a but recognizable at the magnified scale in Fig. 6.36b. The $CH_3CO^{*\cdot}$ intensity is used to estimate the lifetime of an electronically excited state of $CH_3CO(A)^{\cdot}$. The lifetime is approximately 3 μs, considering a flight time of 25 μs. The letter "A" denotes the lowest electronically excited metastable state. Higher discrete excitation states such as B, C, D, and so on are possible; however, it is natural that most molecules are excited to the lowest A state and the abundance of highly excited molecules is negligibly low. The experiment on the acetone precursor $(CH_3)_2CO^{+\cdot}$ results in a lifetime of ~ 80 μs for $(CH_3)_2CO$ (A) [83]. The long lifetime is reasonable because the nonradical neutral acetone is stable. The radical molecule $CH_3CO(A)^{\cdot}$ has the shorter lifetime than the non-radical molecule $(CH_3)_2CO(A)$.

In Fig. 6.36, the identified fragments are $CH_3^{\cdot}$, CH_3^+, and CO. It is intriguing that all the CO fragments are neutral, whereas most CH_3 fragments are ionic with a small amount of the neutral radical $CH_{3\cdot}$. From the peak intensities in Fig. 6.36, the branching ratios of CAD and ETD to the precursor ion yield are determined as

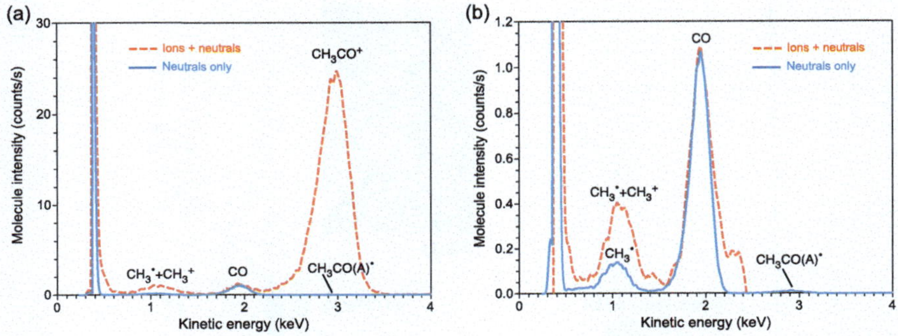

Fig. 6.36 Kinetic-energy-domain STJD-MS-II spectra for precursor acetyle CH_3CO^+ ions [83]: (**a**) raw spectra and (**b**) spectra with the enlarged *y*-axis after subtracting the precursor acetyle CH_3CO^+ events. The acceleration voltage was 3 kV. The red dashed lines represent data acquired without engaging the ion deflector (ions and neutrals), whereas the blue solid lines represent the data with engaging the ion deflector (neutrals only). The largest peak at 3 keV in (**a**) is due to the precursor CH_3CO^+ ions without colliding with Xe. $CH_3CO(A)^{\cdot}$ is the metastable neutral acetyle before dissociation. The letter "A" denotes the lowest electronically excited states. The mark "·" denotes the radical state. All the CO fragments are neutral; a majority of the CH_3 fragments are ionic. The ratio of $CH_3CO(A)^{\cdot}$ to CH_3CO^+ yields an excited-A-state lifetime of ~ 3 μs. A similar experiment for acetone $(CH_3)_2CO$ is reported in Ref. [83]. The branching ratios of the possible reaction paths can be analyzed from the ionic and neutral peak intensities. Reproduced with permission from Ohkubo et al. [83]. Copyright 2011 Elsevier B.V.

follows:

$$CH_3CO^+ + Xe \rightarrow \left[CH_3CO^+(A)\right] \rightarrow CH_3^+ + CO + Xe$$
$$\Delta H = 3.37\,\text{eV (CAD)}\ \ 4\% \qquad (6.7)$$

$$CH_3CO^+ + Xe \rightarrow \left[CH_3CO^+(A)\right] \rightarrow CH_3^{\cdot} + CO^+ + Xe$$
$$\Delta H = 7.57\,\text{eV (CAD)}\ \ 0\% \qquad (6.8)$$

$$CH_3CO^+ + Xe \rightarrow CH_3CO(A)^{\cdot} + Xe^+$$
$$\Delta H = 5.20\ \text{eV (ETD)}\ \ 1\% \qquad (6.9)$$

where H is the heat of formation, representing the change in enthalpy for the formation of the molecules from the constituent elements. The ΔH values are obtained by subtracting H of the left side from H of the right side, ignoring "[$CH_3CO^+(A)$]." The ΔH values also represent the reaction thresholds, which are the minimum energies to initiate the reactions. The ΔH values are consistent with the experimental branching ratios. No Reaction (6.8) with the highest ΔH occurrs, which is consistent with the observation in Fig. 6.36. The CAD reaction is predominant for the acetyle precursor. On the other hand, for the precursor radical acetone ions, $(CH_3)_2CO^{+\cdot}$, the electron transfer neutralization is predominant. This is reasonable because the acetone ions are radicals with an unpaired electron. The dissociation paths in acetone are more complex than acetyle because there are three fragments: CH_3CO, CO, and CH_3; see Ref. [83].

ETD in acetyle and acetone is governed by the lifetime of an excited state, which is different from the prompt dissociation of peptides. Focusing on the electronically excited states due to CAD and ETD, Reactions (6.10) and (6.11) indicate the branching ratios determined experimentally. The energy level of $CH_3CO^+(A)$ is not known, but it should be between 10.2 and 14.4 eV, considering that no CO^+ are detected. A possible range of ΔH is between 0 and −4.20 eV. It is important to note that conventional MS/MS can identify only Reaction (6.10) but cannot detect Reactions (6.9) and (6.11), which exclusively produce the neutral fragments.

$$\begin{aligned} CH_3CO^+(A) \rightarrow CH_3^+ + CO\,(CAD) \quad 70\% \\ \Delta H = 0 \text{ to } -4.20\,\text{eV} \end{aligned} \tag{6.10}$$

$$\begin{aligned} CH_3CO(A)^{\cdot} \rightarrow CH_3^{\cdot} + CO \quad (ETD)\ 30\% \\ \Delta H = -2.13\,\text{eV} \end{aligned} \tag{6.11}$$

6.4.4.3 Interstellar Chemistry of Prebiotic Organic Molecules

The dissociation path indicated by Reactions (6.9) and (6.11) for ETD resembles DR, which is a dissociation path of singly-charged ions that capture an electron (ECD). It is believed that DR is the crucial process for forming prebiotic organic molecules in interstellar space [129]. The dissociation in DR is a process of metastable neutral molecules, similar to Reaction (6.11). The neutral radical $CH_3^{\cdot}$ is one of the sources for a series of chemical reactions to produce more complex organic molecules. These molecules can lead to the formation of amino acids on the surface of dusts in interstellar space, as illustrated in Fig. 6.37. This is attributed to the origin of life on the Earth.

Ground-based DR experiments normally require a huge magnetic synchrotron storage ring in a MeV range [130]. Semiconductor detectors are capable of measuring the kinetic energy of MeV molecules in the same way as the α-particle in Fig. 6.24. Kinetic-energy-domain MS for neutral fragments is executable in an MeV range. However, all the neutral fragments simultaneously arrive the semiconductor detector surface, as observed in PSD fragments in Fig. 6.29 and Fig. 6.30. Semiconductor detectors cover the size of analyte beams. Therefore, the simultaneous impact of all the fragments conveys the same energy as that of the precursor molecule to the detector, which precludes fragmentation analysis. To solve this problem, a mesh technique is employed for distinguishing individual fragments by solving simultaneous equations [131]. On the other hand, an array STJD with 200-μm square pixels can detect individual fragments: the smallness of one STJ pixel is advantageous for avoiding simultaneous impact of multiple fragments. KER at several eV leads

Fig. 6.37 Prebiotic organic molecules in interstellar space. Radicals formed through electron–ion dissociative recombination (DR) probably produce the prebiotic organic molecules such as acetone, acetic acid, and formic acid in Sagittarius B2, and amino acids in a subsequent series of chemical reactions. Night sky photograph courtesy of M. Hamada

to bending the fragment trajectory; therefore, only one of the fragments impinges one pixel at a time. Imaging analysis of ionic and neutral fragments with an array STJD is promising for investigating the entire gas-phase reaction behavior including KER [83].

Recently, the MMC type was deployed on a heavy-ion synchrotron storage ring called CRYRING that was transported from Stockholm to Darmstadt. Imaging detection of neutral fragments is in progress [132, 133]. Another powerful instrumentation is a compact electrostatic storage rings in an energy range of a few 10 keV [134, 135]. However, the detector used for the electrostatic storage ring experiments was based on MCP, not analyzing different neutral fragments, but counting the number of all fragments. Therefore, lifetimes of metastable molecules after capturing an electron were measurable only. In a range of ~ 10 keV, semiconductor detectors cannot measure kinetic energy. Superconductor detectors play a crucial role in electrostatic storage ring facilities to reveal DR reaction paths. Furthermore, all the storage ring experiments are restricted to detecting neutral fragments only, neglecting ionic fragments. The compact STJD-MS/MS instrument in Fig. 6.35 enables the detection of both ionic and neutral fragments; therefore, we can analyze full reaction paths related to DR. The neutral MS eliminating neutral loss contributes to revealing fundamental gas phase reactions on the ground, simulating chemical reaction in interstellar space.

6.4.5 Possible Application to Atom Probe Tomography

Atom probe tomography (APT) also faces the *m/z* overlap problem [136]. APT is a method combing field ion microscopy (FIM) and imaging TOF–MS [137, 138], visualizing atomic-scale structure of materials that are shaped into a nano-scale needle tip form. The needle tips become ion sources. An imaging MS detector is placed at the detector position in Fig. 6.23. A nano-needle tip is cooled at a few 10 K to suppress thermal vibration of atoms, which is favorable to superconductor detectors to avoid blackbody radiation. The atoms on the tip surface are ionized and repelled from the tip under an intense electric field, often assisted by pulse laser shots. This is called field evaporation, which means that APT is a destructive analytical method. A divergent electric field at the tip apex provides a high magnification of a few million [139]. The enlarged pattern of the tip surface structure is imaged by a position sensitive MS detector. It is possible to reconstruct 3D structure of materials at a sub-nanometer resolution from a series of ion-images in the course of the field evaporation. Because of this capability, it is also called 3D-APT. The TOF-MS section operates on the linear mode; the requirement for *m/z* resolution is not as high as that for modern high mass-resolution TOF–MS instruments. However, a flight length is normally shorter than conventional TOF–MS; therefore, the requirements to the imaging MS detectors are demanding. The detectors should have a capability of imaging as well as *m/z* separation. Typical imaging detectors consist of a delay line imager attached to an MCP. The MCP multiplies secondary electrons after ion impact; the location of the electron shower is identified by the delay-line imager. For 3D-APT instrumentation, important specifications include lateral spatial resolution, in-depth spatial resolution, and *m/z* separation [139]. The in-depth resolution in 3D-APT is considerably higher than the lateral resolution. We skip further discussion on the in-depth resolution because it depends on the field evaporation process irrespective of detector performance.

One of the typical *m/z* overlap cases in APT is a combination of $^{28}Si^{2+}$ and $^{14}N^{+}$ at *m/z* 14 [136]. SiC:N is a typical example. XAFS reveals the substitution sites of the N dopant in the SiC lattice structure, as shown in Fig. 6.17 [58]. 3D-APT hopefully visualizes 3D atomic structure of the N dopant inside SiC diodes or transistors; however, it is impossible owing to the *m/z* overlap. The intensity of $^{28}Si^{2+}$ from the SiC matrix is extremely higher than that of $^{14}N^{+}$. The atom probe community is searching for kinetic-energy sensitive MS detectors to overcome the *m/z* overlap [136, 140–142]. As described before, the superconductor detectors solve the *m/z* overlap problem. Let us contemplate whether superconductor detectors can contribute to APT, considering the requirement for imaging MS detectors. This approach facilitates the expansion of analytical application areas, where superconductor detectors overcome the fundamental limits of conventional instruments.

The kinetic-energy resolving power $E_0/\Delta E$ required for 3D-APT is 10 [136]. The 200-μm square STJD has an $E_0/\Delta E$ value of 6–8, as shown in Fig. 6.27b. STJDs with thick Al layers potentially meet the $E_0/\Delta E$ requirement of 10. All the superconductor detector types in Table 5.1 exhibit similar $E_0/\Delta E$ values for ions because the resolving power is predominantly determined by ion-surface collision process rather than the intrinsic energy resolution of the superconductor detectors. Nevertheless, the current resolving power of the superconductor detectors is adequate for distinguishing different charge states. SSPDs also enable charge-state discrimination, although a shaper threshold for the discrimination should be achieved by selecting proper superconductor materials. The cold needle-tip ion-source allows us to skip the IR blocking filter in Fig. 6.25. This is impetus to considering the application to APT.

When we assume a magnification factor of 1 M and a tolerable lateral resolution of 1 nm on needle tips, the imaging MS detector should have a 2D spatial resolution of 1 mm. In general, APT cannot achieve an atomic scale resolution; therefore, this spatial resolution is adequate. Assuming a field of view of ~ 100 nm on needle tips, the required sensitive area is roughly 10×10 cm^2. We also assume a requirement of the 1-ns time resolution for *m/z* separation. These performance figures are 10 times worse than the high-level requirements for a 2D precision of 1/1000 and a time resolution of 100 ps [136]. The 100-ps resolution is very challenging for all the superconductor detector types as well as most non-superconductor solid state detectors. In contrast, vacuum tubes such as EM, MCP, and PMT have a temporal resolution of a few 10 ps. Nevertheless, the above performance figures can cover most practical cases of APT analysis. The required number of pixels is 100×100. With thermal detectors such as TESD or MMCD, we can fabricate an array detector with 10^4 pixels covering 10 cm with multiplex readout. However, the rise time of output pulses of the TES type is inherently in a range of several μs because of thermal nature and multiplexing. The MMC type records a pulse rise time of 70 ns;[143] it is still insufficient. The slow response leads to an unacceptably low *m/z* resolution. The thermal detectors operating at ~ 50 mK also face a problem of IR radiation from the ion source side even when the ion source temperature is a few 10 K.

With quantum detectors such as STJD and MKID, we can reach sub-μs rise time as shown in Fig. 6.32 because electronic excitation and relaxation are faster than thermal processes. The observed $\tau_{\rm rise}$ of the STJD pulses was 400 ns, limited by readout electronics, for X-ray photon in Fig. 3.14 and for ions in Fig. 6.32. When we use constant fraction discrimination, the time resolution can be improved to 90 ns. The time response of Josephson junctions is determined by a time constant of *CR*; it is 40 ps for the 200-μm square junction [144]. This time constant in a range of <100 ps after ion impact is equivalent to the combination of the quasiparticle excitation and relaxation process times. Therefore, the 100-ps requirement can be cleared; however, readout circuitry for kinetic energy measurement is considerably slow. We have an option of another design with an additional Josephson junction for fast-slow coincidence [120]. With respect to the number of pixels, the technical achievement at AIST is 1024-pixels of 200-μm square junctions. For STJDs, each pixel has its own readout channel; it is advantageous to fast time response but disadvantageous to a large-scale array. The detector size of 2.8×2.8 mm^2 with a spatial resolution of 0.2 mm in close-packed pixel arrangement is technically feasible at this moment. However, this is far behind the requirements of 10×10 cm^2 and 10^4 pixels. Measurement of a restricted area on needle tips is one possible approach of array STJDs.

Instead of using STJD, MKID imagers can have 2×10^4 pixels, which meet the requirement [145]. Because both STJD and MKID are quantum detectors with an energy gap, they are tolerant of heat input. The intrinsic time response of MKID imagers is in the same order as that of STJDs. However, it seems that readout electronics with the 1-ns time-resolution for the 10^4-pixel MKID imagers is problematic for frequency-domain multiplexing. In addition, the resonance circuit in Fig. 3.35 for each MKID pixel occupy a large space, leading to a small filling factor, because only inductor segment can detect ions. Nevertheless, MKIDs with a large number of pixels are adequate for APT.

A prototype SSPD 1D-imager has a spatial resolution of 500 nm for X-rays [146]. SSPDs with a practical size record a τ_{rise} value of a few 100 ps with a reachable timing jitter of 80 ps [147]. Promising readout electronics is a superconductor time-to-digital converter using single-flux-quantum (SFQ) digital circuits with a timing jitter less than 100 ps [119]. In addition, the SS type inherently has a strip geometry, which resembles the delay-line imaging scheme of the current APT MS detectors [148]. The 2D delay-line scheme was reported recently [149]. In addition to the delay line scheme, pixel type SSPD is also possible like STJD and MKID [150]. A readout scheme of orthogonal time-amplitude multiplexing can realize a large scale SSPD pixel array with a temporal resolution of several 10 ps [151]. The SSPD imagers accept direct ion impact, which means that no MCP multiplication is needed, leading to a higher detection efficiency. However, in both delay line scheme and pixel one, it is currently problematic to cover 10×10 cm with nanostrips with the 1-ns time-resolution. A detector size of 1×1 cm^2 is currently practical. Even when an SSPD MS imaging detector is fabricated, we cannot measure the charge states of individual ions. Nevertheless, it is possible to measure doubly-charged ions by eliminating singly-charged ion events. Unfortunately, it is impossible to detect singly-charge ions exclusively in a combination of $^{14}N^+$ and $^{28}Si^{2+}$. This restricts sample variety for SSPD-APT.

There are high hurdles for the application of superconductor detectors to 3D-APT. Nonetheless, it is worth deliberating challenging applications, which seem to be impossible at the beginning. Later, a promising scheme will materialize [24].

References

1. K. Birkinshaw, Int. J. Mass Spectrom. **215**, vii (2002)
2. M. Frank, S. Labov, G. Westmacott, W. Benner, Mass Spectrom. Rev. **18**, 155 (1999)
3. S. Friedrich, J. Synchrotron Rad. **13**, 159 (2006)
4. J.N. Ullom, J. Low Temp. Phys. **151**, 746 (2008)
5. J. Ullom, D. Bennett, Supercond. Sci. Technol. **28**, 084003 (2015)
6. W.B. Doriese et al., Rev. Sci. Instrum. **88**, 053108 (2017)
7. S.J. Lee et al., Rev. Sci. Instrum. **90**, 113101 (2019)
8. S. Fretwell, K.G. Leach, C. Bray, G.B. Kim, J. Dilling, A. Lennarz, X. Mougeot, F. Ponce, C. Ruiz, J. Stackhouse, S. Friedrich, Phys. Rev. Lett. **125**, 032701 (2020)
9. S. Friedrich et al., Phys. Rev. Lett. **126**, 021803 (2021)
10. P. Szypryt et al., Rev. Sci. Instrum. **90**, 123107 (2020)

11. G.C. O'Neil et al., Phys. Rev. A **102**, 032803 (2020)
12. T. Okumura et al., Phys. Rev. Lett. **127**, 053001 (2021)
13. Y-H. Kim, W. Cui, H-S. Jo, L. You (eds.), Special Issue: Low Temperature Detectors LTD20. J. Low Temp. Phys. **214**(3–4), **215**(3–4), **216**(1–2) (2024), and the previous proceedings series
14. Y.H. Kim, S.-J. Lee, B. Yang, Supercond. Sci. Technol. **35**, 063001 (2022)
15. NIST X-Ray Transition Energies Database. https://www.nist.gov/pml/x-ray-transition-energies-database
16. Lawrence Berkeley National Laboratory, X-ray data booklet. https://xdb.lbl.gov
17. M. Ohkubo, J. Appl. Phys. **134**, 081101 (2023)
18. X. Llovet, C.J. Powell, F. Salvat, A. Jablonski, J. Phys. Chem. Ref. Data **43**, 013102 (2014)
19. R. Jenkins, R. Manne, R. Robin, C. Senemaud, Pure & Appl. Chem. **63**, 735 (1991)
20. M.O. Krause, J. Phys. Chem. Ref. Data **8**, 307 (1979)
21. J.H. Hubbell et al., J. Phys. Chem. Ref. Data **23**, 339 (1994)
22. D.A. Wollman, K.D. Irwin, G.C. Hilton, L.L. Dulcie, D.E. Newbury, M.M. Martinis, J. Microsc. **188**, 196 (1997)
23. P. Zimmermann, S. Peredkov, P.M. Abdala, S. DeBeer, M. Tromp, C. Muller, J.A. van Bokhoven, Coord. Chem. Rev. **423**, 213466 (2020)
24. D.M. Schlosser et al., Nucl. Instrum. Methods in Phys. Res. A 624, 270 (2010)
25. D.A. Wollman et al., AIP Conf. Proc. **449**, 799 (1998)
26. M. Frank, L.J. Hiller, J.B. de Grand, C.A. Mears, S.E. Labov, M.A. Linderman, H. Netel, D. Chow, A.T. Barfknecht, Rev. Sci. Instrum. **69**, 25 (1998)
27. M. Suga et al., Prog. Sold. State Chem. **42**, 1 (2014)
28. L.J. Bauer, H.A. Mustafa, P. Zaslansky, I. Mantouvalou, Acta Biomat. **109**, 142 (2020)
29. D.A. Wollman, S.W. Nam, G.C. Hilton, K.D. Irwin, N.F. Bergren, D.A. Rudman, J.M. Martinis, D.E. Newbury, J. Microsc. **199**, 37 (2001)
30. K. Tanaka, A. Takano, A. Nagata, S. Nakayama, K. Takahashi, M. Ajima, K. Obara, K. Chinone, Microsc. **69**, 298 (2020)
31. G. Fujii, M. Ukibe, S. Shiki, M. Ohkubo, X-ray Spectrom. **46**, 325 (2017)
32. N. Zen, S. Shiki, M. Ohkubo, IEEE Trans. Appl. Super. **19**, 1008 (2009)
33. P. Zaslansky, J.D. Currey, C. Fleck, Bioinspir. Biomim. **11**, 051003 (2016)
34. R. Osorio et al., J. Dent. **39**, 148 (2011)
35. M. Toledano, M. Yamauti, E. Osorio, R. Osorio, Caries Res. **46**, 201 (2012)
36. R. Osorio, E. Osorio, I. Cabello, M. Toledano, Caries Res. **48**, 276 (2014)
37. O. Thanatvarakorn, M.S. Islam, S. Nakashima, A. Sadr, T. Nikaido, J. Tagami, Dent. Mater. J. **35**, 769 (2016)
38. B.A. Gotliv, A. Veis, Cells Tissues Organs **189**, 12 (2009)
39. R. Yee, C. Holmgren, J. Mulder, D. Lama, D. Walker, W.P. Helderman, J. Dent. Res. **88**, 644 (2009)
40. R. Cantor et al., Microsc. Microanal. **22**(Suppl. 3), 434 (2016)
41. M.H. Carpenter et al., J. Low Temp. Phys. **200**, 437 (2020)
42. M.H. Carpenter et al., IEEE Trans. Appl. Super. **31**, 2102206 (2021)
43. K. Suenaga, T. Okazaki, E. Okunishi, S. Matsumura, Nature Photo. **6**, 545 (2012)
44. T.C. Lovejoy, Q.M. Ramasse, M. Falke, A. Kaeppel, R. Terborg, R. Zan, N. Dellby, O.L. Krivanek, Appl. Phys. Lett. **100**, 154101 (2012)
45. K. Maehata, T. Hara, K. Mitsuda, M. Hidaka, K. Tanaka, Y. Yamanaka, J. Low Temp. Phys. **184**, 5 (2016)
46. K. Yamada, N. Kawakami, T. Moronaga, K. Hayahsi, C. Ichihara, T. Hara, Appl. Phys. Express **13**, 082008 (2020)
47. T. Hayashi, H. Muramatsu, R. Konno, N.Y. Yamasaki, K. Mitsuda, A. Takano, K. Maehata, T. Hara, J. Low Temp. Phys. **199**, 908 (2020)
48. T. Hayashi, R. Miyagawa, Y. Yagi, K. Tanaka, R. Ota, N. Y. Yamasaki, K. Mitsuda, K. Maehata, T. Hara, J. Low Temp. Phys. (2024)
49. J.J. Rehr, R.C. Albers, Rev. Mod. Phys. **72**, 621 (2000)
50. S. Friedrich, T. Funk, O. Drury, S.E. Labov, S.P. Cramer, Rev. Sci. Instrum. **73**, 1629 (2002)

51. M. Ohkubo, P. Fons, A. Kushino, Y. E. Chen, M. Ukibe, Y. Kitajima, Nucl. Instrum. Methods in Phys. Res. A **559**, 731 (2006)
52. S. Shiki, N. Zen, M. Ukibe, M. Ohkubo, A.I.P. Conf, Proc. **1185**, 409 (2009)
53. S. Shiki, M. Ukibe, N. Matsubayashi, N. Zen, M. Koike, Y. Kitajima, M. Ohkubo, J. Low Temp. Phys. **176**, 604 (2014)
54. S. Friedrich, O.B. Drury, T. Funk, D. Sherrell, V.K. Yachandra, S.E. Labov, S.P. Cramer, A.I.P. Conf, Proc. **705**, 985 (2004)
55. C.J. Titus et al., J. Chem. Phys. **147**, 214201 (2017)
56. Y. Il Joe et al., Phys. Rev. Appl. **13**, 034026 (2020)
57. P. Fons et al., Phys. Rev. Lett. **96**, 045504 (2006)
58. M. Ohkubo, S. Shiki, M. Ukibe, N. Matsubayashi, Y. Kitajima, S. Nagamachi, Sci. Rep. **2**, 831 (2012)
59. Y. Maeda et al., J. Phys. Chem. C **124**, 11032 (2020)
60. S. Mitsuya, H. Ishii, J. Kawai, Appl. Phys. Lett. **89**, 134104 (2006)
61. J. Uhlig et al., Phys. Rev. Lett. **110**, 138302 (2013)
62. The FEFF Project. http://feff.phys.washington.edu/feffproject-feff.html
63. H.M. Tsai et al., Appl. Phys. Lett. **79**, 2393 (2001)
64. Y.K. Chang et al., J. Appl. Phys. **86**, 5609 (1999)
65. U. Schwingenschlögl, A. Chroneos, C. Schuster, R.W. Grimes, J. Appl. Phys. **112**, 046101 (2012)
66. S-R.G. Christopoulos, N. Kuganathan, A. Chroneos, Sci. Rep. **10**, 7459 (2020)
67. G.C. O'Neil et al., J. Phys. Chem. Lett. **8**, 1099 (2017)
68. J. Chen, H. Zhang, I.V. Tomov, X. Ding, P.M. Rentzepis, Chem. Phys. Lett. **437**, 50 (2007)
69. M.A. Reis, P.C. Chaves, A. Taborda, X-ray Spectrom. **40**, 141 (2011)
70. M.R.J. Palosaari et al., Phys. Rev. Appl. **6**, 024002 (2016)
71. A. Yamazaki et al., Nucl. Instrum. Methods Phys. Res. B **404**, 92 (2017)
72. S. Shiki, G. Fujii, S. Tomita, K. Sasa, Nucl. Instrum. Methods in Phys. Res. B **554**, 165451 (2024)
73. L. Miaja-Avila, G.C. O'Neil, Y. Joe, K.M. Morgan, J.W. Fowler, W.B. Doriese, B. Ganly, D. Lu, B. Ravel, D.S. Swetz, J.N. Ullom, X-ray Spectrom. **50**, 9 (2021)
74. J.J. Thomson, *Rays of Positive Electricity and Their Application to Chemical Analyses* (Longmans, Green and Co., London, 1913)
75. D.W. Koppenaal, C.J. Barinaga, M.B. Denton, R.P. Sperline, G.M. Hieftje, G.D. Schilling, F.J. Andrade, J.H. Barner IV., Anal. Chem. **77**, 418A (2005)
76. I.S. Gilmore, M.P. Seah, Int. J. Mass Spectrom. **202**, 217 (2000)
77. G.W. Fraser, Int. J. Mass Spectrom. **215**, 13 (2002)
78. G. Westmacott, F. Zhong, M. Frank, S. Friedrich, S.E. Labov, W.H. Benner, Rapid Commun. Mass Spectrom. **14**, 600 (2000)
79. D.Z. Keifer, M.F. Jarrold, Mass Spectrom. Rev. **36**, 715 (2016)
80. E.J. Sternglass, Phys. Rev. **108**, 1 (1957)
81. D. Twerenbold, Nucl. Instrum. Method in Phys. Res. A **370**, 253 (1996)
82. M. Ohkubo, Physica C **468**, 1987 (2008)
83. M. Ohkubo, S. Shiki, M. Ukibe, S. Tomita, S. Hayakawa, Int. J. Mass Spectrom. **299**, 94 (2011)
84. D. Twerenbold, J.-L. Vuilleumier, D. Gerber, A. Tadsen, B. van den Brandt, P.M. Gillevet, Appl. Phys. Lett. **68**, 3503 (1996)
85. H. Hilton, J.M. Martinis, D.A. Wollman, K.D. Irwin, L.L. Dulcie, D. Gerber, P.M. Gillevet, D. Twerenbold, Nature **391**, 672 (1998)
86. R.J. Wenzel, U. Matter, L. Schultheis, R. Zenobi, Anal. Chem. **77**, 4329 (2005)
87. B. Spengler, D. Kirsch, R. Kaufmann, M. Karas, F. Hilenkamp, Rapid Commun. Mass Spectrom. **4**, 301 (1990)
88. B. Seyfried, J. Siekmann, O. Belgacem, R.J. Wenzel, P. Turecek, G. Allmaier, J. Mass Spectrom. **45**, 612 (2010)

89. M. Frank, C.A. Mears, S.E. Labov, W.H. Benner, D. Horn, J.M. Jaklevic, A.T. Barfknecht, Rapid Commun. Mass Spectrom. **10**, 1946 (1996)
90. K.K. Murray, R.K. Boyd, M.N. Eberlin, G.J. Langley, L. Li, Y. Naito, Pure Appl. Chem. **85**, 1515 (2013)
91. S. Shiki, M. Ukibe, R. Maeda, M. Ohkubo, Y. Sato, S. Tomita, Nucl. Instrum. Methods in Phys. Res. A **595**, 391 (2008)
92. S. Shiki, M. Ukibe, R. Maeda, M. Ohkubo, Jpn. J. Appl. Phys. **47**, 3682 (2008)
93. M. Ohkubo, N. Zen, T. Kitazume, M. Ukibe, S. Shiki, M. Koike, Prog. Supercond. **14**, 77 (2012)
94. M. Ohkubo, Y. Shigeri, T. Kinumi, N. Saito, M. Ukibe, Y. E. Chen, A. Kushino, A. Kurokawa, H. Sato, S. Ichimura, Nucl. Instrum. Methods Phys. Res. A **559**, 779 (2006)
95. M. Ohkubo, M. Ukibe, Y. Chen, S. Shiki, Y. Sato, S. Tomita, S. Hayakawa, J. Low Temp. Phys. **151**, 760 (2008)
96. S. Tomita, Y. Sato, M. Ohkubo, M. Ukibe, S. Hayakawa, Appl. Phys. Lett. **91**, 053507 (2007)
97. M. Ohkubo et al., J. Low Temp. Phys. **167**, 943 (2012)
98. M. Ohkubo, M. Ukibe, N. Saito, A. Kushino, S. Ichimura, S. Friedrich, IEEE Trans. Appl. Super. **15**, 932 (2005)
99. A. Kushino, M. Ohkubo, K. Fujioka, Cryogenics **45**, 637 (2005)
100. A. Kushino, S. Kasai, S. Kohjiro, S. Shiki, M. Ohkubo, J. Low Temp. Phys. **151**, 650 (2008)
101. A. Kushino, M. Ohkubo, Y. E. Chen, M. Ukibe, S. Kasai, K. Fujioka, Nucl. Instrum. Methods Phys. Res. A **559**, 654 (2006)
102. A. Kushino, S. Kasai, J. Supercond. Nov. Magn. **28**, 715 (2015)
103. S. Shiki, M. Ukibe, Y. Sato, S. Tomita, S. Hayakawa, M. Ohkubo, J. Mass Spectrom. **43**, 1686 (2008)
104. R. Thissen, O. Witasse, O. dile Dutuit, C. Simon Wedlund, G. Gronoffe, J. Lilenstena, Phys. Chem. Chem. Phys., **13**, 18264 (2011)
105. J. Axelsson, C.T. Reimann, B.U.R. Sundqvist, Int. J. Mass Spectrom. **133**, 141 (1994)
106. D. Keifer, D. Shinholt, M. Jarrold, Anal. Chem. **87**, 10330 (2015)
107. A.B. Kanu, P. Dwivedi, M. Tam, L. Matz, H.H. Hill, J. Mass Spectrom. **43**, 1 (2008)
108. M. Ohkubo, IEICE Trans. Electron. **E90-C**, 550 (2007)
109. M. Ohkubo, A.I.P. Conf, Proc. **1185**, 381 (2009)
110. B. Brehm, G. Defrenes, Int. J. Mass Spectrom. Ion Phys. **26**, 251 (1978)
111. M.J. Besnard, L. Hellner, G. Dujardin, J. Chem. Phys. **89**, 6554 (1988)
112. D., McCammon R., Almy E., Apodaca W., Bergmann Tiest W., Cui S., Deiker M., Galeazzi M., Juda A., Lesser T., Mihara J. P., Morgenthaler W. T., Sanders J., Zhang E., Figueroa-Feliciano R. L., Kelley S. H., Moseley R. F., Mushotzky F. S., Porter C. K., Stahle A. E., Szymkowiak (2002) A High Spectral Resolution Observation of the Soft X-Ray Diffuse Background with Thermal Detectors. The Astrophysical Journal 576(1) 188–203 https://doi.org/10.1086/341727
113. N. Zen, H. Shibata, Y. Mawatari, M. Koike, M. Ohkubo, Appl. Phys. Lett. **106**, 222601 (2015)
114. I. Charaev et al., Nat. Nano. **18**, 343 (2023)
115. R.L. Merino, P. Seifert, J.D. Retamal, R.K. Mech, T. Taniguchi, K. Watanabe, K. Kadowaki, R. H. Hadfield, D.K. Efetov, 2D Mater. **10**, 021001 (2023)
116. K. Suzuki, M. Ohkubo, M. Ukibe, K. Chiba-Kamoshida, S. Shiki, S. Miki, Z. Wang, Rapid Commun. Mass Spectrom. **24**, 3290 (2010)
117. L.D. Plath, H. Abroshan, C. Zeng, H.J. Kim, R. Jin, M.E. Bier, J. Am. Soc. Mass Spectrom. **33**, 521 (2022)
118. L.X. Jiang, L.D. Plath, M.A. Halim, S. Friedrich, M.E. Bier, Anal. Chem. **94**, 5284 (2022)
119. K. Sano, Y. Muramatsu, Y. Yamanashi, N. Yoshikawa, N. Zen, M. Ohkubo, Physica C **504**, 97 (2014)
120. E. Esposito, M. Ejrnaes, S. Pagano, D.P. de Lara, R. Cristiano, Appl. Phys. Lett. **82**, 2109 (2003)
121. N. Zen, A. Casaburi, S. Shiki, K. Suzuki, M. Ejrnaes, R. Cristiano, M. Ohkubo, Appl. Phys. Lett. **95**, 172508 (2009)

122. K. Suzuki, S. Shiki, M. Ukibe, M. Koike, S. Miki, Z. Wang, M. Ohkubo, Appl. Phys. Express **4**, 083101 (2011)
123. E. Sage et al., Nature Commun. **6**, 6482 (2015)
124. S. Domingez-Medina et al., Science **362**, 918 (2018)
125. D.M. Sipe, L.D. Plath, A.A. Aksenov, J.S. Feldman, M.E. Bier, ACS Nano **12**, 2591 (2018)
126. S. Hayakawa, J. Mass Spectrom. **39**, 111 (2004)
127. A. Florescu-Mitchell, J. Mitchell, Phys. Rep. **430**, 277 (2006)
128. L. Snyder, Proc. Natl. Acad. Sci. U.S.A. **103**, 12243 (2006)
129. A.G.G.M. Tielens, Rev. Mod. Phys. **85**, 1021 (2013)
130. R. Thomas, Mass Spectrom. Rev. **27**, 485 (2008)
131. G. Angelova, O. Novotny, J.B.A. Mitchell, C. Rebrion-Rowe, J.L. Le Garrec, H. Bluhme, K. Seiersen, L.H. Andersen, Int. J. Mass Spectrom. **232**, 195 (2004)
132. O. Novotny, S. Allgeier, C. Enss, A. Fleischmann, L. Gamer D. Hengstler, S. Kempf, C. Krantz, A. Pabinger, C. Pies, D.W. Savin, D. Schwalm, A. Wolf, J. Appl. Phys. **118**, 104503 (2015)
133. L. Gamer, D. Schulz, C. Enss, A. Freischmann, L. Gastaldo, S. Kempf, C. Krantz, O. Novotny, D. Schwalm, A. Wolf, J. Low Temp. Phys. **184**, 839 (2016)
134. T. Tanabe, K. Noda, M. Saito, E.B. Starikov, M. Tateno, Phys. Rev. Lett. **93**, 043201 (2004)
135. Y. Ebara, T. Furukawa, J. Matsumoto, H. Tanuma, T. Azuma, H. Shiromaru, K. Hansen, Phys. Rev. Lett. **117**, 133004 (2016)
136. T.F. Kelly, Microsc. Microanal. **17**, 1 (2011)
137. T.F. Kelly, M.K. Miller, Rev. Sci. Instrum. **78**, 031101 (2007)
138. B. Gault, A. Chiaramonti, O. Cojocaru-Mirédin, P. Stender, R. Dubosq, C. Freysoldt, S.K. Makineni, T. Li, M. Moody, J.M. Cairney, Nat. Rev. Methods Primers **1**, 51 (2021)
139. B. Gault, M.P. Moody, F. De Geuser, A. La Fontaine, L.T. Stephenson, D. Haley, S.P. Ringer, Microsc. Microanal.anal. **16**, 99 (2010)
140. M.K. Miller, T.F. Kelly, K. Rajan, S.P. Ringer, Mater. Today **15**, 158 (2012)
141. B. Gault, A. Chiaramonti, O. Cojocaru-Mirédin, P. Stender, R. Dubosq, C. Freysoldt, S.K. Makineni, T. Li, M. Moody, J.M. Cairney, Nat. Rev. Dis. Primers. **1**, 51 (2021)
142. C. Bacchi, G. Da Costa, E. Cadel, F. Cuvilly, J. Houard, C. Vaudolon, A. Normand, F. Vurpillot, Microsc. Microanal. **28**, 1076 (2022)
143. H. Rotzinger, J. Adams, S.R. Bandler, J. Beyer, H. Eguchi, E. Figueroa-Feliciano, W. Hsieh, G.M. Seidel, T. Stevenson, J. Low Temp. Phys. **151**, 351 (2008)
144. M. Maezawa, M. Aoyagi, H. Nakagawa, I. Kurosawa, S. Takada, Appl. Phys. Lett. **66**, 2134 (1995)
145. A.B. Walter et al., Publ. Astron. Soc. Pac. **132**, 125005 (2020)
146. C. Watanabe, M. Ukibe, N. Zen, G. Fujii, K. Makise, M. Ohkubo, Te-Hui Lee, and Di-Jing Huang. IEEE Trans. Appl. Super. **29**, 2400104 (2019)
147. M. Ejrnaes, A. Casaburi, R. Cristiano, N. Martucciello, F. Mattioli, A. Gaggero, R. Leoni, J.-C. Villégier, S. Pagano, Supercond. Sci. Technol. **24**, 035018 (2011)
148. Q.-Y. Zhao, D. Zhu, N. Calandri, A.E. Dane, A.N. McCaughan, F. Bellei, H.-Z. Wang, D.F. Santavicca, K.K. Berggren, Nat. Photo. **11**, 247 (2017)
149. B.G. Oripov, D.S. Rampini, J. Allmaras, M.D. Shaw, S.W. Nam, B. Korzh, A.N. McCaughan, Nature **622**, 730 (2023)
150. E.E. Wollman, V.B. Verma, A.E. Lita, W.H. Farr, M.D. Shaw, R.P. Mirin, S.W. Nam, Opt. Express **27**, 35279 (2019)
151. L.-D. Kong, H. Wang, Q.-Y. Zhao, J.-W. Guo, Y.-H. Huang, H. Hao, S. Chen, X.-C. Tu, L.-B. Zhang, X.-Q. Jia, L. Kang, J. Chen, P.-H. Wu, Nat. Photo. **17**, 65 (2023)

Correction to: Quantum Detection

Correction to:
Chapter 3 in: M. Ohkubo, *Superconductor Detectors*, https://doi.org/10.1007/978-3-031-84193-4_3

The original version of the chapter 3 was published with errors that had inadvertently remained during the editing process. Figure captions 3.21 and 3.32 are updated. The chapter has been updated. The basic facts have not been changed.

The updated version of this chapter can be found at
https://doi.org/10.1007/978-3-031-84193-4_3

M. Ohkubo, *Superconductor Detectors*, https://doi.org/10.1007/978-3-031-84193-4_7

Appendix A
Mathematical Derivation

A.1 Definition of Energy Resolution and Major Probability Distribution Functions

The energy resolution in spectroscopy is defined as the smallest detectable difference between the energy values of two particles [1]. It is denoted by ΔE, which is the FWHM value of a peak due to monoenergetic particle (E_0) incident events on pulse height spectra. Resolving power expressed by $E_0/\Delta E$ is also used for one of the detector performance figures. Figure A.1 shows three major probability distribution functions for quantum radiation detectors: binomial, Gauss, and Poisson distributions. All the functions play important roles in the basic statistical detector physics. Full-energy peaks (or total absorption peaks) to monochromatic particles can be usually expressed by the Gauss distribution; a factor of 2.355 often appears to convert standard deviation to FWHM. The phrase "high energy resolution" means small FWHM values. The variance ratio of the actual distribution for carrier generation to the Poisson distribution, both of which have the same average carrier number, is called the Fano factor denoted by F [2]. The Fano factor is important for describing fluctuation in a variety of fields. Nonetheless, if an exact distribution function is analytically known, F disappears from the expression for the theoretical limit of ΔE as described below. Unfortunately, the analytical formulas of exact distribution functions are usually unknown. Therefore, F always appears in discussion on the energy resolution in quantum detection, even though most quantum detectors have no connection to the Poisson distribution.

In Fig. A.1, it is assumed that the particles have the kinetic energy E_0 that equals to $40 \times E_g$ and the charge generation is governed by the binomial probability distribution function. When 80% of the particle energy is consumed to generate charge carriers ($p = 0.8$), the average number $\overline{N}$ of charge carriers is 32; 20% of the particle energy is lost without generating charge carriers. The average charge carrier number of 32 is sufficient for the binomial distribution to be well approximated by the Gauss distribution. The charge carrier number in practical quantum detectors is normally very large, such as ~ 6000 for Si and ~ 2,000,000 for Nb; therefore, the Gauss

M. Ohkubo, *Superconductor Detectors*, https://doi.org/10.1007/978-3-031-84193-4

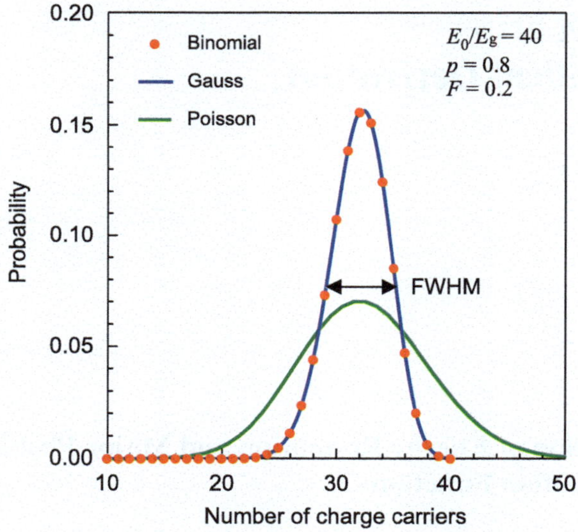

Fig. A.1 Major probability distribution functions: binomial distribution (red dots), Gauss distribution (blue solid line), and Poisson distribution (green solid line) for quantum detectors with energy gap E_g. Assumed parameters are the particle energy E_0 ($= 40E_g$) and the Fano factor ($F = 0.2$); the average value of the number of charge carriers is 32. The energy resolution is defined as full width at half maximum (FWHM) of the peak. Each distribution curve is normalized such that its integral equals 1.0. The definition of ΔE is also applied to thermal detectors. Reproduced with permission from Ohkubo [11]. Copyright 2023 AIP Publishing LLC

distribution can replace the binomial distribution. FWHM values estimated from experimental data are often narrower than the binomial distribution as well as the Poisson distribution that is the extreme case that the energy consumption to generate charge carriers in the course of collision cascades is negligibly small. It seems that no exact analytical treatment is possible for the actual probability function considering all physical processes of absorption, excitation, relaxation, and charge collection. A single statistical model cannot describe all the physical processes. One possible way is to utilize Monte Carlo simulation with reasonable hypotheses, although we need to deliberate on the hypotheses.

When all the particle energy is consumed to generate charge carriers, $\overline{N}$ equals to 40. This situation represents $F = 0$ because no fluctuation of the charge carrier number occurs, leading to zero ΔE. This never happens in practical cases. Unavoidable energy loss results in a situation with $\overline{N} < 40$; correspondingly, a fluctuation of the carrier number arises. Since the exact probability distribution function is unknown, the Fano factor is often determined from experimental data of the output charge and the peak width of full-energy absorption peaks on pulse height spectra or from calculated data with Monte Carlo simulation. F and ε are often treated as independent parameters; however, they should be interdependent according to the definitions of F and ε, as discussed in the text.

A.2 Energy Resolution in Binomial Function for Quantum Detection

The binomial distribution function is closer to the exact distribution function rather than the Poisson distribution for charger carrier generation in actual detectors; the Poisson distribution cannot be a physical basis for practical quantum detectors. The binomial distribution is expressed by

$$P(x) = \frac{n!}{(n-x)!x!}p^x(1-p)^{n-x}, \tag{A.1}$$

where n is the number of trials to generate one electron–hole (e–h) pair in semiconductors or one quasielectron-quasihole pair in superconductors, p is the success probability for each trial, and x is the number of the exact number of success times. Note that "pair" does not signify a bound state such as Cooper pairs, but simultaneous generation. The average values of the success times $\overline{x}$ and the standard deviation σ are expressed by

$$\overline{x} = np = \overline{N} \text{ and} \tag{A.2}$$

$$\sigma = \sqrt{np(1-p)} = \sqrt{\overline{N}(1-p)}. \tag{A.3}$$

To adjust the binomial distribution to quantum detection, we replace n with E_0/E_g, $\overline{x}$ with the average carrier number $\overline{N}$, and p with the fraction of the energy consumed to generate e–h pairs. The other fraction $1-p$ is dissipated in loss processes without contributing to generating carriers. It should be noted that Eqs. (A.2) and (A.3) are based on the purely statistical model, not considering the details of physical processes such as X-ray absorption, subsequent electron emission, energy down-conversion cascades of an energetic photoelectron or an Auger electron, Compton recoil electrons, and other processes. Nonetheless, they can express a statistical essence of the quantum detection. It is natural that there is a constraint on relation between $\overline{N}$ and p or F and ε, which is described below. The experimental determination of F ignores this constraint, missing the physical basis, although there seems to be no better practical way.

The constraint in the binomial distribution is such that $\overline{N}$ is expressed by

$$\overline{N} = p\frac{E_0}{E_g} = \frac{E_0}{\epsilon}. \tag{A.4}$$

The $\varepsilon(= E_g/p)$ is always larger than E_g because some energy losses always arise without exception in practical quantum detectors. The energy resolution represents the FWHM values of probability distributions. At an extremely small p, the distribution function is approximated by the Poisson distribution, of which variance σ_0^2 is equal to $\overline{N}$. The standard deviation is given by

$$\sigma_0 = \sqrt{\overline{N}} = \sqrt{\frac{E_0}{\epsilon}}. \tag{A.5}$$

The variance ratio of the binomial distribution to the Poisson distribution corresponds to the Fano factor, which is expressed by [3]

$$F = \left({}^{\sigma}/_{\sigma_0} \right)^2 = 1 - p = 1 - \frac{E_g}{\varepsilon}. \tag{A.6}$$

In the extreme limit at $p \cong 0$ or $F \cong 1$, each carrier generation event along an electron track after X-ray absorption is almost completely uninfluenced by the previous generation or collision. In other word, the carrier generation collisions are independent. Correspondingly, almost no charge carriers are generated; this is useless for practical EDS detectors. When only a very small fraction of E_0 is consumed to generate charge carriers, the N distribution or the peak shape is well approximated by the Poisson distribution. This means that a detector has a poor energy resolution as depicted by the green line in Fig. A.1. Contrarily, when almost all the energy E_0 is consumed to generate charge carriers ($p \cong 1$ or $F \cong 0$), the collisions along the track are almost completely dependent each other. The standard deviation becomes as close as zero, and an extremely sharp peak appears at 40 in Fig. A.1.

For Monte Carlo simulation with m events, F is obtained by

$$F = \frac{\frac{1}{m}\sum_{i=1}^{m}\left(N_i - \frac{1}{m}\sum_{j=1}^{m}N_j\right)^2}{\frac{1}{m}\sum_{j=1}^{m}N_j} = \frac{\overline{N^2} - \overline{N}^2}{\overline{N}}. \tag{A.7}$$

Eq. (A.7) is also valid for any analytical probability distribution functions for carrier generation.

By using Eqs. (A.4) and (A.6), Eq. (A.3) is translated to

$$\sigma = \sqrt{FE_0/\varepsilon}. \tag{A.8}$$

The standard deviation in a unit of eV is written as

$$\sigma_{\text{eV}} = \varepsilon \times \sigma = \sqrt{FE_0\varepsilon}. \tag{A.9}$$

When E_0/E_g is large enough as is in the case of 40, the binomial distribution is approximated by Gauss distribution. This results in a conversion factor of $2\sqrt{2\ln 2}$ (= 2.355) from standard deviation to FWHM:

$$E = 2.355\sqrt{FE_0\varepsilon}. \tag{A.10}$$

This is the famous formula for quantum detectors with energy gaps. When the binomial distribution with p is valid, we can skip F, and Eq. (A.10) is translated to

$$E = 2.355\sqrt{\left(\frac{1}{p} - 1\right)E_0 E_g}. \tag{A.11}$$

No Fano factor appears in Eq. (A.11). Another derivation route is that Eqs. (A.3) and (A.4) directly yield Eq. (A.11) after adding 2.355. Afterward, by substituting Eq. (A.6) that is the Fano factor definition for Eq. (A.11), Eq. (A.10) is obtained. This derivation is a straightforward way, which means that we need not consider the Fano factor when we know the exact distribution function. When other analytical distributions are valid, the Fano factor also disappears. This is natural since the Fano factor is a variance ratio of an exact distribution to a Poisson distribution that is invalid for practical quantum detectors. However, no analytical distribution function is known; therefore, F serves as a convenient factor to represent the degree of deviation from the Poisson distribution with the same $\overline{N}$. The Poisson distribution is the reference to a hypothetical quantum detector with the worst energy resolution. Consequently, it is difficult to estimate the theoretical limit of ΔE because of indeterminacy on both F and ε, although Eq. (A.10) is absolutely correct.

We still need further experimental and theoretical investigations on statistical detector physics. Nevertheless, the recent ε–E_g relation data on AlGaAs [4] together with the classical data on Si and Ge approaches a model with the binominal distribution, provided that the phonon loss in semiconductor detectors is subtracted from E_0, as discussed in Chap. 3, Sect. 3.1.4.

A.3 Energy Resolution in Canonical Ensemble for Thermal Detection

The fundamental fluctuation of thermal detectors is determined by energy fluctuation in an absorber with heat capacity C, which is treated as the canonical ensemble. The probability p_s for the specific state E_s of the absorber as the subsystem in Fig. 4.1b should be proportional to the number of states of the heat bath with $\Omega(E_b)$. Because the total system energy E_t is constant, the situation is expressed by $E_b = E_t - E_s$. By applying normalization factor such that $\sum_s p_s = 1$, p_s is given by

$$p_s = \frac{\Omega(E_t - E_s)}{\sum_s \Omega(E_t - E_s)}. \tag{A.12}$$

The entropy of the heat bath is expressed by $S(E_t - E_s) = k_B \ln \Omega(E_t - E_s)$. Eq. (A.12) is rewritten as

$$p_s = \frac{\exp\left(\frac{S(E_t - E_s)}{k_B}\right)}{\sum_s \exp\left(\frac{S(E_t - E_s)}{k_B}\right)}. \tag{A.13}$$

The subsystem and the heat bath is in thermal equilibrium, so that the probable energy of the subsystem is very smaller than that of the bath regardless of non-zero probability that the subsystem has an energy comparable to the bath. Therefore, the Taylor expansion of the entropy in $E_\mathrm{t} \gg E_s$ is given by

$$S(E_\mathrm{t} - E_s) = S(E_\mathrm{t}) - E_s \frac{\partial S(E_\mathrm{t})}{\partial E_\mathrm{b}} + \frac{1}{2} E_s^2 \frac{\partial^2 (E_\mathrm{t}^2)}{\partial E_\mathrm{b}^2} + \cdots \tag{A.14}$$

Here, we introduce the temperature of the bath in a condition that the energy exchange is too small to change the temperature:

$$\frac{\partial S(E_\mathrm{t})}{\partial E_\mathrm{b}} = 1/T_\mathrm{b}. \tag{A.15}$$

T_b is practically constant, so that the third term and higher partial differential terms in Eq. (A.14) are negligible, Eq. (A.14) is then transformed to

$$S(E_\mathrm{t} - E_s) = S(E_\mathrm{t}) - \frac{E_s}{T_\mathrm{b}}. \tag{A.16}$$

By substituting Eq. (A.16) for Eq. (A.13), the probability for the system with E_s is given by

$$p_s = \frac{\exp(-E_s/k_\mathrm{B} T_\mathrm{b})}{\sum_s \exp(-E_s/k_\mathrm{B} T_\mathrm{b})} = \frac{1}{Z} e^{-\beta E_s}, \tag{A.17}$$

where

$$Z = \sum_s e^{-\beta E_s} \text{ and} \tag{A.18}$$

$$\beta \equiv \frac{1}{k_\mathrm{B} T_\mathrm{b}}. \tag{A.19}$$

Eq. (A.17) is called the Boltzmann distribution or the canonical distribution. It is apparent that as E_s increases, p_s decreases exponentially: a high temperature is required for a high energy state of the subsystem. The normalization factor of Eq. (A.18) is called the partition function.

We now know the probability of the subsystem with E_s; therefore, it is possible to estimate the mean energy $\overline{E}$ and the standard deviation σ:

$$\overline{E} = \sum_s p_s E_s = \frac{1}{Z} \sum_s E_s e^{-\beta E_s} \text{ and} \tag{A.20}$$

$$\sigma^2 = \overline{(E - \overline{E})^2} = \overline{E^2} - \overline{E}^2. \tag{A.21}$$

Calculating the derivative of ln Z with d(ln x)/dx = 1/x, we can obtain

$$\frac{\partial \ln Z}{\partial \beta} = \frac{\partial \ln Z}{\partial Z}\frac{\partial Z}{\partial \beta} = \frac{1}{Z}\frac{\partial Z}{\partial \beta} = -\frac{1}{Z}\sum_s E_s e^{-\beta E_s}. \quad (A.22)$$

$$\text{Thus, } \overline{E} = -\frac{\partial \ln Z}{\partial \beta}. \quad (A.23)$$

The second order partial derivative of ln Z is obtained as follows, by using $(g/f)' = (g'f - f'g)/f^2$:

$$\frac{\partial^2 \ln Z}{\partial \beta^2} = -\frac{\partial}{\partial \beta}\left(\frac{1}{Z}\sum_s E_s e^{-\beta E_s}\right) = \frac{1}{z}\sum_s E_s^2 e^{-\beta E_s} - \frac{1}{Z^2}\left(\sum_s E_s e^{-\beta E_s}\right)^2 = \overline{E^2} - \overline{E}^2. \quad (A.24)$$

Thus, considering that the heat capacity of the subsystem is given by $C = \frac{\partial \overline{E}}{\partial T}$ and $T = T_b$ in thermal equilibrium, we obtain

$$\sigma^2 = \frac{\partial^2 \ln Z}{\partial \beta^2} = -\frac{\partial \overline{E}}{\partial \beta} = -\frac{dT}{d\beta}\frac{\partial \overline{E}}{\partial T} = \frac{C}{k_B \beta^2} = k_B T^2 C \text{ and} \quad (A.25)$$

$$\Delta E = 2.355\sigma = 2.355\sqrt{k_B T^2 C}. \quad (A.26)$$

Eq. (A.26) is the famous formula for the energy resolution limit of thermal detectors in FWHM. Nonetheless, we can exceed this thermostatistical limit by narrowing a readout bandwidth or using TES with ETF as a highly sensitive thermometer [5].

Appendix B
IUPAC Notation of Characteristic X-Ray Lines and Interaction with Detector Materials

The traditional Siegbahn notation is widely used for X-ray emission spectral lines at present. On the other hand, international union of pure and applied chemistry (IUPAC) recommends new notation based on electron transition between electron shells of atoms since 1991 [6]. This article follows the IUPAC notation because the IUPAC notation is systematic and advantageous to deal with such characteristic X-rays as L lines and M lines as well as K lines. The Siegbahn notation follows the IUPAC notation in parentheses for reference. The IUPAC notation is based on an inner shell location for a hole and an outer shell from which an electron fills the hole. The hole can be created by an X-ray photon, electrons, protons, and even atoms. The electron transition from an upper shell to a lower shell leads to the emission of a characteristic X-ray photon. The conventional Siegbahn notation of "$K\alpha_1$" due to the electron transition from L_3 ($2p_{3/2}$) to K(1 s) is replaced by "K-L_3" in the IUPAC notation. When the X-ray line for the transition from L_2 ($2p_{1/2}$) to K(1 s) experimentally overlap with the K-L_3 line, they are collectively denoted as K-$L_{2,3}$ ($K\alpha_{1,2}$). We can refer to the electron configuration and binding energies in Ref. [7] to calculate the energy of an X-ray emission line from the energy difference between two shells. Under the IUPAC recommendation, it is straightforward to recognize radiative transition paths from an upper orbital to a lower one and to denote any possible emission lines systematically. This is beneficial to discuss the L-line analysis that is material analysis using characteristic L-lines instead of K-lines. The advantage of the L-line analysis with electron probes is described in Chap. 6, Sect. 6.3.1.1.

Let us consider what happens when a hole is created on one of the electron shells. If an X-ray photon with E_0 is absorbed by an electron on the K shell of an atom, a photoelectron is emitted from the K shell and thus the atom is ionized to singly charged positive ion when E_0 exceeds the ionization potential. We need to consider the shake-up and shake-off effects to be precise. The photoelectron energy E_{eK} is expressed by.

$$E_{eK} = E_0 - E_K \tag{A.27}$$

M. Ohkubo, *Superconductor Detectors*, https://doi.org/10.1007/978-3-031-84193-4

where E_K is the electron binding energy of the K shell. When E_0 is lower than E_K, an electron on the L shell is emitted and the photoelectron energy is given by

$$E_{eL} = E_0 - E_L. \tag{A.28}$$

Figure 6.2 illustrates these processes. Similar ionization processes continue to higher shells of M, N, O, and P, depending on the atomic number Z.

As an example, when an Nb atom absorbs a 5.9 keV photon, which is normally used as a reference X-ray source with ^{55}Fe, and a hole is created on the L_1 orbital, E_{eL1} is equal to 3.3 keV. No hole on the K shell is created because 5.9 keV is lower than the Nb K-edge energy of 18.985 keV. After emitting the photoelectron, the L vacancy is filled with a transition from an electron on the M shell. The transition prompts the emission of a characteristic X-ray photon with an energy of E_L–E_M (L_1–M_4, L_1–M_5, L_2–M_4, L_2–M_5, L_3–M_4, L_3–M_5, etc.) at radiative transition probabilities as shown in Fig. 6.4. This X-ray can excite another Nb atom when it has sufficient energy to creat a hole. These processes continue until no shell is available.

The photoelectron with $E_{eL1} = 3.3$ keV can also excite another Nb atom, creating a hole on the L, M, or N shell. Besides the photoelectron, Auger electrons due to the non-radiative transition have an energy in a range between a few 100 eV and a few keV, depending on the location of the initial hole. Similarly, the Auger electron can create a hole on the L, M, or N shell of other Nb atoms. The hole is filled by an electron transition from an upper shell, emitting a characteristic X-ray or an Auger electron, similar to the primary X-ray due to the radiative transition. These processes also continue until no shell is available.

The X-ray absorption is utilized for X-ray absorption spectroscopy (XAS) or X-ray absorption fine structure (XAFS). On the other hand, X-ray emission and electron emission after creating a hole on a shell are also utilized for material analysis. When characteristic X-rays escaping from materials are analyzed, it is called X-ray fluorescence (XRF) analysis with an X-ray probe, electron probe microanalysis (EPMA) with an electron probe, particle induced X-ray analysis (PIXE) with a proton or helium ion probe. The characteristic X-ray line-shape analysis is called X-ray emission spectroscopy (XES) to investigate electronic structure or chemical states. Analysis of photoelectrons is called X-ray photoelectron spectroscopy (XPS) or electron spectroscopy for chemical analysis (ESCA) for measuring electronic structure of outer shells of surface layers within a few nm. When Auger electrons are analyzed, it is called Auger electron spectroscopy (AES) for qualitative and quantitative analysis of elements in surface layers. Naturally, for measuring the energy of a primary X-ray or particle, we must minimize the escape of characteristic X-rays, photoelectrons, and Auger electrons such that all the energy of the incoming particle is transferred to the absorber.

The above-mentioned introduction to the IUPAC notation and X-ray absorption in Nb reminds us of what happens in Nb-STJDs. For 1-keV X-ray incidence, which is in a range that Nb-STJDs perform well, the energy range of photoelectrons is between 533 and 798 eV for a hole on one of the M_1, M_2, M_3, M_4, or M_5 shells. The photoelectron energy range is far larger than the binding energy of the N shell. Therefore, the photoelectrons can create another hole on the N shell. This vacancy is filled with an electron from an upper level, emitting either X-ray photon or Auger electron. Therefore, X-ray absorption is converted to energy loss processes of the energetic electrons in the Nb absorber. The photoelectrons and Auger electrons behave equally: the energetic electrons generate e–h pairs or collide with free electrons. This is called the collision cascade. The primary and secondary electrons lose their energies mainly through electron–electron inelastic scattering or phonon emission. Before stopping inside the Nb absorber, they generate e–h pairs in an eV range. The electrons and holes create a phonon bubble, and finally break Cooper-pairs in a meV range when Nb is in a superconducting state, as shown in Fig. 3.3b. In contrast to semiconductors, Debye phonons excited by electrons contribute to generating quasiparticles because the Debye energy is higher than $2\Delta_g$.

Particularly for superconductors, there is a possibility of local superconductor-norma (S–N) transition due to keV energy absorption along electron collision cascade tracks [8–10]. When the S–N transition occurs, it causes a large loss of quasiparticles before contributing to generating an output pulse. The electrons forming Cooper pairs return to normal electrons, not generating quasiparticles, as described in Chap. 3, Sect. 3.2.3.1. Therefore, we have to utilize superconductor quantum detectors in a low particle energy range below ~ 2 keV. When the S–N transition occurs, the number of quasiparticles is drastically reduced.

When the characteristic X-rays of Nb L_3-$M_{4,5}$ ($L\alpha_{1,2}$) at 2.16 keV are not absorbed by the detector material and escape for the Nb surface, an escape peak appears at a location of E_0–E_{eL}, as shown in Fig.3.24. In addition to the escape peak, continuum events appear in a range lower than E_0 because some photoelectrons or Auger electrons escape from the Nb surface, as shown in Fig. 14 in Ref. [11]. For the continuum events, the absorption events should occur extremely near the absorber surface. When all the secondary characteristic X-rays and secondary electrons convey their energies to the absorber, one count is registered near E_0 on pulse height spectra. The word "near" means that there is a statistical fluctuation, as discussed in Appendix A. The collection of numerous events forms a full-energy absorption peak (or total absorption peak), the centroid of which is E_0 with fluctuation represented by FWHM.

Appendix C
Terminology and Nomenclature

C.1 International Standards

Superconductor detector types are categorized according to the international electrotechnical committee (IEC) standard: Superconductivity—Part 22-1: Superconducting electronic devices—Generic specification for sensors and detectors (IEC 61788-22-1) [12, 13]. The standard defines terms, nomenclature, graphical symbols for diagrams based on the international consensus under Technical Committee 90—Working Group 14 (IEC-TC90/WG14). The WG14 is responsible for superconductor electronic devices in collaboration with IEEE [14]. Standards for measurement methods were published for the normal state resistance and the critical current of Josephson junctions (IEC 61788-22-2) [15, 16] and for dark count rate of SSPDs (IEC 61788-22-3) [17]. A standard method for SQUID noise measurement, which can affect other quantum magnetic sensors such as optically pumped magnetometers (OPMs), flux gates (FGs), magnetic tunnel junctions (MTJs), and diamond nitrogen-vacancy (NV) center magnetometers, is under discussion [20].

The readers may notice the unfamiliar graphical symbol for Josephson junctions (JJs) and related components in addition to the conventional switches, resistors, battery in Figs. 4.2 and 4.5. The JJ symbol is different from the simple cross symbol used commonly. Regrettably, the cross symbol is assigned to the functional element or the attribute for "magnetic field effect or dependence" (S00123) under "IEC 60617—Graphical Symbols for Diagrams" [18], which contains all symbols for circuit components. Because of this conflict, we cannot register the cross symbol in the IEC 60617 database. With respect to the standard Josephson junction symbol (S01926), the two dots symbolizing a Cooper pair are added to the cross to minimize the change from the simple cross. The magnetic field dependence attribute on the Cooper pair symbol is logical for Josephson junctions. In addition to the magnetic field dependence, other logical reasons appear in Ref. [13]. For superconductor electronics, there are two additional standard symbols: normal-superconducting boundary (S01925) and superconducting region with one superconducting connection (S01924), which can be used for circuits consisting of superconducting parts and normal-conducting parts. The

M. Ohkubo, *Superconductor Detectors*, https://doi.org/10.1007/978-3-031-84193-4

IEC-IEEE collaboration will add new symbols as a common language for superconductor electronics, collecting opinions of the community. The IEC-IEEE standardization group recommends using the standard symbols and measurement methods for fair trade and scientific papers. The graphical symbols for diagrams, publications, and glossary can be searched on IEC Products & Services Portal [19]. Note that the Cooper pair (two dots) symbolizes superconductor devices, while the vertical bar used for the symbols of diodes, transistors, diodes, and thyristors symbolizes semiconductor devices.

C.2 Detector or Sensor

We have to define a distinction between "sensor" and "detector," which helps to read this book. The boundary between the two terms is ambiguous; they are often used interchangeably. Nevertheless, the term "sensor" tends to be used for measuring stationary or rather slowly varying fields and quantities such as magnetic field and temperature. On the other hand, the term "detector" is often used for measuring single particles or quanta such as photons. In other words, sensors operate on the current mode, whereas detectors operate on the pulse mode, following Chap. 2. Some MS detectors can seamlessly change the detection mode from the pulse mode with single-ion sensitivity to the current mode in a wide dynamic range against ion flux, which enables the measurement from trace amounts of impurities to main components.

Typical superconductor sensors installed in hospitals and research institutes are superconductor quantum interference device (SQUID) magnetic sensors, which enable magnetocardiography (MCG), magnetoencephalography (MEG), and magnetospinography (MSG). The SQUID magnetic sensors measure an extremely small magnetic field produced by heart motion or neural function. Magnetograph is a typical example exhibiting the superiority of superconductor sensors. The frequency range of magnetography is ~ 100 Hz to 10 kHz; SQUIDs operating in a MHz range easily cover that range [20]. SQUIDs are also utilized as current–voltage conversion amplifiers for superconductor detectors; they are indispensable for X-ray spectroscopy with TESDs and MMCDs.

In this book, "detector" is used throughout the text because we focus on the applications to single particle detection. The collective usage of "sensor" and "detector" is exemplified by the term "TES detector" (TESD). It means a detector using a transition edge temperature sensor. More precisely, "TES X-ray detector (TESXD)" can also be used by adding "X-ray" according to the IEC standards for nomenclature [12]. By skipping "X-ray," the acronyms become TESD, MMCD, STJD, and MKID in this book, considering the nomenclature consistency and the distinction between "sensor" and "detector." Note that we do not deny the conventional usages of TES, MMC, SSPD, STJ, and MKID.

C.3 Superconductor, Superconducting, or Superconductive

IEC-TC90/WG14 is in the middle of the transition from "superconducting electronics" to "superconductor electronics" by collecting opinions widely. In this book, we use the phrase "superconductor detector" instead of the conventional "superconducting detector." This usage unifies two adjectives, "superconducting" and "superconductive." There is a long history for using the two adjective "superconducting" or occasionally "superconductive.". Under the IEC standards, "superconducting" qualifies a material in a superconducting state below T_c [21], whereas "superconductive" qualifies a material that can exhibit superconductivity under a certain condition [22]. Superconducting detectors means that detectors are always in a superconducting state, whereas superconductive detectors can be in either superconducting state or normal-conducting state. This is the precise differentiation. Strictly speaking, STJD can be called "superconducting detector," while TESD or SSPD should be called "superconductive detector" because they are not in a fully superconducting state. However, both adjectives often imply that a material or a device is in a superconducting state in the community.

The above-mentioned complication on two adjectives arise only in English. German has the single adjective "der Supraleitende Detektor." We need to have a wider view of different languages; French, Japanese, Chinese, Korean, and others, in which the noun corresponding to "superconductor" is used as an adjective. This noun usage is also possible in English. Translation of both "superconducting detector" and "superconductive detector" into the other languages results in "superconductor detector." In addition, "semiconductor detector" is the terminology [23]. Considering many languages in addition to English, this book adopts the term "superconductor detector," which indicates that detectors consist of a superconductor and that superconductivity plays an important role, irrespective of whether the superconductor is in a superconducting or normal-conducting state. When we need to indicate that devices or states are surely superconducting, "superconducting" should qualify the words, such as "superconducting state" and "superconducting gap."

C.4 Nanostrip or Nanowire

The superconductor strip (SS) type was initially called "superconducting single photon detector (SSPD)" [24]. Associated with the term of "semiconductor single photon detector" in the semiconductor community, the SSPD is reasonable. However, other types of superconductor detectors also enable single photon detection in an infrared to γ-ray range. The SSPD is not a unique term for the SS type. In addition, a variety of nomenclatures for the SS type have appeared in the literature. Therefore, the requests for terminology standardization from the experts were natural; this was one of the important triggers for the cooperative standardization efforts between IEC and IEEE.

The SS type is often called superconducting nanowire single photon detector (SNSPD). However, in most cases, the dimensions of superconductor strips are unsuitable to the definition of "nanowire" in the ISO standard for nano-objects [25]. Under ISO, "nanowire" is defined as a conducting or semiconducting nanofiber, which shall have two external dimensions in the nanoscale, and the third dimension significantly larger. The nanoscale is a length scale approximately between 1 and 100 nm. The dimensions of most SSPDs do not satisfy the condition of nanowire. Moreover, superconductor strips are frequently and significantly wider than the sense of wires that have a diameter or two similar transverse dimensions and a very long longitudinal dimension.

More importantly, the detector physics of the SS type is treated within a 2D framework. The term "nanowire" should be used for 1D physics, on which the transverse extent of long superconductors should be neglected [26]. Superconductor nanowires fall into the 1D-system category, providing that the diameter is less than a physical quantity to be concerned, and thus we can assume the physical quantity or current density to be uniform across the wire cross-section. On the other hand, for SSPDs, we should consider 2D detection mechanisms such as hot spot formation, vortex-antivortex pair (VAP) depairing and single vortex crossing (SVC) [27]. In most theoretical papers, the term "strip" is frequently used to discuss the photon detection mechanisms of superconductor strips, of which thicknesses are in the same order as ξ or less and widths are considerably larger than ξ and even comparable to or large than λ. It has recently been found that ultrawide strips with a width of several-μm are still capable of detecting infrared single photons near 1 eV [28, 29]. It is unreasonable to call the ultrawide strips "microwires" because it is far from our common sense of wires. Note that the IEC definition of wire is "the cross-section of a wire may have any shape, but the term wire is not generally used for ribbons or tapes" [30]. Under ISO, nano-ribbon and nano-tape mean thin objects that have a wide transverse dimension and a very long longitudinal dimension. Therefore, the terms "ribbons" and "tapes" are equivalent to "strips" in the field of electronics: "stripline" is one of the typical terms using "strip."

The SS type is also used for detecting X-ray photons, ions, neutral particles, electrons, and biomacromolecules [31]. Therefore, we use the term "superconductor strip particle detector (SSPD)" in this book, keeping the same acronym as the original SSPD [24]. One of the pioneers of SSPDs used the word "strip" for α-particle detection [32].

C.5 Bolometer or Calorimeter

Thermal detectors are referred to as either "bolometer" or "calorimeter." The "bol" means "ray" in Greek, while the "calor" means heat in Latin. The term "bolometer" originally represented resistive thermometers for measuring photon flux [33]. Currently, the term "bolometer" is widely used for radiation measurement devices with any types of thermometers. It is common that bolometers measure particle flux or radiation power, which is equivalent to the current mode.

The term "calorimeter" is generally assigned to devices for measuring the amount of heat produced by chemical or physical processes such as burning or melting. In elementary particle physics, on the other hand, "calorimeter" means a complex detector system to measure single-particle energy. The operation principle of the calorimeters for elementary particle physics is not a temperature rise, but a complex of several types of detectors such as scintillator, semiconductor, and gaseous detectors. Nonetheless, the calorimeters measure the energy of each particle; it is equivalent to the heat induced by each particle incidence event. Therefore, the term "calorimeter" is commonly used for single particle detection on the pulse mode. There is an overlap between "bolometer" and "calorimeter" in the literature. However, it is a fair distinction that bolometers measure particle flux or radiation power, whereas calorimeters measure the energy of each particle through a temperature rise.

In quantum detectors, there is no nomenclature corresponding to bolometer and calorimeter. When distinction is necessary in quantum detection, we use "current mode" or "analog mode" for the flux measurement and "pulse mode" or "pulse counting mode" for single particle detection.

References

1. IEC Electropedia "energy resolution": https://www.electropedia.org/iev/iev.nsf/display?openform&ievref=395-03-106
2. U. Fano, Phys. Rev. **72**, 26 (1947)
3. D. Durnford, Q. Arnaud, G. Gerbier, Phys. Rev. D **98**, 103013 (2018)
4. M.D.C. Whitaker, G. Lioliou, A.B. Krysa, A.M. Barnett, Semicond. Sci. Technol. **35**, 095026 (2020)
5. K.D. Irwin, G.C. Hilton, D.A. Wollman, J.M. Martinis, J. Appl. Phys. **83**, 3978 (1998)
6. R. Jenkins, R. Manne, R. Robin, C. Senemaud, Pure. Appl. Chem. **63**, 735 (1991). https://doi.org/10.1351/pac199163050735
7. G.P. Williams, X-ray data booklet, Electron binding energies for the elements. https://xdb.lbl.gov/Section1/Table_1-1.pdf
8. D. Van Vechten, K.S. Wood, Phys. Rev. B **43**, 12852 (1991)
9. K.S. Wood, D. Van Vechten, Nucl. Instrum. Methods Phys. Res. A, **314**, 86 (1992)
10. A.G. Kozorezov, A.F. Volkov, J.K. Wigmore, A. Peacock, A. Poelaert, R. den Hartog, Phys. Rev. B **61**, 11807 (2000)
11. M. Ohkubo, J. Appl. Phys. **134**, 081101 (2023)
12. IEC 61788-22-1, Superconductivity—Part 22-1: Superconducting electronic devices—Generic specification for sensors and detectors. https://webstore.iec.ch/publication/26674
13. M. Ohkubo, IEEE CSC SNF **11**(40), ST575 (2017). https://snf.ieeecsc.org/files/ieeecsc/2023-05/edOhkub0M_Abstract_042017_0.pdf
14. Guide to IEC/IEEE Cooperation. https://www.iec.ch/basecamp/guide-iecieee-cooperation
15. IEC 61788-22-2, Superconductivity—Part 22-2: Normal state resistance and critical current measurement—High-T_C Josephson junction. https://webstore.iec.ch/en/publication/63379
16. M. Ohkubo, G. Uehara, J. Beyer, M. Mimura, H. Tanaka, K. Ehara, S. Tanaka, T. Noguchi, E.E. Mitchell, C.P. Foley, R.L. Fagaly, Supercond. Sci. Technol. **35**, 045002 (2022)
17. IEC 61788-22-3: Superconductivity—Part 22-3: Superconducting strip photon detector—Dark count rate. https://webstore.iec.ch/publication/63471
18. IEC 60617—Graphical Symbols for Diagrams. http://std.iec.ch/iec60617
19. IEC Products & Services Portal. https://products.iec.ch/home
20. M. Ohkubo, Rev. Sci. Instrum. **94**, 111501 (2023)
21. IEC Electropedia, Superconducting. https://www.electropedia.org/iev/iev.nsf/display?openform&ievref=815-20-03
22. IEC Electropedia, Superconductive. https://www.electropedia.org/iev/iev.nsf/display?openform&ievref=815-20-04
23. IEC Electropedia, Semiconductor detector. https://www.electropedia.org/iev/iev.nsf/display?openform&ievref=395-03-35
24. G.N. Gol'tsman, O. Okunev, G. Chulkova, A. Lipatov, A. Semenov, K. Smirnov, B. Voronov, A. Dzardanov, C. Williams, R. Sobolewski, Appl. Phys. Lett. **79**, 705 (2001)

M. Ohkubo, *Superconductor Detectors*, https://doi.org/10.1007/978-3-031-84193-4

25. ISO 80004-1:2023 Nanotechnologies—Vocabulary—Part 1: Core vocabulary. https://www.iso.org/standard/79525.html
26. H. Bartolf, A. Engel, A. Schilling, K. Il'in, M. Siegel, H. -W. Hübers, A. Semenov, Phys. Rev. B **81**, 024502 (2010)
27. A. Semenov, Supercond. Sci. Technol. **34**, 054002 (2021)
28. D.Y. Vodolazov, Phys. Rev, Appl. **7**, 034014 (2017)
29. Y.P. Korneeva, D.Y. Vodolazov, A.V. Semenov, I.N. Florya, N. Simonov, E. Baeva, A.A. Korneev, G.N. Gol'tsman, T.M. Klapwijk, Phys. Rev. Appl. **9**, 064037 (2018)
30. IEC Electropedia, Wire. https://www.electropedia.org/iev/iev.nsf/display?openform&ievref=151-12-28
31. I.E. Zadeh, J. Chang, J.W.N. Los, S. Cyger, A.W. Elshaari, S. Steinhauer, S.N. Dorenbos, V. Zwiller, Appl. Phys. Lett. **118**, 190502 (2021)
32. D.H. Andrews, R.D. Fowler, M.C. Williams, Phys. Rev. **76**, 154 (1949)
33. D.H. Andrews, W.F. Brucksch, W.T. Ziegler, E.R. Blanchard, Rev. Sci. Instrum. **13**, 281 (1942)

Index for Elements and Compounds

M. Ohkubo, *Superconductor Detectors*, https://doi.org/10.1007/978-3-031-84193-4

Index

A

B

M. Ohkubo, *Superconductor Detectors*, https://doi.org/10.1007/978-3-031-84193-4

F

G

N

O

P

T

The manufacturer's authorised representative in the EU is Springer Nature Customer Service Centre GmbH, Europaplatz 3, 69115 Heidelberg, Germany. If you have any concerns regarding our products, please contact ProductSafety@springernature.com

Printed and bound by CPI Group (UK) Ltd, Croydon, CR0 4YY
05/10/2026
02228133-0002